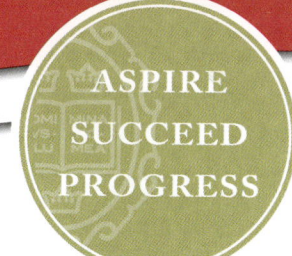

W0051274

exam success

in Cambridge International AS & A Level

Physics

Second Edition

Darren Forbes

Hossam Attya
Camille Pervenche
Jaykishan Sharma

Oxford excellence for Cambridge AS & A Level

OXFORD
UNIVERSITY PRESS

OXFORD
UNIVERSITY PRESS

Great Clarendon Street, Oxford, OX2 6DP, United Kingdom

Oxford University Press is a department of the University of Oxford.
It furthers the University's objective of excellence in research,
scholarship, and education by publishing worldwide. Oxford is a
registered trade mark of Oxford University Press in the UK and in
certain other countries

British Library Cataloguing in Publication Data
Data available

978-1-38-200553-1

1 3 5 7 9 10 8 6 4 2

Paper used in the production of this book is a natural, recyclable
product made from wood grown in sustainable forests.
The manufacturing process conforms to the environmental
regulations of the country of origin.

Printed in the UK by Bell and Bain Ltd, Glasgow

Acknowledgements
The publisher and authors would like to thank the following for
permission to use photographs and other copyright material:

Cover photo: Frank Fox/Science Photo Library.

p168: paintings/Shutterstock.

All artwork by QBS Learning and Aptara.

Every effort has been made to contact copyright holders of
material reproduced in this book. Any omissions will be rectified in
subsequent printings if notice is given to the publisher.

Contents

Access your support website for the answers to Try this and Exam-style questions here: www.oxfordsecondary.com/caie-al-sci-exam-success

Introduction

Exam Success in Physics will help you to reach your highest potential and achieve the best possible grade. The guide is designed to help you learn the content of the AS & A Level Physics syllabus (9702), to help you develop techniques about how to revise and to answer the questions in the exam papers.

The first part of the syllabus (Units 1 to 11) is examined at AS Level by Papers 1, 2 and 3. You can gain an AS Level qualification in Physics by taking these three papers. The second part of the syllabus (Units 12 to 25) is examined by Papers 4 and 5. You can only gain an A Level qualification in Physics if you take these two papers and the AS papers.

The AS **and** A Level papers can be taken together in the same series. Alternatively, you can follow the staged assessment by taking all three AS papers in one series and the A Level papers in a later series. If your grade for the AS papers was not as good as you hoped, you may retake the papers in another series. However, if you retake you have to take all three papers in the same series.

The table gives a summary of the five examination papers:

Level of assessment	Exam paper	Units	Style of questions	Weighting
AS	Paper 1	1 to 11	40 multiple choice questions (MCQs)	31% of the AS Level 15.5% of the A Level
AS	Paper 2	1 to 11	5 or 6 structured questions	46% of the AS Level 23% of the A Level
AS	Paper 3	1 to 11	2 or 3 questions on practical Physics	23% of the AS Level 11.5% of the A Level
A Level	Paper 4	12 to 25	9 structured questions	38.5% of the A Level
A Level	Paper 5	1 to 25	2 or 3 structured questions	11.5% of the A Level

Assessment objectives

The papers are written to match the three assessment objectives.

- AO1: Knowledge and understanding
- AO2: Handling, applying and evaluating information
- AO3: Experimental skills and investigations

You can find details of these three objectives in the syllabus. You should appreciate that AO1 and AO2 have equal weighting in Papers 1, 2 and 4. In other words, 50% of the marks in those papers are to test what you know (and can recall) and 50% assesses how you can use your knowledge and understanding to analyse and interpret information. AO3 is assessed in Papers 3 and 5.

Know your syllabus

Download the syllabus from the CAIE website (https://www.cambridgeinternational.org) and use it all the time during your course. The subject content of the syllabus is divided into 25 topics. In this book they are known as Units. Each topic is divided into learning outcomes. These tell you exactly what you need to learn and what you should be able to do with the knowledge you acquire.

Command words

Each exam question has a command word. This is usually at the beginning of the question as in '**Describe** the difference between a scalar quantity and a vector quantity'. Sometimes command words are not at the beginning as in 'Use the graph to **determine** the acceleration of the car.'

You can find brief explanations of the command words in the syllabus.

Do your own marking

Mark schemes for the exam-style questions in this book are available on the support website (www.oxfordsecondary.com/caie-al-sci-exam-success). Find out what examiners are looking for by using the mark schemes to mark answers written by your peers. However, do not learn mark schemes in the hope that the same questions will be set when you take the exams. Candidates that try this usually gain very few marks because they are not answering the questions set.

Key features of the book

The following features in the guide will help you develop the skills you need to learn effectively and use your knowledge in unfamiliar contexts.

Knowledge check

This feature recaps the skills and knowledge you should already have before working through the unit.

Worked Example

These give examples of questions, and show you how best to answer them.

Remember

These include key information that you must remember if you are to achieve a high grade.

Link

These show where in the book you can find more information about a topic.

Key terms

These give easy-to-understand definitions of important terms.

Exam tip

These provide advice on how to master your exam technique and help you understand exactly what examiners are looking for.

Try this

These feature gives you an opportunity to test out the skills and knowledge you've learnt with a practice question. Answers to Try this and the end-of-chapter Exam-style questions are available on the OUP support website: www.oxfordsecondary.com/caie-al-sci-exam-success.

Practical Skills

These describe practical skills that you might be tested on and are intended as reminders of work you may have already done in the lab.

Maths Skills

These remind you of the vital mathematical skills that you need in order to answer exam questions in chemistry.

↑ Raise your grade

Here, you can read model answers that achieved maximum marks, as well as find out how to improve answers where marks were lost and learn how to avoid common errors in exams.

1 Physical quantities and units

Knowledge check

You should be able to:

- know some of the SI base units
- describe the difference between vectors and scalars
- use standard form to represent numbers.

Physical quantities

Physical quantities, such as kinetic energy, electric current, and temperature, are expressed by a number (the 'magnitude') and a unit.

> **Key term**
>
> **Solar constant:** the average amount of energy reaching the Earth per square metre each second.

Making estimates

Being able to estimate a quantity is an important skill that can be improved with practice. Start with quantities you know, or can estimate reasonably accurately, and combine them to estimate a value for an unknown quantity.

Worked Example

The **solar constant** (the average amount of energy reaching the Earth per square metre each second) is $1.4\,\text{kW m}^{-2}$ (see Figure 1.1). The radius of the Earth $\approx 6.4 \times 10^{3}\,\text{km}$. Estimate the energy received per day on Earth from the Sun.

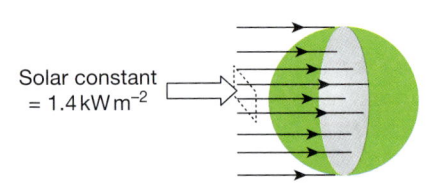

Solar constant $= 1.4\,\text{kW m}^{-2}$

Answer

energy per day = 'area' of Earth × solar constant × time

$$= \pi \times (6.4 \times 10^{6})^{2} \times (1.4 \times 10^{3}) \times (24 \times 60 \times 60) = 10^{22}\,\text{J}$$

▲ **Figure 1.1** Solar constant

Table 1.1 gives the value of some physical quantities that are useful to know in order to estimate other quantities.

▼ **Table 1.1** Useful quantities

Quantity	Value	Quantity	Value
Diameter of a nucleus	$10^{-14}\,\text{m}$	Atmospheric pressure	$1.0 \times 10^{5}\,\text{Pa}$
Diameter of an atom	$10^{-10}\,\text{m}$	One day	$8.64 \times 10^{4}\,\text{s}$
Wavelength of visible light	400–$700\,\text{nm}$	One year	$3.1 \times 10^{7}\,\text{s}$
Radius of the Earth	$6.4 \times 10^{6}\,\text{m}$	Speed of a car	20–$30\,\text{m s}^{-1}$
Mass of an electron	$9.1 \times 10^{-31}\,\text{kg}$	Speed of sound in air	$330\,\text{m s}^{-1}$
Mass of a proton	$1.7 \times 10^{-27}\,\text{kg}$	Speed of light in a vacuum	$3 \times 10^{8}\,\text{m s}^{-1}$
Mass of a postage stamp	$5 \times 10^{-5}\,\text{kg}$	Energy of an alpha particle	$5\,\text{MeV}$
Mass of an apple	$0.1\,\text{kg}$	Freezing point of water	$0\,^{\circ}\text{C}$
Mass of an adult	$70\,\text{kg}$	Boiling point of water	$100\,^{\circ}\text{C}$
Density of air	$1.3\,\text{kg m}^{-3}$	Specific heat capacity of water	$4200\,\text{J kg}^{-1}\,^{\circ}\text{C}^{-1}$
Density of water	$10^{3}\,\text{kg m}^{-3}$	Charge on an electron e	$1.6 \times 10^{-19}\,\text{C}$
Gravitational field strength, g	$9.81\,\text{N kg}^{-1}\,(9.81\,\text{m s}^{-2})$	Charge transferred by a lightning flash	$10\,\text{C}$

SI units

The SI system of units (Système International d'Unités) is an internationally agreed system of units, built on seven base units (see Table 1.2).

▼ **Table 1.2** SI base units

Unit	Symbol	Base unit
metre	m	length
kilogram	kg	mass
second	s	time
ampere	A	electric current
kelvin	K	temperature
mole*	mol	amount of a substance
candela**	cd	light intensity

*A Level only ** not part of the A Level course.

Expressing derived units in terms of base units

The familiar units for quantities such as force (newtons, N), and energy (joules, J) can be 'broken down' into their base units by using an equation which links these quantities to quantities whose units are known.

Worked Example

Determine the base units of force.

Answer

The base units of force can be found using $F = ma$.

units of force (N) = units of mass (kg) × units of acceleration ($m\,s^{-2}$)

The SI base units of force are $kg\,m\,s^{-2}$.

Table 1.3 lists the derived units that you will encounter at A and AS Level.

▼ **Table 1.3** Derived units

Quantity	Symbol	Unit	SI base units
force	F	newton (N)	$kg\,m\,s^{-2}$
pressure	p	pascal (Pa)	$kg\,m^{-1}\,s^{-2}$
energy	E	joule (J)	$kg\,m^2\,s^{-3}$
power	P	watt (W)	$kg\,m^2\,s^{-1}$
frequency	f	hertz (Hz)	s^{-1}
electric charge	Q	coulomb (C)	$A\,s$
potential difference (p.d.)	V	volt (V)	$kg\,m^2\,s^{-3}\,A^{-1}$
electrical resistance	R	ohm (Ω)	$kg\,m^2\,s^{-3}\,A^{-2}$
capacitance	C	farad (F)	$A^2\,kg^{-1}\,m^{-2}\,s^4$
magnetic flux density*	B	tesla (T)	$kg\,s^{-2}\,A^{-1}$
magnetic flux*	Φ	weber (Wb)	$kg\,m^2\,s^{-2}\,A^{-1}$

*A Level only

Some quantities have no units (e.g., the refractive index of glass). These quantities are called dimensionless quantities.

Using SI base units to check equations

For any equation to be correct, a necessary condition is that the equation must be **homogeneous**; the units of the left-hand side of the equation must match those on the right-hand side.

Worked Example

A student is studying how the period of oscillation of a simple pendulum depends on its length l. She cannot recall whether the theoretical equation for the period T of the pendulum is:

$$T = 2\pi\sqrt{\frac{g}{l}} \quad \text{or} \quad T = 2\pi\sqrt{\frac{l}{g}}$$

Which equation could be correct?

Answer

The units of the first equation are: $\dfrac{[\mathrm{ms}^{-2}]^{\frac{1}{2}}}{\mathrm{m}^{\frac{1}{2}}} = \mathrm{s}^{-1}$

The units of the second equation are: $\dfrac{[\mathrm{m}]^{\frac{1}{2}}}{[\mathrm{ms}^{-2}]^{\frac{1}{2}}} = \mathrm{s}$

and so only the second equation can be correct. (This method does not prove the equation – it cannot show whether the 2π in the equation is correct).

★ Exam tip

Checking the homogeneity of an equation does not prove it is correct – it only shows whether an equation could be correct.

Key term

Homogeneous: for an equation, the units of the left-hand side match those on the right-hand side.

Using prefixes

Physics is often concerned with very small numbers, such as the diameters of atomic nuclei, and very large numbers, such as the masses of stars and planets. These can be expressed in standard form (a number between 1.0 and 10.0 multiplied by a multiple of 10); for example, the radius of the Earth is 6.37×10^6 m.

Another way of expressing such numbers is to use prefixes (see Table 1.4).

▼ **Table 1.4** Prefixes

Prefix	Symbol	Multiple	Example
pico	p	10^{-12}	pF (picofarad)
nano	n	10^{-9}	nC (nanocoulomb)
micro	µ	10^{-6}	µA (microamp)
milli	m	10^{-3}	mV (millivolt)
centi	c	10^{-2}	cm (centimetre)
deci	d	10^{-1}	dB (decibel)
kilo	k	10^{3}	kg (kilogram)
mega	M	10^{6}	MΩ (megaohm)
giga	G	10^{9}	GJ (gigajoule)
tera	T	10^{12}	TW (terawatt)

Tables and graphs

Measurements from an experiment should always be recorded in a neat table. You should make sure the table has enough columns to include any repeated readings, averages, and calculated values that you use later to analyse your results.

Each column heading should have the quantity being recorded, and the unit it is measured in, separated by a solidus (/); for example m/g or T/s. Alternatively, the units can be in brackets.

Scalar and vector quantities

Scalar quantity: A scalar quantity only has magnitude (size); for example mass, speed, distance, and temperature.

Vector quantity: A vector quantity is one which has both magnitude and a direction; for example, displacement, velocity, force, and momentum.

Adding and subtracting vectors

A vector such as velocity can be represented by an arrow, the length of the arrow indicating the *magnitude of the quantity* (the speed) and the direction of the arrow indicating the *direction of travel*. Vectors are usually indicated by bold type (**a**), or with an arrow ($\vec{a}$).

Two vectors **a** and **b** (see Figure 1.2) are added by placing one of the vectors so that it begins at the end of the other vector. The sum of the two vectors is a straight line drawn from the beginning of the first vector to the end of the second.

▲ **Figure 1.2** Adding vectors

The same procedure can be used to subtract one vector from another.

To find **a** – **b**, first draw the vector **–b** by drawing the vector **b**, but pointing in the opposite direction to **b**, and then add this to **a** (see Figure 1.3).

▲ **Figure 1.3** Subtracting vectors

Representing a vector as two perpendicular components (resolving vectors)

It is often useful to separate a vector, such as a force or velocity, into two components at right angles to each other.

In Figure 1.4 the horizontal component of the force F is $F\cos\theta$; the vertical component is $F\sin\theta$. The **resultant vector** (the sum of the two components) is F.

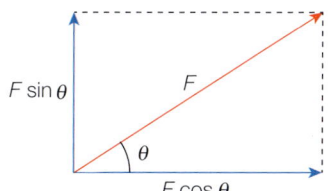
▲ **Figure 1.4** Components of a vector

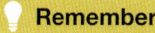 **Remember**

The component adjacent to the angle θ is always the cosine component.

↑ Raise your grade

1 The energy released by the fission of one uranium nucleus is 3.2×10^{-11} J.
State this energy in pJ. [1]

energy = ...32.... pJ ✔

3.2×10^{-11} J $= 32 \times 10^{-12}$ J;

p = pico = 10^{-12}

2 The orbital period of Jupiter is 0.37 Gs. Express this time in years. [2]

$0.37 \, Gs = 0.37 \times 10^9 \, s = 3.7 \times 10^8$ ✔

Correct prefix of 'G' = giga = 10^9

time = ...3.7×10^8.... years ✗

The answer needs to be converted from seconds to years. The correct answer is 11.9 years
[No. of seconds in a year = 3.1×10^7].

3 The speed v of sound waves in air is given by the equation:

$$v = \sqrt{\frac{\gamma p}{\rho}}$$

where p is the pressure of the air, ρ is the density of the air, and γ is a dimensionless constant.

Show that this equation is homogeneous. [2]

units of $v = ms^{-1}$ ✔   units of $\sqrt{\frac{\gamma p}{\rho}} = \left(\frac{Nm^{-2}}{kgm^{-3}}\right)^{\frac{1}{2}} = \left(\frac{kgms^{-2}m^{-2}}{kgm^{-3}}\right)^{\frac{1}{2}} = (m^2 s^{-2})^{\frac{1}{2}} = ms^{-1}$ ✔

Same units, so equation is homogeneous.   Working shown clearly.

4 A ship is travelling due east at a speed of $5\,km\,h^{-1}$. A patrol vessel is 12 km due south of
the ship, moving with a speed of $13\,km\,h^{-1}$, and wishes to intercept the ship.

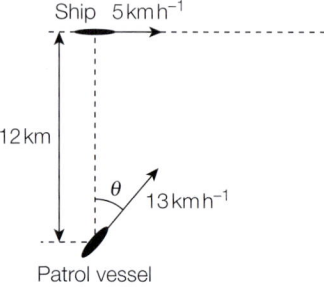

Ship $5\,km\,h^{-1}$

12 km

θ

$13\,km\,h^{-1}$

Patrol vessel

(a) At what angle θ should the patrol vessel sail in order to intercept the ship?

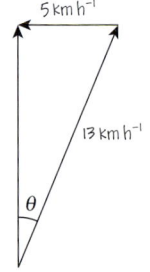

$5\,km\,h^{-1}$

$13\,km\,h^{-1}$

θ

$\sin\theta = \frac{5}{13}$; $\theta = \sin^{-1}\left(\frac{5}{13}\right) = 22.6°$ ✔

Correct answer with
working shown clearly.

$\theta = $ ....$22.6°$....

(b) How much time elapses before the patrol vessel reaches the ship? [3]

velocity in direction of ship $= 13\cos 22.6 = 12\,km\,h^{-1}$ ✔   Correct method.

time taken to travel 12 km $= \frac{12}{12} = 1$ hour ✔   Correct calculation.   time = ...1.0 hour....

Errors and uncertainties

Systematic and random errors

Systematic errors

Systematic errors cause all the recordings of a measurement to be displaced one way or the other from the true or accurate value (see Figure 1.5). Causes include zero errors in instruments, incorrectly calibrated scales, or changes in environmental conditions such as temperature. Using a micrometer with a zero error to measure the diameter of a resistance wire, for example, will give a very precise value (to the nearest 0.01 mm) but not a very accurate value if no allowance is made for the zero error.

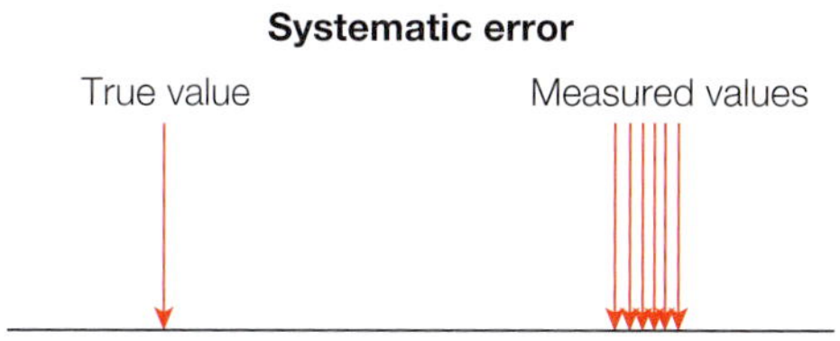

Systematic errors can be reduced or eliminated by, for example, checking for any zero error in a micrometer or vernier calipers, or using two ammeters in series to check they read the same value.

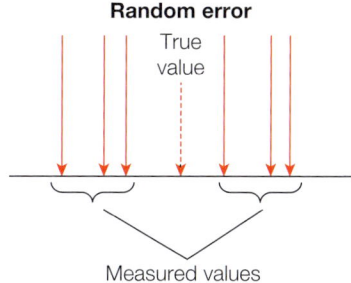
▲ **Figure 1.5** Precise but inaccurate

Random errors

Random errors occur principally because of the limitations of the experimenter; for example, in judging the start and finish of an oscillation. A random error means that the values of a measurement are scattered in a random fashion (see Figure 1.6). The error can be reduced by taking several values and calculating a mean. If a range of values are recorded, a reasonable estimate of the absolute uncertainty of the measurement is half the range.

▲ **Figure 1.6** Accurate but imprecise

Precision and accuracy

Measurements should ideally be both precise and accurate, with all the readings grouped closely around the 'true' value. Vernier calipers (see Figure 1.7) can be used to measure distances from a few millimetres up to 10 cm or more. If several values are close together and one value is significantly different from all the others (an **outlier**), it can usually be rejected as an **anomalous result**.

Precision

A precise measurement is the degree to which the measurement is **repeatable**. The **precision** of an instrument is the smallest non-zero reading (the smallest division) that can be measured by the instrument or the size of the smallest division on a measuring instrument. A micrometer (see Figure 1.8) is a more precise instrument than a metre rule because it can measure to a precision of ± 0.01 mm, whereas it is only possible to record values to the nearest millimetre using a metre rule.

▲ **Figure 1.7** Vernier calipers

Accuracy

Accuracy is how close the value(s) are to the true value. It is a measure of the confidence an experimenter has in a measurement. An accurate measurement can be obtained using a correctly calibrated instrument skilfully (e.g., by avoiding parallax errors). Accuracy is expressed by the absolute or percentage uncertainty in a measurement.

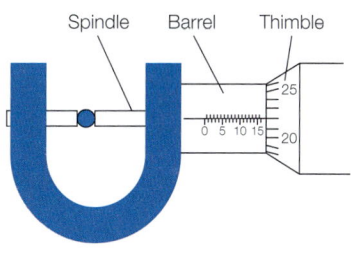
▲ **Figure 1.8** Micrometer

Key terms

Accuracy: how close the measurement is to the true value.

Anomalous result: if several values are close together and one value is significantly different from all the others, it is usually rejected as an anomalous result.

Precision: the smallest division that can be measured by an instrument.

Random errors: errors caused by the experimenter, e.g., judging the start and finish of an oscillation.

Systematic errors: errors that affect all the readings, e.g., incorrectly calibrated scales.

Uncertainty

Suppose five measurements of the diameter d of a glass marble are recorded: 2.52 cm, 2.49 cm, 2.48 cm, 2.51 cm and 2.49 cm.

The mean value is 2.498 cm and the range of values (largest to smallest) is 0.04 cm. The diameter of the marble should be recorded as 2.50 ± 0.02 cm.

The mean value is only given to the nearest 0.01 cm because the uncertainty in the value is 0.02 cm. The value is stating that:

- d lies between 2.48 cm and 2.52 cm

- the uncertainty in the value is 0.02 cm (half the range).

The **percentage uncertainty** in the value of d can be found from:

$$\text{percentage uncertainty} = \frac{\text{absolute uncertainty}}{\text{mean value}} \times 100\%$$

In this example, the percentage uncertainty in d is:

$$\frac{0.02}{2.50} \times 100 = 0.8\%$$

To calculate the total percentage uncertainty in a quantity that depends on several measurements, add the individual percentage uncertainties, but first you need to multiply each value by the measurement's index (power). For example, if a quantity p is given by:

$$p = \frac{x^3 y^{\frac{1}{2}}}{z^4}$$

the percentage uncertainty in p is equal to:

$$3 \times \% \text{ uncertainty in } x \quad + \quad \tfrac{1}{2} \times \% \text{ uncertainty in } y \quad + \quad 4 \times \% \text{ uncertainty in } z$$

Uncertainties and significant figures

Measurements taken during an experiment may vary when the experiment is repeated. Suppose, for example, five measurements of the time t taken for an object to fall a fixed distance through a viscous liquid are measured as 7.1 s, 7.4 s, 6.8 s, 6.7 s, and 7.5 s. The average time is 7.1 s with a range (largest value – smallest value) of 0.8 s. The value of t can be expressed as 7.1 ± 0.4 s.

As well as random variations, uncertainties arise both from the limitations of the instrument being used (both its accuracy and its precision) and the difficulty in judging the measurement (e.g., deciding when an oscillation of a pendulum is completed). If a metre rule is calibrated in millimetres, it is tempting to judge the uncertainty in a measurement made with the rule as ± 1 mm, but if it is used to measure the length of a resistance wire, for example, there may be kinks in the wire. The actual overall uncertainty in the measurement may be 2–5 mm.

When calculating a quantity using measured values, the calculated quantity also has an uncertainty. For example, the resistance of a resistance wire is found by measuring the current I through the wire and the p.d. V across it.

$$I = 0.15 \pm 0.01 \text{ A} \qquad V = 1.54 \pm 0.02 \text{ V}$$

The nominal resistance of the wire, $R = \dfrac{V}{I} = \dfrac{1.54}{0.15} = 10.2667 \, \Omega$

$$\text{largest value of } R = \frac{V_{max}}{I_{min}} = \frac{1.56}{0.14} = 11.1 \, \Omega$$

$$\text{smallest value of } R = \frac{V_{min}}{I_{max}} = \frac{1.52}{0.16} = 9.5 \, \Omega$$

The large range of possible values of R (between 9.5 Ω and 11.1 Ω) means the resistance cannot be known as precisely as 10.2667 Ω! Instead, the value of R should be given as: $R = 10.3 \pm 0.8 \, \Omega$

Key term

Percentage uncertainty: can be found by dividing the absolute uncertainty by the mean value and multiplying it by one hundred.

★ **Exam tip**

If a calculation involves a number of stages, or uses several different measured values, do not 'round' any intermediate values in the calculation. Instead, only round the final answer.

💡 **Remember**

As a general rule, **express a calculated value to the same number of significant figures as, or one more than, the significant figures of the least precise value used in the calculation.**

In this example, the current I is expressed to only two significant figures, and so the resistance R should be expressed to two or three significant figures.

↑ Raise your grade

An experiment is performed to measure the Young modulus E of copper using a long, thin copper wire. The diameter d and length ℓ_0 of the wire are measured, and then a tensile force F is applied to the wire. The extension x of the wire is then recorded. The measurements made are shown in the first two columns of the table.

Measurement	Value	Percentage uncertainty
Length of wire ℓ_0	0.953 ± 0.002 m	$\dfrac{0.002}{0.953} \times 100 = 0.2\%$
Diameter of wire d	0.21 ± 0.01 mm	$\dfrac{0.01}{0.21} \times 100 = 4.8\%$
Tensile force F	5.15 ± 0.05 N	$\dfrac{0.05}{5.15} \times 100 = 1.0\%$
Extension of wire x	1.2 ± 0.1 mm	$\dfrac{0.1}{1.2} \times 100 = 8.3\%$

✔✔ Method of calculation and values are correct

(a) State a suitable instrument for measuring: [2]

> A micrometer measures to a precision of 0.01 mm.

> A ruler can only measure to the nearest mm. A travelling microscope or other vernier scale is needed.

(i) d .... micrometer ✔ ....

(ii) x .... 30 cm ruler ✗ ....

(b) Complete the table by calculating the percentage uncertainties in the measurements. [2]

(c) The Young modulus is found from the equation:

$$E = \frac{4F\ell_0}{\pi d^2 x}$$

✔ Correct substitutions including conversion of mm to m

(i) Calculate the value of E.

$$E = \frac{4Fl_0}{\pi d^2 x} = \frac{4 \times 5.15 \times 0.953}{\pi \times (0.21 \times 10^{-3})^2 \times (1.2 \times 10^{-3})} = 1.18 \times 10^{11}$$

✔ Correct value.

$E =$ .... 1.18×10^{11} .... N m^{-2}

(ii) Calculate the percentage uncertainty in E.

> The percentage uncertainty in d^2 is twice the percentage uncertainty in d. (If there had been a d^3 term the percentage uncertainty would be three times as much as that in d.)

Total percentage uncertainty $= 0.2 + 4.8 + 1.0 + 8.3 = 14.3\%$ ✗

> The correct value is:
> $0.2 + 2 \times 4.8 + 1.0 + 8.3 = 19.1\%$.

Percentage uncertainty = ... 14.3% ....

(iii) State the value of E and its uncertainty to the appropriate number of significant figures. [4]

14.3% of $1.18 \times 10^{11} = 0.169 \times 10^{11}$

✔ The value is correct allowing for 'error carried forward' from (c) (ii); only one decimal place is justified because of the large uncertainty.

$E =$ .... $(1.2 \pm 0.2) \times 10^{11}$ .... N m^{-2}

(d) Suggest a possible cause of systematic error in the measurements made. [1]

.... 'zero error' in the micrometer .... ✔ A valid answer. ....

Exam-style questions

1. The friction force F on a sphere of radius r falling with velocity v through a liquid is given by the equation:

$$F = 6\pi r\eta v$$

where η is the viscosity of the liquid.

What are the SI base units of viscosity?

A $kg\,m\,s$ B $kg\,m^{-1}\,s$ C $kg\,m\,s^{-1}$ D $kg\,m^{-1}\,s^{-1}$ [1]

2. The stress needed to fracture a material with a crack is given by the equation:

$$\sigma = k\sqrt{\frac{\gamma E}{d}}$$

where E is the Young modulus, d the width of the crack, and k is a dimensionless constant. For the equation to be homogeneous, what are the units of γ?

A N B J C $N\,m^{-2}$ D $J\,m^{-2}$ [1]

3. What is the best estimate of the kinetic energy of an Olympic 100 m runner, running at top speed?

A 0.4 kJ B 4 kJ C 40 kJ D 400 kJ [1]

4. Which one of the following physical quantities is a vector?

A work B mass C momentum D power [1]

5. Two vectors, **a** and **b** are as shown below.

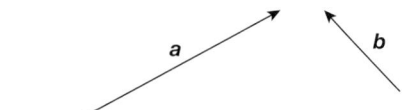

Which vector represents **a** – **b**?

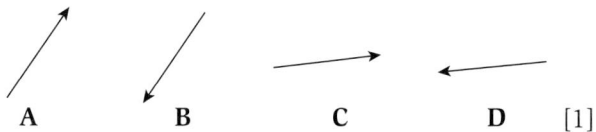

A B C D [1]

6. One light-year is the distance travelled by light in one year. The diameter of the Milky Way galaxy is approximately 100 000 light-years.

What is the best estimate of this distance in metres?

[The speed of light is $3 \times 10^8\,m\,s^{-1}$.]

A $10^{15}\,m$ B $10^{17}\,m$ C $10^{19}\,m$ D $10^{21}\,m$ [1]

7. (a) Express the following in standard form:
   (i) 470 kΩ (ii) 1000 μF (iii) 0.05 nm [3]

   (b) Express the following using an appropriate prefix:
   (i) $6.4 \times 10^6\,m$
   (ii) 0.0075 A
   (iii) $3.0 \times 10^8\,m\,s^{-1}$ [3]

8. (a) Describe the difference between a *scalar* quantity and a *vector* quantity. [1]

   (b) In the following list, underline all the **vector** quantities.
   weight acceleration stress power work [2]

9. The diagram shows a car travelling up a hill at constant speed.

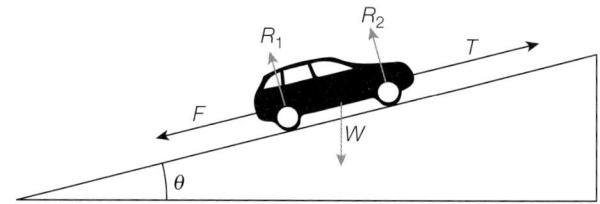

   (a) State:
   (i) the horizontal component of the friction force F
   (ii) the vertical component of the engine force T
   (iii) the component of the weight force W acting down the slope. [3]

   (b) By resolving forces along the slope, write down an equation relating F, T, and W. [1]

10. The density of brass was found by measuring the diameter d and the mass m of a brass sphere.

d	16.5 ± 0.1 mm
m	19.7 ± 0.1 g

The density ρ can be found using the equation:

$$\rho = \frac{6m}{\pi d^3}$$

What is the percentage uncertainty in the value of the density?

A 0.5% B 0.6% C 1.1% D 2.3% [1]

You should be able to:

- describe displacement, velocity, average speed and acceleration and how these are represented by graphs of motion

- use the equations for velocity $\left(v = \dfrac{\Delta s}{\Delta t} \right)$ average speed $\left(v = \dfrac{\text{total distance}}{\text{time taken}} \right)$ and acceleration $\left(a = \dfrac{\Delta v}{\Delta t} \right)$

- describe the difference between vectors (e.g., speed, distance,) and scalars (e.g., velocity, acceleration, displacement).

Using graphs

Displacement–time graphs

The **gradient** of the displacement–time graph (see Figure 2.1) is the **velocity** $\left(\dfrac{\Delta s}{\Delta t} \right)$.

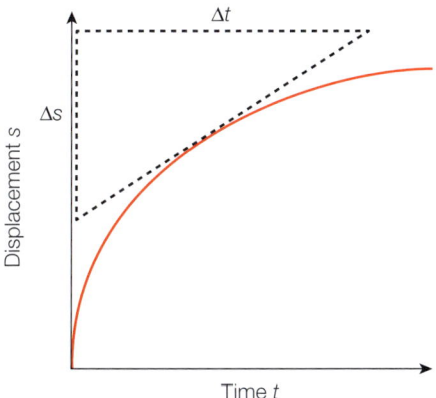

▲ **Figure 2.1** Velocity is the gradient of a displacement–time graph

> ★ **Exam tip**
>
> When calculating a velocity by finding the gradient, draw a tangent, and complete a large triangle as shown in Figure 2.1. Δs can be read directly from the axis of the graph.

Changes in the gradient of a displacement–time graph give information about acceleration as well as velocity (see Figure 2.2).

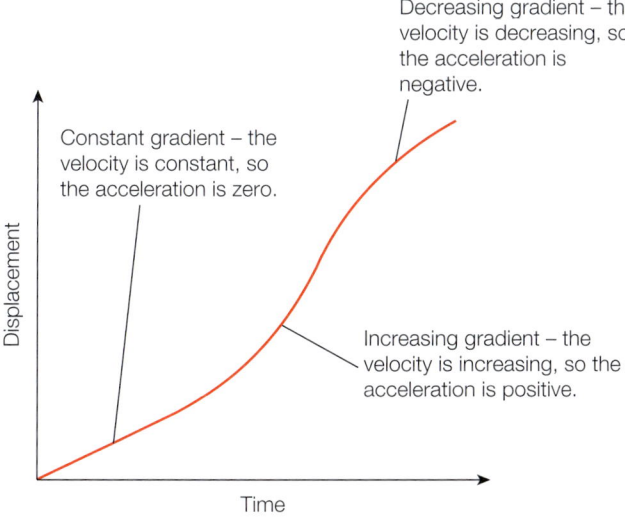

Constant gradient – the velocity is constant, so the acceleration is zero.

Decreasing gradient – the velocity is decreasing, so the acceleration is negative.

Increasing gradient – the velocity is increasing, so the acceleration is positive.

▲ **Figure 2.2** Acceleration in displacement–time graphs

Velocity–time graphs

Velocity–time graphs are very useful. The gradient of a velocity–time graph is the acceleration, and the area under a velocity–time graph between any two points is the displacement (see Figure 2.3).

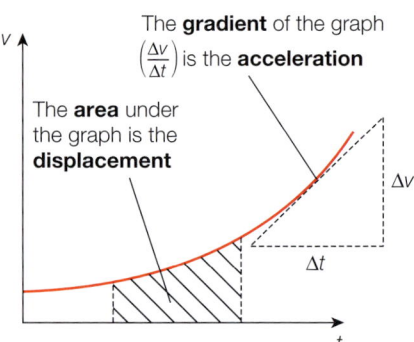

▲ **Figure 2.3** Acceleration and displacement can be calculated from a velocity–time graph

Worked Example

Figure 2.4 shows the velocity–time graph for a car starting from rest.

(a) Determine the car's acceleration at $t = 4\,\text{s}$.

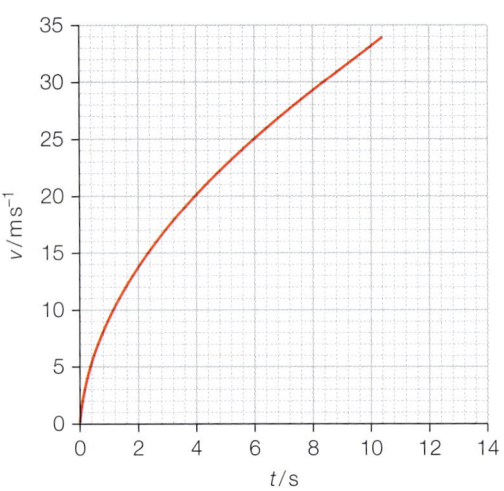

▲ **Figure 2.4**

(b) Estimate the distance travelled in the first 6 s.

> 💡 **Remember**
>
> Don't forget to include the units in your answer.

Answer

(a) The acceleration at $t = 4\,\text{s}$ is the gradient at $t = 4\,\text{s}$ (see Figure 2.5).

$$a = \frac{\Delta v}{\Delta t} = \frac{32.0 - 11.0}{8.0} = 2.6\,\text{ms}^{-2}$$

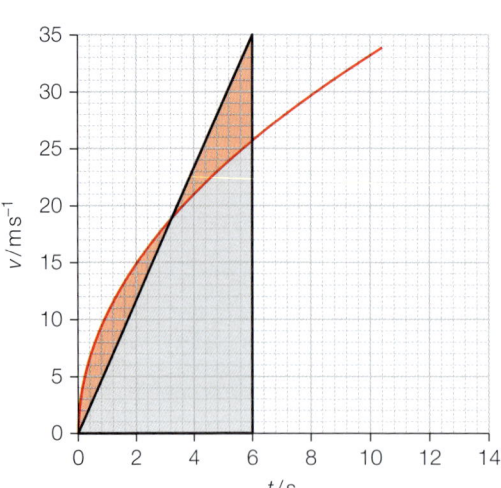

▲ **Figure 2.5**

(b) The distance travelled is the area under the graph (see Figure 2.6) from $t = 0\,\text{s}$ to $t = 6\,\text{s}$.

distance travelled $\approx \frac{1}{2} \times 6 \times 35 = 105\,\text{m}$

The part of the triangle above the line roughly matches the part of the area beneath the line not covered by the triangle.

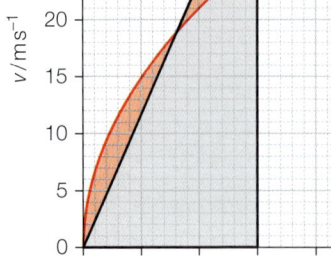

▲ **Figure 2.6**

> 📐 **Maths Skills**
>
> For more on estimating areas see Appendix: Maths skills.

Equations of motion

If an object is travelling with **constant acceleration**, the **equations of motion** may be used to analyse its motion.

A velocity–time graph for an object moving with constant acceleration a, starting with velocity u, and reaching a velocity v in t seconds, shows how these equations arise.

From Figure 2.7:

$$\text{acceleration} = \text{gradient}$$

$$a = \frac{v - u}{t}$$

so $\qquad v = u + at \qquad$ (equation 1)

The displacement, s, of an object can be found by working out the area under the graph (see Figure 2.8):

$$\text{displacement} = \text{area under the graph}$$

$$s = ut + \tfrac{1}{2}(v - u)t$$

from equation 1, $v - u = at$ so:

$$s = ut + \tfrac{1}{2}(at)t$$

$$s = ut + \tfrac{1}{2}at^2 \qquad \text{(equation 2)}$$

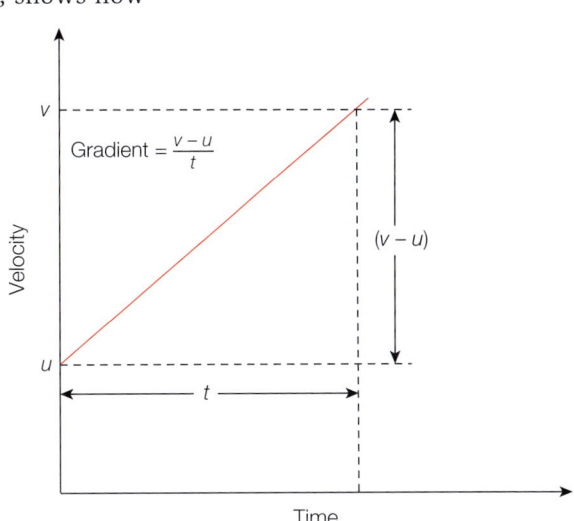

▲ **Figure 2.7** The acceleration is given by the gradient of the velocity–time graph

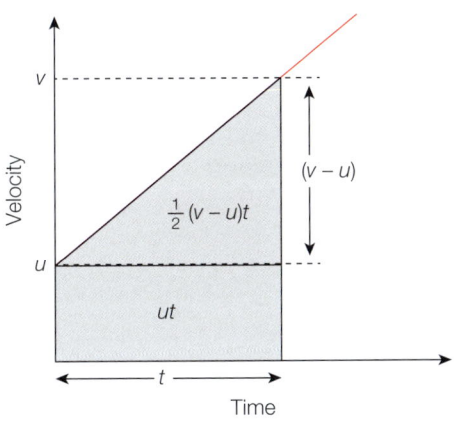

▲ **Figure 2.8** The displacement is the area under the graph

The third equation can be derived from equations 1 and 2:

First, 'square' equation 1:

$$v^2 = (u + at)^2 = u^2 + 2uat + a^2t^2$$

$$= u^2 + 2a(ut + \tfrac{1}{2}at^2)$$

so $\qquad v^2 = u^2 + 2as \qquad$ (equation 3)

> 💡 **Remember**
>
> These equations only apply when the acceleration or deceleration is **constant**, including zero acceleration (constant velocity).

> 💡 **Remember**
>
> The four equations of motion: $v = u + at$
>
> $$s = ut + \tfrac{1}{2}at^2$$
>
> $$v^2 = u^2 + 2as$$
>
> $$s = \frac{u + v}{2}$$

> ★ **Exam tip**
>
> You need to be able to recall these equations and you may be asked to derive them.

Using the equations of motion

Worked Example

An aeroplane touches down at the end of a runway travelling at a speed of $72\,\mathrm{m\,s^{-1}}$ (see Figure 2.9). It decelerates uniformly at a rate of $3\,\mathrm{m\,s^{-2}}$.

▲ **Figure 2.9**

Calculate:

(a) the speed of the aeroplane 8 s after touchdown

(b) the distance travelled along the runway before coming to rest.

Answer

(a) Using $v = u + at$:

$$v = 72 + (-3) \times 8 = 48\,\mathrm{m\,s^{-1}}$$

(b) Using $v^2 = u^2 + 2as$ with $v = 0$ when the aeroplane comes to rest:

$$0^2 = 72^2 + 2 \times (-3) \times s$$

$$s = 864\,\mathrm{m}$$

> 💡 **Remember**
>
> The acceleration is negative because the aeroplane is decelerating.

Motion under gravity

At low speeds air resistance has a negligible effect on falling objects and can be ignored. The equations of motion can be used to solve problems where $a = g$, the acceleration of free fall.

> ⭐ **Exam tip**
>
> The value of g is provided in Exam Papers 1, 2, and 4.

Worked Examples

1 A stone is thrown vertically upwards with a velocity of $30.0\,\mathrm{m\,s^{-1}}$ and falls back down to the ground (Figure 2.10).

Calculate:

(a) the velocity of the stone after 4.0 s

(b) the maximum height reached by the stone

$a = g = 9.8\,\mathrm{m\,s^{-2}}$

$30\,\mathrm{m\,s^{-1}}$

▲ **Figure 2.10**

Answer

(a) Using $v = u + at$:

$$v = 30.0 + (-9.81) \times 4.0 = -9.24\,\mathrm{m\,s^{-1}}$$

The velocity is $9.24\,\mathrm{m\,s^{-1}}$ **downwards**.

(b) Using $v^2 = u^2 + 2as$ with $v = 0$ when the stone reaches its highest point:

$$0^2 = 30.0^2 + 2 \times (-9.81) \times s$$

$$s = 45.9\,\mathrm{m}$$

> 💡 **Remember**
>
> Take care deciding on the **sign** of g. An object thrown into the air, for example, is always accelerating downwards, whether the object happens to be moving upwards or downwards.

2 A hot-air balloon (see Figure 2.11) is ascending at a constant speed of 3.0 m s⁻¹. A sandbag is dropped from the balloon and hits the ground after 5.0 s.

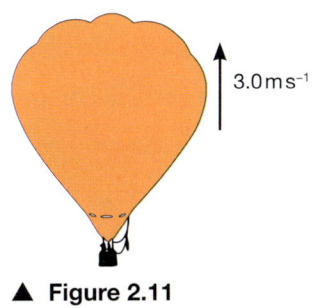

▲ **Figure 2.11**

(a) Calculate the height of the balloon when the sandbag was released.

(b) Draw the graph of velocity against time for the sandbag, from the moment the sandbag is released until it hits the ground. Ignore air resistance.

Answer

(a) Using $s = ut + \frac{1}{2}at^2$ from the moment the sandbag is released until it hits the ground:

$$s = 3.0 \times 5.0 + \frac{1}{2} \times (-9.81) \times 5.0^2$$

$$= -108 \text{ m}$$

> 💡 **Remember**
>
> s is negative as the displacement is downwards.

(b) See Figure 2.12.

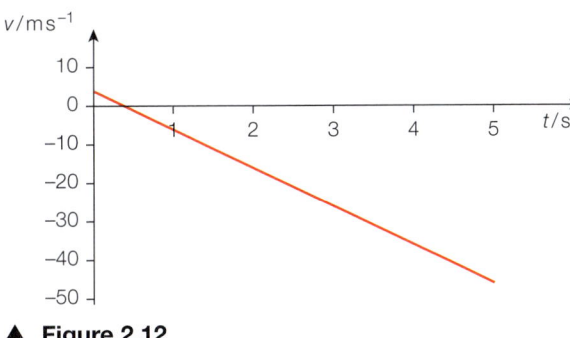

▲ **Figure 2.12**

Projectiles

The movement of any object travelling through the air (see Figure 2.13) can be described in terms of its horizontal and vertical components. If air resistance can be ignored, the components are:

- constant acceleration g vertically downwards

- constant velocity horizontally (as no forces act horizontally).

The equations of motion can be applied separately in the horizontal and vertical directions to solve problems.

If the initial velocity is V, making an angle θ to the horizontal (see Figure 2.14), then:

- the horizontal velocity is $V \cos \theta$

- the initial vertical velocity is $V \sin \theta$.

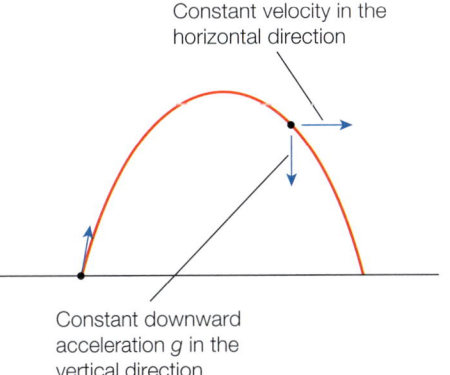

▲ **Figure 2.13** Projectile motion

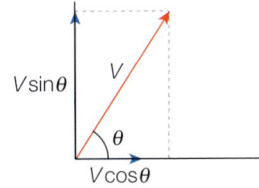

▲ **Figure 2.14** Horizontal and vertical components of velocity

> 📐 **Maths Skills**
>
> For more on resolving vectors in two directions, see Appendix: Maths skills.

Figure 2.15 summarises some facts about the projectile's motion.

- At its maximum height H:
  - the vertical component of velocity is zero
  - the horizontal component of the velocity stays constant at $V\cos\theta$.
- When the projectile has travelled its full range R, the vertical displacement is zero (the net distance travelled vertically is 0).

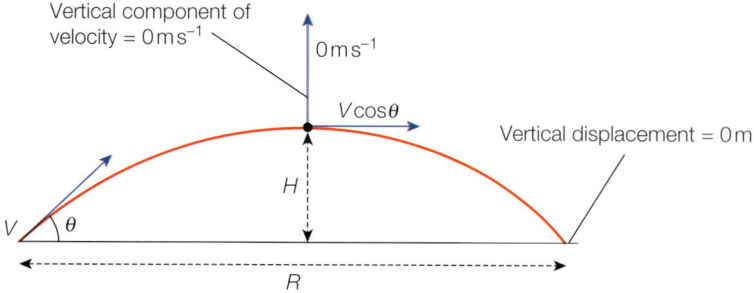

▲ **Figure 2.15** Projectile motion

> 💡 **Remember**
>
> Applying the equations of motion to projectiles, it can be shown that:
> $$H = \frac{V^2\sin^2 q}{2g}$$
> and
> $$R = \frac{V^2\sin 2q}{g}$$

Worked Example

A cannon fires a cannonball with an initial velocity of $12\,\text{m s}^{-1}$ at an angle of 50° to the horizontal.

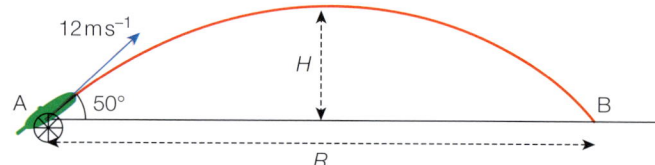

▲ **Figure 2.16**

Show that:

(a) the initial horizontal component of the cannonball's velocity is $7.7\,\text{m s}^{-1}$

(b) the maximum height H reached is $4.3\,\text{m}$.

> ⭐ **Exam tip**
>
> 'Show' means you are expected to derive the answer given, showing all your working.

Answer

(a) Initial horizontal velocity $= 12\cos 50°$

$$= 7.7\,\text{m s}^{-1}$$

(b) Using $v^2 = u^2 + 2as$ applied vertically from the starting point A to the highest point:

$$0^2 = (12\sin 50)^2 + 2 \times (-9.81) \times H$$

$$H = \frac{(12\sin 50)^2}{2 \times 9.8} = 4.3\,\text{m}$$

> 💡 **Remember**
>
> The vertical component of the velocity at the highest point is zero.

↑ Raise your grade

The graph shows how the velocity of a drag-racing car changes with time. The graph can be divided into three separate stages, as shown below.

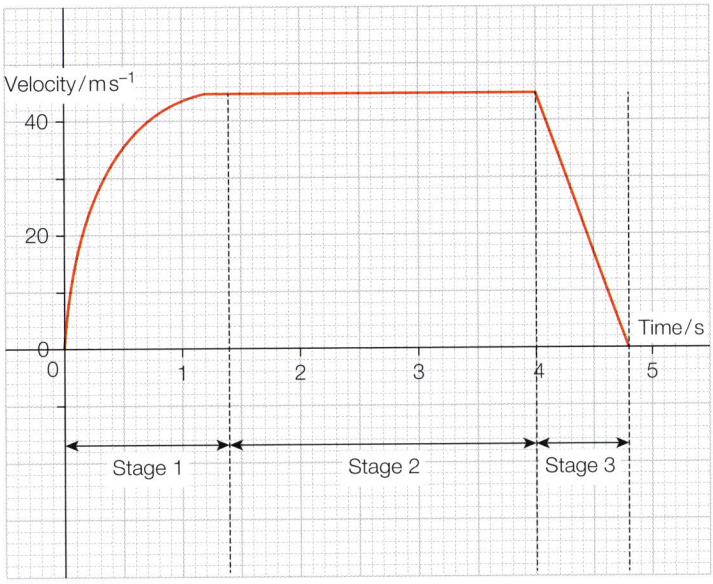

(a) Describe the motion of the car during the three separate stages. [3]

> The car accelerates at first, then travels at constant speed. After 4 seconds it starts to slow down

✔ ✗ ✗

> The statement is correct, but lacks sufficient detail. A better answer would be: *In stage 1 the car accelerates from rest. As the velocity increases, the acceleration decreases (the gradient of the graph decreases). In stage 2 the velocity is constant at 45 m s⁻¹. After 4 s, at the beginning of stage 3, the car decelerates quickly and at a uniform rate, coming to rest after 4.8 s.*

(b) Use the graph to determine the acceleration of the car.

(i) after 0.5 s ✗ ✗ [2]

$$\text{Acceleration} = \frac{velocity}{time} = \frac{35}{0.5} = 70$$

> Acceleration is **change** in velocity/time. A tangent should be drawn at $t = 0.5$ s and the gradient of the tangent calculated.
>
> A good answer would be:
>
> acceleration = gradient of graph at $t = 0.5$ s
> $$= \frac{55 - 22}{1.2} = 27.5 \text{ m s}^{-2}$$

acceleration = $\underline{70}$ m s⁻²

(ii) after 4.5 s. [2]

$$\text{acceleration} = \frac{45}{5.0 - 4.0} = 45 \quad ✗ ✗$$

> The candidate has misread the second time value (it should be 4.8 s not 5.0 s). The answer should also be negative (the car is decelerating).
> A good answer would be $a = \frac{45 - 0}{4.0 - 4.8} = -56.3 \text{ m s}^{-2}$

acceleration = $\underline{45}$ m s⁻²

(c) Estimate the total distance travelled by the car. [2]

✔   ✗

$$\text{Total distance} = \text{area under graph} = 45 \times 4 + \tfrac{1}{2} \times 0.8 = 198$$

> The correct method has been used to find the distance travelled for one mark, but the calculation hasn't taken into account the area under the curve of the graph between $t = 0$ s and $t = 1.4$ s.
>
> A better answer: distance = $51 \times 5.0 \times 0.20 + 45 \times 2.6 + \tfrac{1}{2} \times 45 \times 0.80 = 186$ m
>
> 51 small squares under the curved part of the graph

distance = $\underline{198}$ m

Exam-style questions

1   A cyclist travels from one town to the next at an average speed of 40 km h⁻¹. She completes the return journey at an average speed of 20 km h⁻¹.

   What was her average speed for the whole journey?

   A 25 km h⁻¹   **B** 27 km h⁻¹

   C 30 km h⁻¹   **D** 33 km h⁻¹   [1]

2   The graph shows the distance travelled by a car in the first 20 s of a journey.

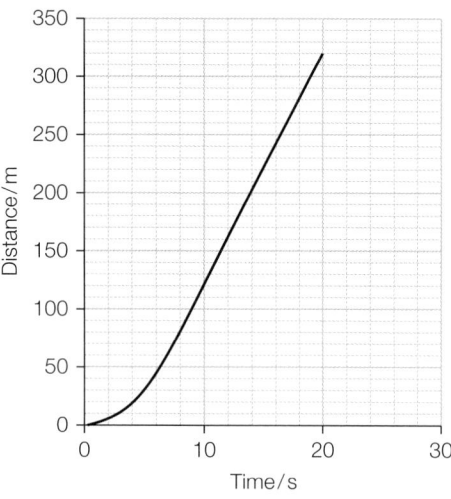

   What is the best estimate of the speed of the car after 10 s?

   A 8 m s⁻¹   **B** 12 m s⁻¹   **C** 16 m s⁻¹   **D** 20 m s⁻¹   [1]

3   A stone is thrown vertically upwards with a speed of 20 m s⁻¹ near the edge of a cliff and falls down to hit the beach below the cliff 6.0 s later.

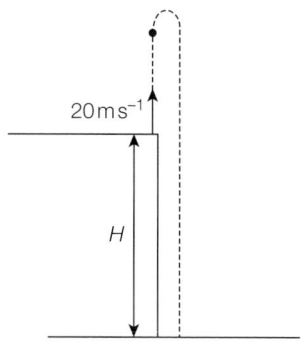

   What is the height *H* of the cliff?

   A 56.4 m   **B** 86.4 m   **C** 176 m   **D** 296 m   [1]

4   **(a)** Explain the difference between a *scalar* quantity and a *vector* quantity.   [1]

   (b) Underline the vector quantities in the list below:
   speed displacement acceleration velocity   [1]

   (c) A tennis player hits a tennis ball horizontally with a speed of 60 m s⁻¹. The ball is initially at a height of 1.40 m above the ground and 11.90 m from the net. Air resistance is negligible.

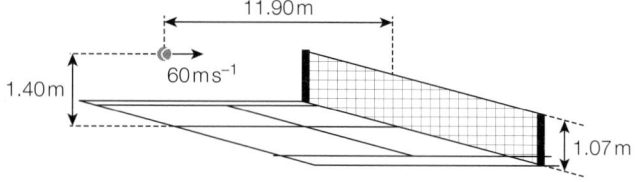

   (i) Calculate the time it takes for the ball to reach the net.

   (ii) Show that the ball passes over the net if the net is 1.07 m high.   [4]

   (d) Calculate the distance the ball is from the net when it lands on the other side of the court.   [2]

   (e) The distance the ball moves is different from its displacement. Explain why.   [1]

5   A basketball player throws a basketball into the hoop of the basket.

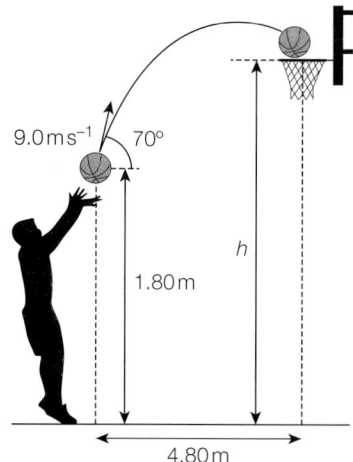

   (a) Calculate:

   (i)   the initial horizontal component of the velocity of the basketball

   (ii) the time it takes for the ball to reach the basketball

   (iii) the height *h* of the basket above the ground.   [5]

   (b) Determine the velocity of the ball as it reaches the basket.   [3]

Knowledge check

You should be able to:

- explain how forces affect the motion of an object by accelerating it

- describe how objects interact with each other through pairs of forces

- calculate the momentum of objects using $p = mv$.

Dynamics is concerned with the affect forces have on the movement of masses as described by Newton's three laws of motion.

Momentum and Newton's laws of motion

Mass and inertia

When a car brakes suddenly we feel we are 'thrown forward' – in reality we are just trying to carry on moving in a straight line at constant speed. The resistance of an object to change its motion (to speed up, slow down or change direction) is called its **inertia**. The mass of an object is an indication of its inertia and is measured in kilograms (kg).

The greater the mass of an object, the greater the resistance to change (see Figure 3.1).

Newton's first law of motion

Newton's first law of motion states that a **resultant force** is needed to accelerate or decelerate an object.

> **Key term**
>
> **Newton's first law:** An object will remain stationary, or continue at constant speed in a straight line, unless acted on by an external resultant force.

The first law appears to contradict our daily experience of forces and movement – to make something move at constant speed it has to be continually pushed. But this ignores the friction forces that oppose the motion. A parachutist falling at constant speed has balanced forces of weight acting downwards and air resistance acting upwards – the resultant force on the parachutist is zero (see Figure 3.2).

In Figure 3.3, the driving force D from the engine is initially greater than the air resistance and friction forces F, and so the car accelerates. At a certain speed the friction forces will equal the driving force and the car will travel at constant speed. If the accelerator is released and the driving force decreases, the friction force is bigger then the driving force so the car decelerates.

a Large mass – difficult to start moving

b Large mass – difficult to stop moving

▲ **Figure 3.1** Large masses – difficult to start and stop

Air resistance force F

Weight force W

▲ **Figure 3.2** The forces are balanced ($F = W$) so the parachutist falls at a constant speed

> **Key terms**
>
> **Dynamics:** The effect forces have on the movement of masses as described by Newton's three laws of motion.
>
> **Inertia:** The resistance of an object to change its motion.
>
> **Resultant force:** When two or more forces act on an object, the resultant force can be found from the vector sum of the individual forces.

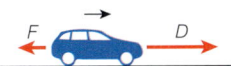  

a $D > F$: accelerates as forces are unbalanced

b $D = F$: moves at a constant speed as forces are balanced

c $D < F$: decelerates as forces are unbalanced

▲ **Figure 3.3** Balanced and unbalanced forces

Momentum

The **momentum** p of an object is its mass m multiplied by its velocity v.

$$p = mv$$

Newton's second law of motion

If a resultant force acts on an object, it speeds up, slows down or changes direction; that is, its momentum changes. Newton's second law links the size of the force applied to the change in momentum.

> **Key term**
>
> **Newton's second law:** The rate of change of momentum of an object is proportional to the resultant force on it.

Consider an object of mass m moving with speed u acted on by a constant force F for a time t (Figure 3.4):

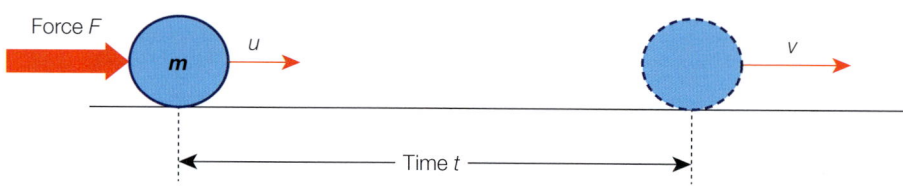

▲ **Figure 3.4** Force = rate of change of momentum

Using Newton's second law:

$$F \propto \frac{\text{change in momentum}}{\text{time taken}} = \frac{mv - mu}{t} = \frac{m(v - u)}{t} = ma$$

$$F \propto ma$$

where a is the acceleration of the object. By defining the unit of force (the newton) as that force which gives a mass of 1 kg an acceleration of $1\,\text{m s}^{-2}$, we can write:

$$F = ma$$

Mass and weight

When an object of mass m is held above the Earth's surface and then released, it accelerates downwards at $9.81\,\text{m s}^{-2}$ (ignoring air resistance) because there is an unbalanced force acting on it (its weight W). Using $F = ma$:

$$W = mg$$

where g is the acceleration of free fall.

Newton's third law of motion

When you push a door, the door pushes you – forces always act in pairs. Newton's third law is often quoted as 'action = reaction', but it can be stated more formally.

Action and reaction forces are often misunderstood. The key points to remember are:

- Forces do not exist individually, but in pairs.

- The forces are of the same type; for example, both gravitational.

- The two forces act on different objects.

- The third law applies to every situation.

> **Remember**
>
> Momentum is a vector. It has units of kg m s^{-1}.

> **Key term**
>
> **Momentum:** defined as mass × velocity, with units of kg m s^{-1}.

> **Remember**
>
> Force = rate of change of momentum
>
> $F = ma$

> **Key term**
>
> **Newton's third law:** When two objects interact, they exert equal and opposite forces on each other.

> **Remember**
>
> Newton's third law refers to equal and opposite forces acting on two different objects.

Conservation of linear momentum

Newton's second law can be written:

$$F = \frac{\Delta(mv)}{\Delta t}$$

so

$$F\Delta t = \Delta(mv)$$

When two objects collide, the same force F acts for the same time Δt on both objects, and so the magnitude of the change in momentum will be the same for both objects.

In Figure 3.7, the momentum of one of the masses will decrease, but the momentum of the other mass increases by an equal amount, and so the **total** momentum of the two masses is the same as before the collision. There is no change in the total overall momentum. This is an example of the principle of conservation of momentum.

Each ball is in contact
with the other for time Δt

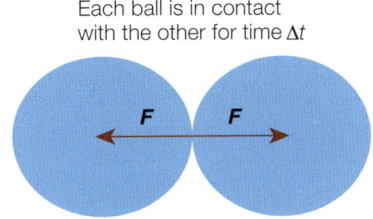

▲ **Figure 3.7** Impulse $= F\Delta t$

Key term

Principle of conservation of momentum: For a system of interacting objects, the total momentum remains constant provided no external resultant force acts on the system.

This is why momentum is such a useful quantity to calculate. Although the momentum of individual objects changes, the total momentum in any interaction (e.g., a collision or explosion) remains constant provided no external force acts.

Worked Example

A ^{235}U nucleus has a mass of 3.9×10^{-25} kg. It decays by emitting an alpha particle, of mass 6.6×10^{-27} kg, with a speed of $1.6 \times 10^7\,\mathrm{m\,s^{-1}}$. What is the recoil velocity of the nucleus?

Answer

The total momentum before and after the alpha particle is emitted is zero. Let the velocity of the nucleus after the alpha particle is emitted be v.

$$\text{momentum of decayed nucleus} + \text{momentum of alpha particle} = 0$$

$$(3.9 \times 10^{-25} - 6.6 \times 10^{-27})v + 6.6 \times 10^{-27} \times 1.6 \times 10^7 = 0$$

so

$$v = \frac{-6.6 \times 10^{-27} \times 1.6 \times 10^7}{(3.9 \times 10^{-25} - 6.6 \times 10^{-27})} = -2.8 \times 10^5\,\mathrm{m\,s^{-1}}$$

Collisions

Elastic collisions

When objects collide, the total momentum of the objects remains constant. If kinetic energy is also conserved in a collision, it is called an **elastic collision**. For elastic collisions, the velocity at which the objects separate is always the same as the velocity of approach. The collisions of molecules in an ideal gas are considered elastic, and the collisions of snooker balls are almost elastic.

Inelastic collisions

Most collisions are **inelastic collisions** – some of the initial kinetic energy is 'lost' (usually transferred as heat or sound energy). A car colliding with a tree, or the collisions of electrons in a gas in which the atoms of the gas become excited, are both examples of inelastic collisions. The principle of conservation of momentum still applies to inelastic collisions.

Key terms

Elastic collision: when kinetic energy is conserved in a collision.

Inelastic collision: when some of the initial kinetic energy is lost, e.g., as heat or sound.

Worked Example

A dynamics trolley of mass 0.90 kg, travelling at a speed of 2.50 m s^{-1}, collides head-on with another trolley of mass 1.80 kg, travelling in the opposite direction with a speed of 1.40 m s^{-1}, as shown in Figure 3.8. The two trolleys stick together after the collision.

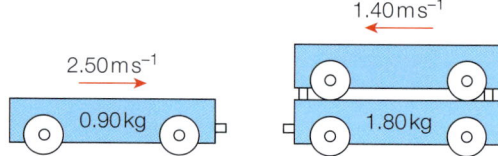

▲ **Figure 3.8** Inelastic collisions

(a) Calculate their common speed, v, after the collision.

(b) Determine in which direction they both move after the collision.

(c) Calculate the total kinetic energy E_k:

    (i) before the collision   (ii) after the collision.

(d) Why are your answers to (c) (i) and (c) (ii) not the same?

Answer

(a) Total momentum before collision = total momentum after collision:

$$0.90 \times 2.50 - 1.80 \times 1.40 = (1.80 + 0.90) \times v$$

$$v = \frac{(0.90 \times 2.50 - 1.80 \times 1.40)}{(1.80 + 0.90)} = -0.10 \text{ m s}^{-1}$$

(b) The final velocity is negative; the trolleys move in the same direction as the 1.80 kg trolley is moving before the collision.

(c) (i) E_k before collision = ½ × 0.90 × 2.50^2 + ½ × 1.80 × (−1.40)^2

              = 4.58 J

   (ii) E_k after collision = ½ × 2.7 × (−0.1)^2 = 0.014 J

(d) Kinetic energy is lost (as heat and sound). As the trolleys 'stick' there is a friction force acting on each trolley accelerating/decelerating it (the friction forces are doing work).

Collisions in two dimensions

The principle of conservation of momentum can be applied to collisions in two dimensions. The total momentum in any direction must remain constant.

In Figure 3.9, a mass M is travelling at speed u towards a mass m which is at rest. When they collide, the line joining their centres makes an angle θ with the original direction of M.

Applying the principle of conservation of momentum:

In the x direction: $Mu = Mv_1 \cos\phi + mv_2 \cos\theta$

In the y direction: $Mv_1 \sin\phi = mv_2 \sin\theta$

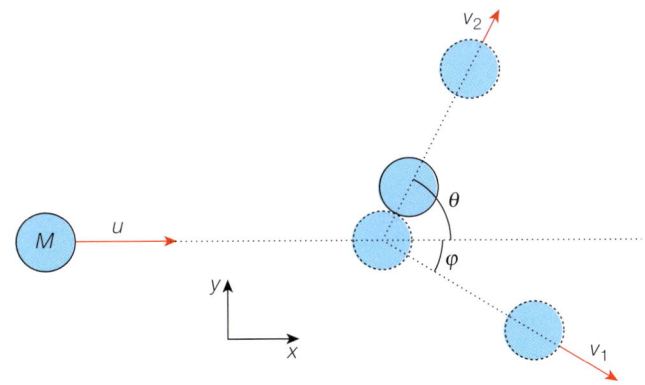

▲ **Figure 3.9** Conservation of momentum in two dimensions

Worked Example

A snooker ball, travelling at a speed of $2.0\,\text{m s}^{-1}$, has an off-centre collision with an identical ball which is at rest, as shown in Figure 3.10. After the collision the first ball moves at an angle of $60°$ to its original direction at a speed of $0.8\,\text{m s}^{-1}$.

(a) Calculate the speed v and direction θ of the second ball after the collision.

(b) State whether the collision is elastic or inelastic. Justify your answer.

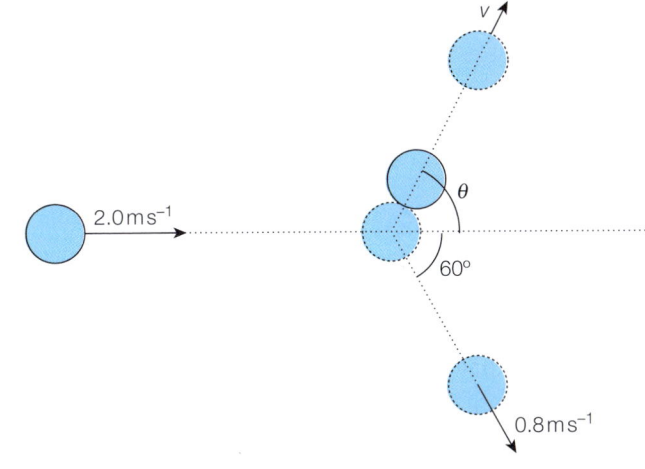

▲ **Figure 3.10** Conservation of momentum in two dimensions

Answer

(a) Let the mass of each ball be m. Using the principle of conservation of momentum:

In the initial direction of travel of the first ball:

$$m \times 2.0 = (m \times 0.8 \cos 60) + (mv \cos\theta)$$

$$v \cos\theta = 1.6 \qquad \text{(equation 1)}$$

At $90°$ to the initial direction of the first ball:

$$0 = mv \sin\theta - m \times 0.8 \times \sin 60$$

$$v \sin\theta = 0.693 \qquad \text{(equation 2)}$$

Equation 2 divided by equation 1:

$$\frac{\sin\theta}{\cos\theta} = \frac{0.693}{1.6} = 0.4333$$

so $\theta = \tan^{-1}(0.433) = 23.4°$

Substituting this value into equation 1:

$$v = \frac{1.6}{\cos 23.4°} = 1.74\,\text{m s}^{-1}$$

(b) Change in $E_k = \tfrac{1}{2} m \times 2.0^2 - \tfrac{1}{2} m \times 0.8^2 - \tfrac{1}{2} m \times 1.74^2 = 0.166\,m$

The collision is inelastic as some kinetic energy is lost.

> **Maths Skills**
>
> $\dfrac{\sin\theta}{\cos\theta} = \tan\theta$

> ★ **Exam tip**
>
> 'Justify' means 'give some evidence for' – in this case some calculation to show whether the collision is elastic or not.

Non-uniform motion

Free fall and air resistance

Figure 3.11 shows different stages after a skydiver jumps from an aeroplane, and Figure 3.12 summarises her motion in a graph.

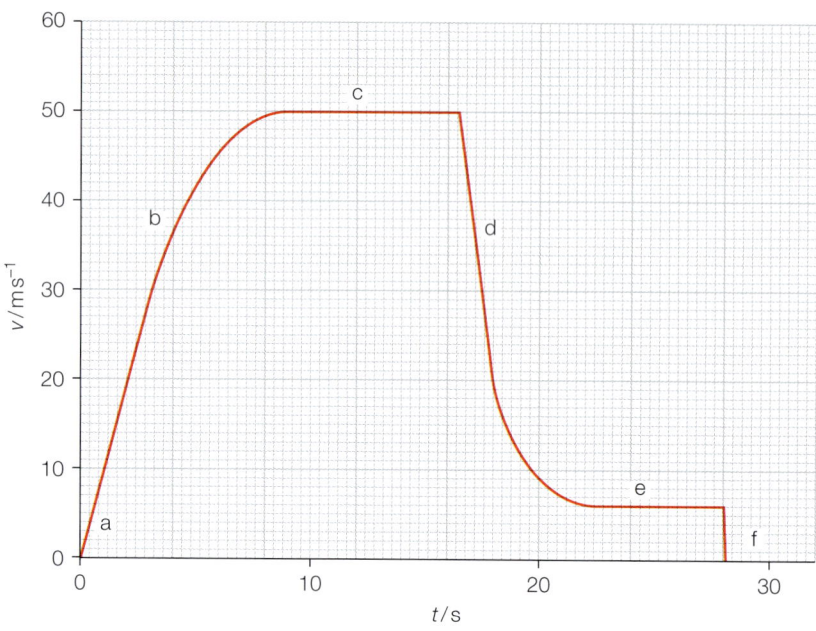

▲ **Figure 3.12** Velocity–time graph for a skydiver

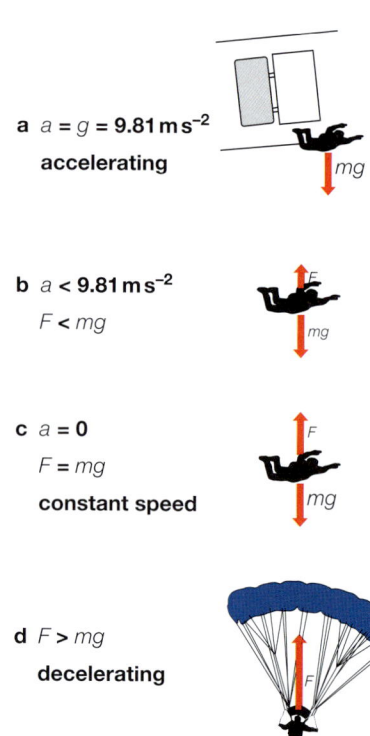

a $a = g = 9.81\,\text{m s}^{-2}$
accelerating

b $a < 9.81\,\text{m s}^{-2}$
   $F < mg$

c $a = 0$
   $F = mg$
   constant speed

d $F > mg$
   decelerating

e $F = mg$
   constant speed

f

▲ **Figure 3.11** Forces on a skydiver

When she first leaves the aeroplane (Figure 3.11a) there is only one force acting on her – her weight mg, and so she accelerates towards the ground at $9.81\,\text{m s}^{-2}$.

As her speed increases, she is hitting more air molecules every second, and so the upward air resistance force F increases (Figure 3.11b). The net, or resultant, force is still downward, and so she continues to accelerate downwards, but at a slower rate.

Eventually the skydiver is falling at such a high speed (about $50\,\text{m s}^{-1}$) that her weight mg and the air resistance force F exactly balance each other (Figure 3.11c), and so there is no resultant force on her. She continues to fall at this speed, known as the **terminal velocity**. Terminal velocity is constant and is the maximum velocity the skydiver will fall at.

When the skydiver opens her parachute, the air resistance force is suddenly greatly increased (Figure 3.11d). The upwards force is now greater than the weight force, and so she begins to decelerate.

After a short time she will reach a much slower speed (about $6\,\text{m s}^{-1}$), where the air resistance force and the weight force balance again (Figure 3.11e), and she will continue to fall at this slower speed until she hits the ground (Figure 3.11f).

Key term

Terminal velocity: When the weight of an object and the air resistance force on an object balance, so there is no resultant force.

⬆ Raise your grade

(a) State the principle of conservation of momentum. [2]

The total momentum of a system remains ✔ ✘
constant.

> The statement is correct but insufficient – it is only true if there is **no resultant force** acting on the system.

(b) Explain what is meant by an *elastic* collision. [1]

Energy is conserved.

> ✘ An elastic collision is one in which the total **kinetic** energy remains constant.

(c) A ball of mass m is travelling with speed u in a straight line when it collides elastically with a stationary ball of mass $2m$, as shown below. After the collision, the smaller mass is moving with a velocity v_1 and the larger mass with a velocity v_2. [4]

Before the collision   After the collision

(i) Show that $v_2 = \dfrac{2u}{3}$.

Conservation of momentum: $mu = mv_1 + 2mv_2 \rightarrow u = v_1 + 2v_2$ (1) ✔

Elastic collision: velocity of separation = velocity of approach

$$v_2 - v_1 = u \qquad\qquad (2) ✔$$

Adding equations (1) and (2): $\quad 3v_2 = 2u \qquad \rightarrow \qquad v_2 = \dfrac{2u}{3}$ ✔

(ii) Calculate v_1.

From equation (1): $\quad v_1 = u - 2v_2 = u - 2 \times \dfrac{2u}{3} = -\dfrac{u}{3}$ ✔

(d) The ball of mass $2m$ then collides with a wall and bounces back with a speed of $\dfrac{u}{3}$. The collision with the wall lasts t seconds. [2]

Calculate:

(i) the change in momentum of the ball

Change in momentum $= 2m \times \dfrac{2u}{3} - 2m \times \dfrac{u}{3} = \dfrac{2mu}{3}$

> ✘ Momentum is a vector – the 'rebound' momentum is negative. The change in momentum is:
> $$2m \times \dfrac{2u}{3} - 2m \times \left(-\dfrac{u}{3}\right) = \dfrac{6mu}{3} = 2mu$$

(ii) the average force acting on the wall during the collision.

$$F = \dfrac{\Delta(mv)}{\Delta t} = \dfrac{\frac{2mu}{3}}{t} = \dfrac{2mu}{3t}$$

> ✔ Method is correct (error carried forward)
> (correct answer is $\dfrac{2mu}{t}$)

(e) The momentum of the ball has changed. Explain how the principle of conservation of momentum still applies. [2]

The wall (and the rest of the Earth it is connected to) gains momentum in the opposite
direction – the total momentum of the ball and wall 'system' remains the same. ✔ ✔

> A good answer.

Exam-style questions

1   An electron in an electron 'gun' is accelerated from rest to a speed of $4.2 \times 10^7\,\text{m s}^{-1}$ over a distance of 15 mm. What is the force on the electron?

   $[m_e = 9.11 \times 10^{-31}\,\text{kg}]$

   A $5.4 \times 10^{-17}\,\text{N}$

   B $1.1 \times 10^{-16}\,\text{N}$

   C $5.4 \times 10^{-14}\,\text{N}$

   D $1.1 \times 10^{-13}\,\text{N}$ [1]

2   A lift of mass $1.4 \times 10^3\,\text{kg}$ is ascending with an acceleration of $1.6\,\text{m s}^{-2}$. What is the tension in the cables supporting the lift?

   A 2.2 kN   **B** 11.5 kN

   C 13.7 kN   **D** 16.0 kN   [1]

3   A box of mass 40 kg is pulled by a force of 600 N at an angle of 40° to the horizontal, as shown below. The friction force F is equal to half the weight of the box.

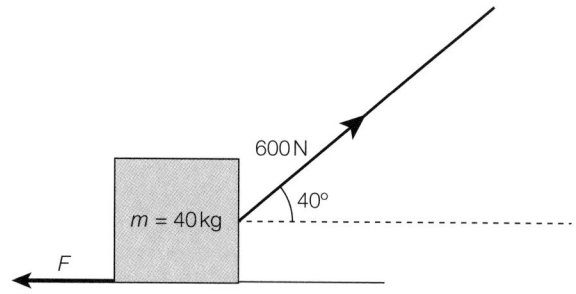

   What is the acceleration of the box?

   A $4.7\,\text{m s}^{-2}$ **B** $6.6\,\text{m s}^{-2}$

   C $9.2\,\text{m s}^{-2}$ **D** $11.0\,\text{m s}^{-2}$ [1]

4   A mass of 1.4 kg moving with a velocity of $0.7\,\text{m s}^{-1}$ collides with a mass of 0.7 kg moving in the opposite direction with a speed of $0.2\,\text{m s}^{-1}$. After the collision, the 1.4 kg mass is moving with a speed of $0.3\,\text{m s}^{-1}$ in the same direction as before.

   What is the speed of the 0.7 kg mass after the collision?

   A $0.6\,\text{m s}^{-1}$ **B** $1.0\,\text{m s}^{-1}$

   C $1.8\,\text{m s}^{-1}$ **D** $2.2\,\text{m s}^{-1}$ [1]

5   Two dynamics trolleys P and Q are initially at rest. The mass of trolley P is 0.90 kg; the mass of the trolley Q is unknown. They are 'exploded apart' by the release of a spring-loaded plunger at the front of one of the trolleys, and move off in opposite directions as shown below.

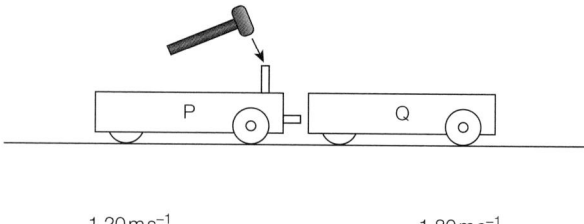

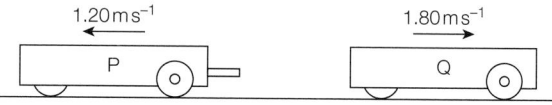

   What is the mass of trolley Q?

   A 0.60 kg   **B** 0.74 kg   **C** 1.35 kg   **D** 1.67 kg   [1]

6   **(a)** State the principle of conservation of momentum.   [2]

   (b) Curling is a sport in which players slide circular 'stones' on a sheet of ice towards a target. The stones each have a mass of 18 kg and a diameter of 19.0 cm.

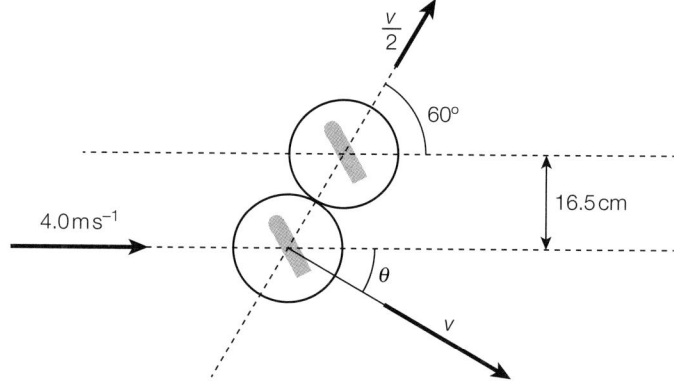

   A stone is travelling at a speed of $4.0\,\text{m s}^{-1}$ when it collides with a second identical but stationary stone, as shown above.

   (i) Calculate the momentum of the first stone before the collision.

   (ii) Show that the stationary stone moves off at an angle of 60° to the original direction of travel of the first stone.   [3]

   (c) The speed v of the first stone after the collision is twice the speed of the second stone.

   Calculate:

   (i) θ, the angle the first stone is deflected

   (ii) v.   [4]

Forces, density, and pressure

4

Types of force

Gravitational forces

Any mass in a gravitational field experiences a force. The **gravitational field strength** g is defined as the force on unit mass (1 kg). The force F on a mass m is:

$$F = mg$$

Near the Earth's surface the gravitational field (see Figure 4.1) is uniform – constant in both **magnitude** and **direction** – and has a value of $9.81\,\text{N kg}^{-1}$.

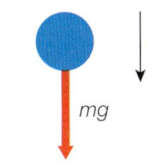

▲ **Figure 4.1** Gravitational field strength

Electrical forces

An electrically charged object in an electric field experiences a force. The **electric field strength** E is defined as the force per unit positive charge (1 C) on a stationary point charge.

The force on a charge q (see Figure 4.2) is:

$$F = Eq$$

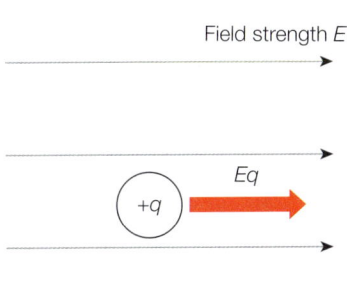

▲ **Figure 4.2** Electric field strength

Upthrust (buoyancy) forces

An object immersed in a fluid (liquid or gas) experiences an upward force, an **upthrust**, equal to the weight of the fluid that has been displaced.

Frictional forces

Friction is a force which always opposes motion. If a book resting on a table is gently pushed one way, there is a friction force in the opposite direction and the book will remain at rest (see Figure 4.3a).

a In equilibrium (at rest)

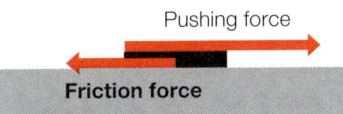

b Beginning to accelerate

▲ **Figure 4.3** Frictional forces on a book

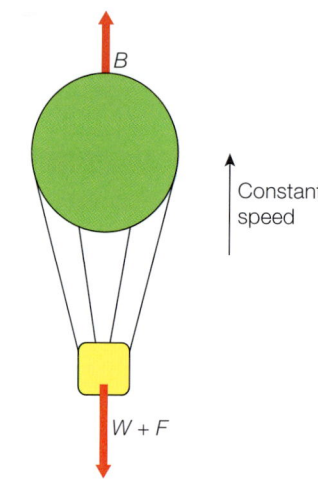

Worked Example

A hot-air balloon ascends at constant speed (Figure 4.4).

(a) State the forces acting on the balloon and their directions.

(b) Derive an equation relating the forces you have identified.

Answer

(a) The weight of the balloon W and the drag or friction force F both act downwards. The upthrust or buoyancy force B acts upwards.

(b) As the balloon is rising at constant speed, the net force on the balloon must be zero.

$$W + F = B$$

▲ **Figure 4.4** Drag and buoyancy

Centre of gravity

The **centre of gravity** of a body (an **object**) is the point where all the weight of the body can be considered to act. Knowledge of the position of the centre of gravity is helpful in assessing the stability of an object.

In Figure 4.5a the object is stable – the weight force is trying to rotate the object anticlockwise, returning it to an upright position.

In Figure 4.5b the object has been tilted further so that the line of action of the weight force is outside the right-hand edge of the object. The weight force is trying to rotate the object clockwise, causing it to topple over.

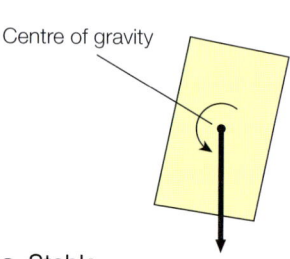

a Stable

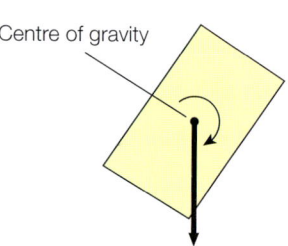

b Unstable

▲ **Figure 4.5** Centre of gravity

The turning effects of forces

The turning effect of a force is called the **moment** of the force.

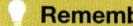 **Remember**

The moment of a force about a point is the force multiplied by the **perpendicular** distance from the line of action of the force to the point.

The SI units of the moment of a force are N m.

For a force F, a perpendicular distance d from a point P, the moment about P is Fd (see Figure 4.6).

The principle of moments

If an object is subjected to a number of forces, but is in equilibrium, the turning effects (moments) of each of the forces must balance out. This statement is known as the **principle of moments**.

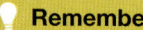 **Remember**

Sum of the clockwise moments = sum of the anticlockwise moments

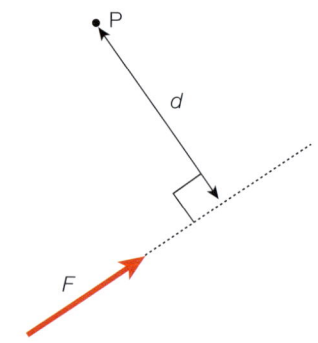

▲ **Figure 4.6** Moment of a force

Key terms

Centre of gravity: the point in a body where all the weight can be considered to act.

Moment: the turning effect of a force.

Principle of moments: if an object is subjected to a number of forces, but is in equilibrium, the turning effects (moment) of each of the forces must balance out.

Worked Example

A horizontal beam is hinged at one end and supports a light fitting. The beam is held in place by a rope, as shown in Figure 4.7. Calculate the tension T in the rope.

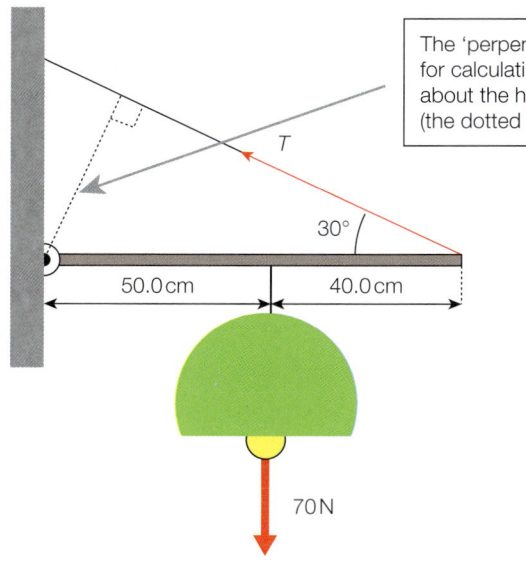

The 'perpendicular distance' for calculating the moment of T about the hinge is $90.0 \times \sin 30°$ (the dotted line).

▲ **Figure 4.7**

Answer

Using the principle of moments about the hinge:

sum of the anticlockwise moments = sum of the clockwise moments

$$T \times (90.0 \times \sin 30°) = 70 \times 50.0$$

$$T = 77.8\,\text{N}$$

Torque and couple

A **couple** is a pair of equal and opposite forces acting on a body, but not along the same line. A couple can only cause a body to rotate.

The **torque** of a couple is the total moment of the couple, and so has the same units as moments (N m).

The torque about point P is $F \times d$ (see Figure 4.8). (The torque is the same value, regardless of the position of P).

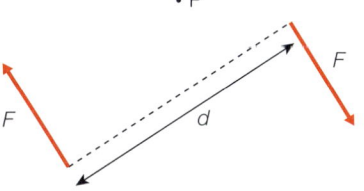

▲ **Figure 4.8** Torque of a couple

Equilibrium of forces

For a body to be in equilibrium, two conditions must be satisfied:

- The resultant force acting in any direction must be zero.
- The resultant torque about any point must be zero.

Remember

Torque of couple = Fd

Key terms

Couple: a pair of equal and opposite forces acting on a body, but not along the same line.

Torque: the total moment of the couple.

Worked Example

A uniform ladder AB, of length 6.0 m and weight 100 N, rests against a smooth wall. The base of the ladder rests on the floor and is 2.0 m from the wall, as shown in Figure 4.9.

Calculate:

(a) the angle the ladder makes with the wall

(c) the force on the ladder from the wall

(c) the size and direction of the force on the ladder from the floor.

Answer

The ladder is uniform, so the centre of gravity of the ladder is halfway along the ladder.

(a) $\sin\theta = \dfrac{2.0}{6.0} = \dfrac{1}{3}$ so $\theta = \sin^{-1}\left(\dfrac{1}{3}\right) = 19.5°$

(b) Taking moments about point B:

$$S \times 6\cos\theta = 100 \times 3\sin\theta$$

$$S = 50\tan\theta = 50\tan 19.5 = 17.7\,\text{N}$$

(c) Resolving vertically: $P = 100\,\text{N}$

Resolving horizontally: $F = S = 17.7\,\text{N}$

The force at B has two components (see Figure 4.10):

The resultant force $R = \sqrt{(100^2 + 17.7^2)}$

$$= 101.6\,\text{N}$$

angle $\phi = \tan^{-1}\left(\dfrac{17.7}{100}\right) = 10.0°$

★ **Exam tip**

One way of approaching many problems on equilibrium is to resolve forces in two perpendicular directions and use the principle of moments about a suitable point: 'resolve, resolve and take moments.'

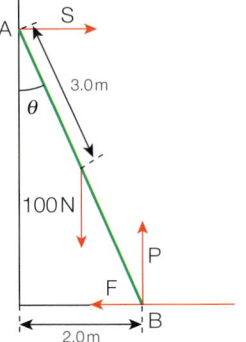

▲ **Figure 4.9**

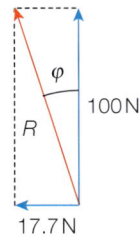

▲ **Figure 4.10**

Equilibrium under three forces

If an object acted on by three forces is in equilibrium, as shown in Figure 4.11a, the resultant of the three forces must be zero. The three forces, drawn as vectors, must form a triangle (see Figure 4.11b).

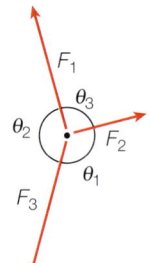

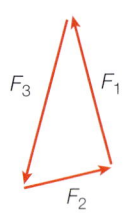

a Free-body diagram   **b** Vector triangle of forces

▲ **Figure 4.11** Equilibrium under three forces

'Three force' questions can be approached in two ways:

* by resolving forces in two directions and taking moments about a suitable point

* drawing a scale diagram of the triangle of forces.

Worked Example

A block of weight W rests on a rough slope making an angle θ with the horizontal, as shown in Figure 4.12. Calculate:

(a) the friction force F **(b)** the normal force N.

Method 1: Resolving forces

(a) Resolving forces along the slope: $F = W \sin\theta$

Resolving forces perpendicular to the slope: $N = W \cos\theta$

Method 2: Drawing a scale diagram of the triangle of forces

See Figure 4.13.

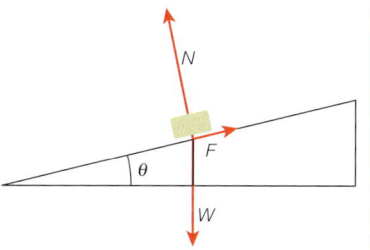

▲ **Figure 4.12** Equilibrium under three forces

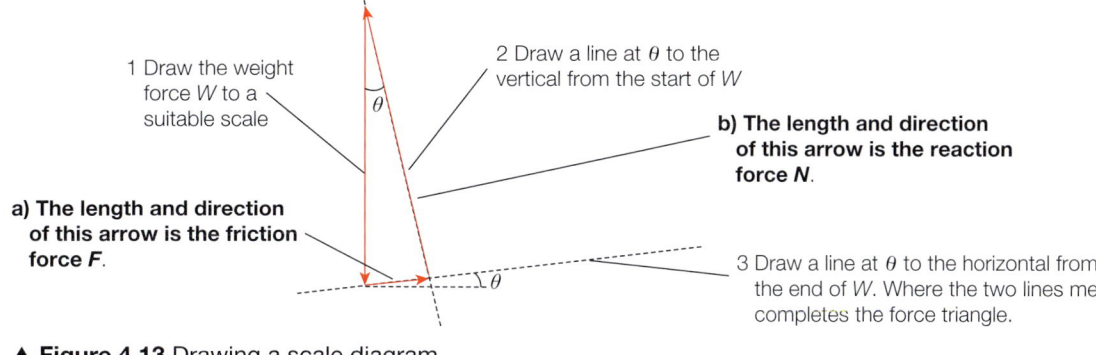

1 Draw the weight force W to a suitable scale

2 Draw a line at θ to the vertical from the start of W

b) The length and direction of this arrow is the reaction force **N**.

a) The length and direction of this arrow is the friction force **F**.

3 Draw a line at θ to the horizontal from the end of W. Where the two lines meet completes the force triangle.

▲ **Figure 4.13** Drawing a scale diagram

Density and pressure

Density

The **density** ρ of a material is defined as the mass per unit volume. For an object of mass m and volume V: $\rho = \dfrac{m}{V}$

Worked Example

The density of air is $1.29\,\text{kg}\,\text{m}^{-3}$. Estimate the mass of air in a school hall.

Answer

Estimated volume of school hall $= 35\,\text{m} \times 20\,\text{m} \times 8\,\text{m} = 5600\,\text{m}^3$

Mass m of air $= \rho V = 1.29 \times 5600 = 7200\,\text{kg}$

Pressure

The **pressure** p is defined as the normal force per unit area (see Figure 4.14):

$$p = \frac{F}{A}$$

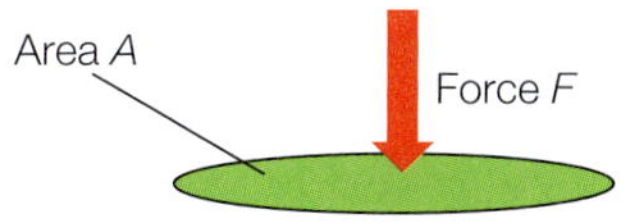

Area A Force F

▲ **Figure 4.14** Pressure

> **Remember**
>
> $$\text{density} = \frac{\text{mass}}{\text{volume}}$$
>
> SI units are $\text{kg}\,\text{m}^{-3}$
>
> Some densities that are useful to know:
>
> air: $1.29\,\text{kg}\,\text{m}^{-3}$
>
> water: $1.0 \times 10^3\,\text{kg}\,\text{m}^{-3}$

> **Remember**
>
> $$p = \frac{F}{A}$$
>
> SI units for pressure are $\text{N}\,\text{m}^{-2}$ or **pascal** (Pa)

> **Key terms**
>
> **Density:** the mass per unit volume.
>
> **Pressure:** the force per unit area.

Worked Example

A bar of gold has dimensions $16.0\,\text{cm} \times 5.0\,\text{cm} \times 2.5\,\text{cm}$ and a mass of $3.86\,\text{kg}$.

(a) What is the density of gold?

(b) What is the **maximum** pressure the bar can exert when placed on a table?

Answer

(a) $\rho = \dfrac{m}{V} = \dfrac{3.86}{0.160 \times 0.050 \times 0.025} = 19.3 \times 10^3\,\text{kg}\,\text{m}^{-3}$

(b) See Figure 4.15 for the orientation for maximum pressure.

$$p_{\text{max}} = \frac{F}{A} = \frac{3.86 \times 9.81}{0.050 \times 0.025} = 30.3\,\text{kPa}$$

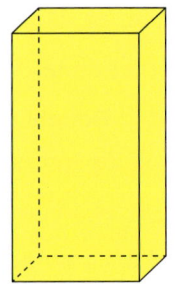

▲ **Figure 4.15** Orientation for maximum pressure

Pressure in liquids and gases

The pressure on the surface of a column of liquid is the atmospheric pressure p_0. Below the surface, the pressure increases with depth due to the weight of liquid above, as shown in Figure 4.16. For a column of liquid of depth Δh, cross-sectional area A and density ρ, the increase in pressure is Δp. For equilibrium:

$$\Delta p A = \rho (A\,\Delta h)\,g$$

$$\Delta p = \rho g \Delta h$$

The net upwards force due to the increased pressure at depth Δh is the buoyancy force discussed earlier.

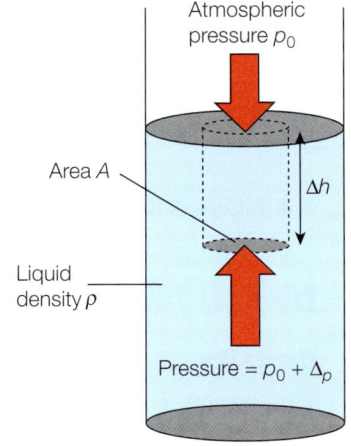

▲ **Figure 4.16** Pressure in liquids

Worked Examples

1   Submersibles used for exploring the deepest oceans can dive to a depth of $6000\,\text{m}$. What is the **total** pressure on the outside of them?

[density of water $= 1.00 \times 10^3\,\text{kg}\,\text{m}^{-3}$, atmospheric pressure $= 1.0 \times 10^5\,\text{Pa}$]

Answer

Hydrostatic pressure p due to the weight of water:

$$p = \rho g \Delta h = 1.00 \times 10^3 \times 9.81 \times 6000 = 5.89 \times 10^7\,\text{Pa}$$

total pressure $=$ hydrostatic pressure $+$ atmospheric pressure

$$= 5.89 \times 10^7 + 1.0 \times 10^5 = 5.90 \times 10^7\,\text{Pa}$$

[Note: the atmospheric pressure is very small compared to the hydrostatic pressure, and can be ignored for most calculations.]

2   The atmospheric pressure at a height of $10\,\text{km}$ is $2.8 \times 10^4\,\text{Pa}$.

  (a) A window of an aeroplane has an area $900\,\text{cm}^2$. Calculate the force on the window when the aeroplane is flying at this height, assuming that the air pressure inside the aeroplane is $1.0 \times 10^5\,\text{Pa}$.

  (b) In which direction is this force?

Answer

(a) Net force exerted on window $= \Delta p \times A$

$$= (1.0 \times 10^5 - 2.8 \times 10^4) \times 900 \times 10^{-4}$$

$$= 6.5\,\text{kN}$$

(b) The force acts **outwards** from inside the aeroplane.

> 💡 **Remember**
>
> $$\Delta p = \rho g \Delta h$$
>
> You need to be able to derive this equation.

↑ Raise your grade

(a) State the two conditions necessary for a body to be in equilibrium. [2]

The overall force acting on the body in any direction must be zero. ✔ ✘

> The candidate has stated the first condition for equilibrium correctly, but should also have stated that the total moment (or torque) acting on the body must be zero.

(b) A uniform beam AB, of length 1.400 m and weight 34 N, is attached by a hinge to a wall at A, as shown in the diagram. The beam is kept horizontal by a wire attached to the beam at C and the wall at D. A lantern of weight 26 N hangs from end B.

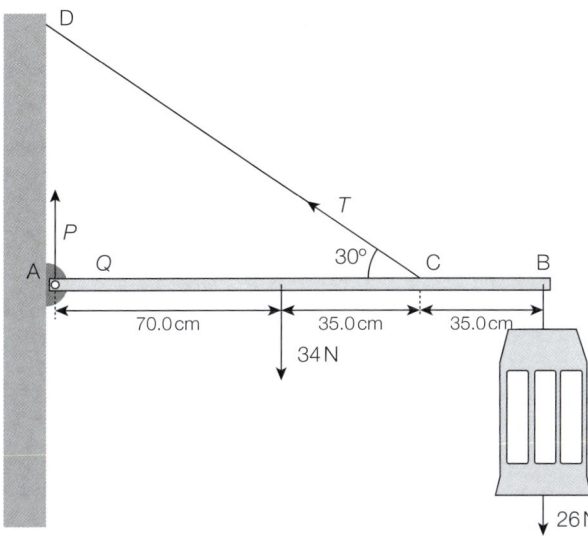

(i) Take moments about A to show that the tension T in the wire is 115 N. [2]

$$T \times 1.05 \sin 30° = 34 \times 0.70 + 26 \times 1.40$$ ✔ ✔

> The candidate has calculated the clockwise and anticlockwise moments correctly, and found the correct value for *T*.

$$T = 115 \text{ N}$$

(ii) Calculate the vertical force *P* exerted by the hinge. [2]

> The method for finding *P* is correct for the first mark, but the calculation of *P* is incorrect – the value of *P* should be 60 – 115 sin 30 = 2.5 N.

Resolving forces vertically: $P + T \sin 30° = 34 + 26$ ✔

$$P = 60 + 115 \sin 30°$$ ✘

$$P = 117.5 \text{ N}$$ vertical force *P* = ...117.5.... N

(iii) Show that the horizontal force *Q* exerted by the hinge is 100 N. [1]

Resolving forces horizontally: $Q = T \cos 30°$ ✔

> A valid method for finding *Q*. An alternative method is to 'take moments' about point D.

$$= 115 \cos 30° = 99.6 \text{ N}$$

(iv) Find the size and direction of the resultant force exerted by the hinge. [2]

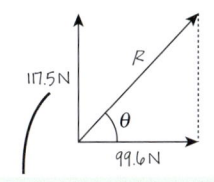

e.c.f. (should be 2.5 N)

$$R^2 = 117.5^2 + 99.6^2$$ ✔

$$R = 154.0 \text{ N}$$

> A valid <u>method</u> for finding *R*, allowing error carried forward (e.c.f.) – the correct value is 99.6 N.

$$\tan \theta = \frac{117.5}{99.6} = 1.18$$ ✔

$$\theta = 49.7°$$

> A valid <u>method</u> for finding θ, again allowing for e.c.f. – the correct value is 1.4°.

Exam-style questions

1 Four forces act on a point.

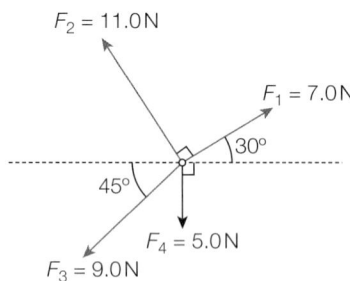

Which statement is **incorrect**?

A The vertical component of F_1 is 3.5 N.

B The horizontal component of F_2 is 9.5 N.

C The vertical component of F_3 is 6.4 N.

D The horizontal component of F_4 is 0.0 N. [1]

2 A load of 400 N is supported by two cables, as shown below. One cable is pulled horizontally with a force of 170 N.

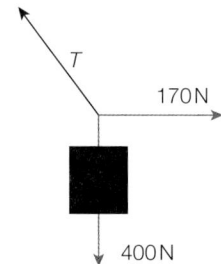

What is the best estimate of the tension in the other cable?

A 165 N   **B** 230 N   **C** 400 N   **D** 435 N   [1]

3 A book is held upright by gripping it in the corner, between thumb and forefinger, as shown below.

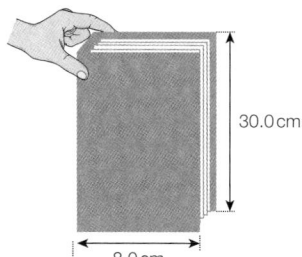

The book weighs 12.0 N. What is the torque applied by the hand to the book?

A 48.0 N cm clockwise

B 48.0 N cm anticlockwise

C 180.0 N cm clockwise

D 180.0 N cm anticlockwise [1]

4 A measuring cylinder contains oil to a depth of 6 cm floating on water. The depth of the water is 9 cm. The density of the oil is 900 kg m⁻³ and the density of water is 1000 kg m⁻³.

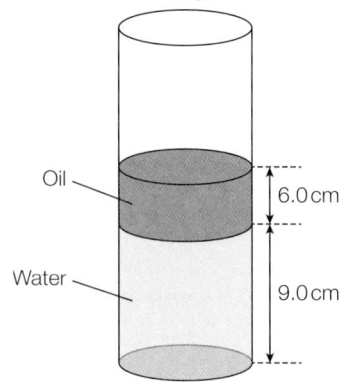

What is the pressure at the bottom of the cylinder due to the liquids?

A 1.38 kPa   **B** 1.41 kPa   **C** 1.87 kPa   **D** 2.21 kPa
[1]

5 **(a)** Define the *moment* of a force. [1]

(b) State the *principle* of moments. [1]

(c) The drawing shows a beam QT, of length 1.5 m and negligible mass, supporting a load of 4.8 kN. The beam is held in equilibrium by cord PR. The cord is at an angle of 70° to the vertical and length TR is 1.0 m.

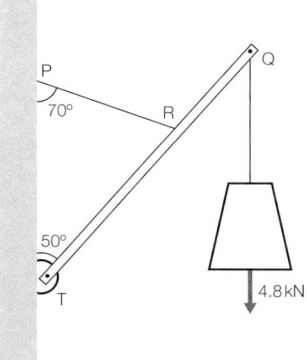

Calculate the tension in the cord. [3]

6 A cable-car is at rest supported by a cable, as shown below.

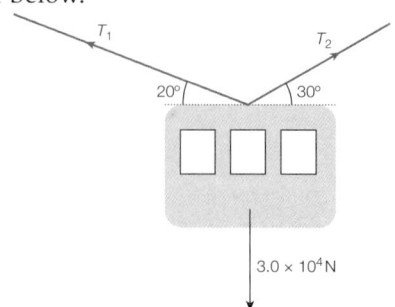

By drawing the triangle of forces to a suitable scale, find the tensions T_1 and T_2 in the cable. [3]

Work, energy, and power 5

Knowledge check

You should be able to:

- describe a range of energy transfers and the processes which cause them
- calculate changes in kinetic energy using $E_k = \frac{1}{2}mv^2$ and changes in gravitational potential energy using $\Delta E_p = mg\Delta h$
- describe power as the rate of transfer of energy.

Energy transfer and conservation

Energy

Energy is transferred when objects move, when their temperature changes or when they change state and so, energy is sometimes described as the ability to do work or change temperature. Energy can be transferred by forces, heating or electric currents. Some energy stores are described as **potential energy**, including elastic potential energy (e.g., of a stretched rubber band), gravitational potential energy (the energy a mass has by virtue of its position in a gravitational field), and electrical potential energy (energy a charged particle has because of its position in an electric field).

Remember

Energy cannot be created or destroyed, only converted from one form to another.

Principle of conservation of energy

An electric motor transfers electrical energy to movement, together with some heat and sound energy. A candle transfers chemical energy into light and heat. A solar cell transfers light energy into electrical energy.

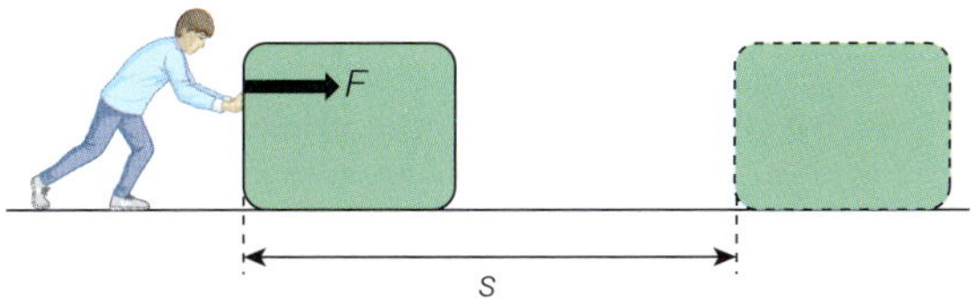

Remember

work done = force × distance moved in the direction of the force

$$W = Fs$$

Work and energy transfer

Work

Work is done when an object moves in the direction of the force (Figure 5.1). When work is done, energy is transferred, perhaps to kinetic energy, gravitational potential energy, heat or sound.

The work done (in joules) is the force (in newtons) multiplied by the distance moved in the direction of the force (in metres).

▲ **Figure 5.1** Work done and energy transferred

Types of energy

Table 5.1 shows different stores of energy and their descriptions.

▼ **Table 5.1** Describing energy

Energy	Description
Gravitational potential energy	Energy a mass has due to its position in a gravitational field
Electrical potential energy	Energy a charged object has due to its position in an electric field
Elastic potential (strain) energy	Energy stored in an object or material due to deformation (e.g., stretching or compressing a spring)
Kinetic energy	The energy a mass has due to its speed
Internal energy	The combined kinetic and potential energies of all the particles in a body
Chemical and nuclear energy	The energy that can be released during chemical or nuclear reactions

Link

See Unit 6 *Deformation of solids* for more about elastic potential energy.

See Unit 16 *Thermodynamics* for more on internal energy and the first law of thermodynamics.

See Unit 18 *Electric fields* for more on electrical potential energy.

Potential energy and kinetic energy

Gravitational potential energy

When a load is lifted, work is done on the load, and the load gains gravitational potential energy. The force needed to just lift a mass m is mg. If the mass is lifted a vertical height Δh (see Figure 5.2):

change in gravitational potential energy of the mass ΔE_p = work done
$$= mg \times \Delta h$$

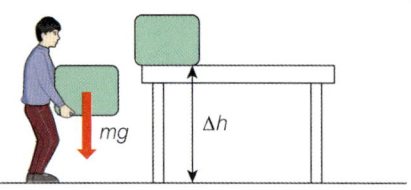

▲ **Figure 5.2** Gravitational potential energy

Kinetic energy

If work is done on an object that is free to move, the object will accelerate and gain kinetic energy E_k (Figure 5.3).

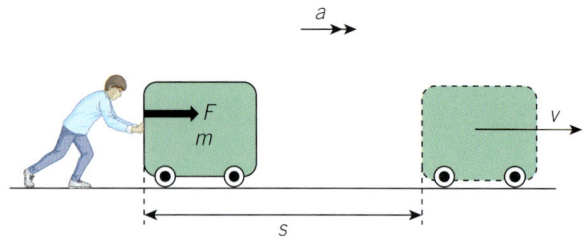

▲ **Figure 5.3** Kinetic energy

$$\text{work done} = E_k = Fs$$

using Newton's second law, $F = ma$:

$$E_k = (ma)s$$

For uniform acceleration:

$$v^2 = u^2 + 2as.$$

So for an object starting from rest ($u = 0$):

$$as = \frac{v^2}{2}$$

so

$$E_k = m\left(\frac{v^2}{2}\right) = \frac{1}{2}mv^2$$

> **Remember**
>
> Change in gravitational potential energy
>
> $$\Delta E_p = mg\Delta h$$

> **Remember**
>
> When calculating work done, kinetic energy, or change in gravitational potential energy, remember that quantities such as mass m, change in height Δh and speed v must all be in SI units for the calculated value to be in joules.

> **Remember**
>
> $$E_k = \frac{1}{2}mv^2$$
>
> You need to be able to derive this equation.

Worked Example

[Take g, the acceleration of free fall, as $9.81\,\text{m s}^{-2}$.]

A child pulls a sledge a horizontal distance of 50 m with a force of 240 N at an angle of 30° to the horizontal, as shown in Figure 5.4.

How much work does the child do?

Answer

work done = $(240 \cos 30°) \times 50 = 10.4\,\text{kJ}$

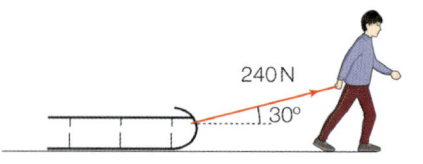

▲ **Figure 5.4** Child pulling a sledge

Efficiency

Energy is transferred by forces or electrical currents doing work. These processes have heating effects and so the transfers that take place often 'waste' energy through this heating. For example, an electric motor may only transfer 80 J usefully for every 100 J provided electrically by the current (see Figure 5.5).

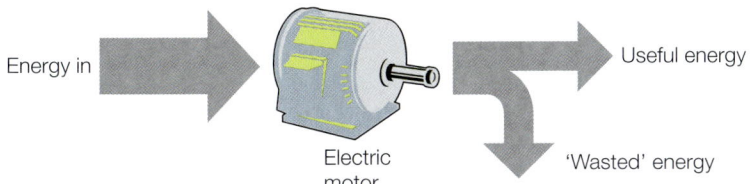

▲ **Figure 5.5** Efficiency of an electric motor

Remember

The **efficiency** of a system is defined as:

$$\text{efficiency} = \frac{\text{useful energy output}}{\text{total energy input}} \times 100\%$$

Key terms

Efficiency: of a system is the useful energy output divided by the total energy input multiplied by 100.

Power: is the rate of doing work (or transferring energy).

Work: is done when an object moves in the direction of the force.

Worked Example

A hydraulic jack lifts one wheel of a car, applying a force of 3.6 kN to to one of the wheels when the handle of the jack is pressed down with a force of 150 N. Each time the handle is moved down 20.0 cm, the car wheel rises 0.5 cm, as shown in Figure 5.6.

(a) Calculate the efficiency of the hydraulic jack.

(b) Explain why the hydraulic jack is not 100% efficient.

Answer

(a) $\text{Efficiency} = \dfrac{\text{useful energy output}}{\text{total energy input}} \times 100\%$

$= \dfrac{3.6 \times 10^{3} \times 0.5 \times 10^{-2}}{150 \times 20.0 \times 10^{-2}} \times 100\%$

$= 60\%$

(b) Energy is 'lost' as heat and sound energy caused by friction between the moving parts of the jack. The platform of the hydraulic jack has mass so gains some potential energy when the wheel is lifted.

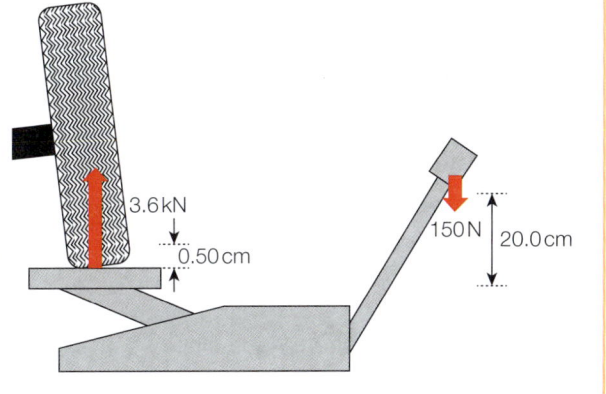

▲ **Figure 5.6** Hydraulic jack

Change in potential energy in a gravitational field

A mass m experiences a force mg when it is in a gravitational field, where g is the gravitational field strength, as shown in Figure 5.7.

In order to move the mass a small distance Δx in the opposite direction to the gravitational field (small enough for the gravitational field strength g not to change significantly), an amount of work $mg\,\Delta x$ must be done on the mass. The gravitational potential energy of the mass increases by the same amount:

$$\Delta E_{p} = mg\,\Delta x$$

If the gravitational field is **uniform** (g is constant):

$$\Delta E_{p} = mg\,\Delta h$$

where Δh is the total distance moved in the opposite direction to the gravitational field.

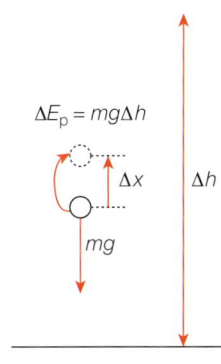

▲ **Figure 5.7** Gravitational field and potential energy

Change in potential energy in an electric field

A charge $+q$ experiences a force Eq when placed in an electric field of strength E, as shown in Figure 5.8.

In order to move the charge a small distance Δx in the opposite direction to the electric field (small enough for the electric field strength E not to change significantly), an amount of work $Eq\Delta x$ must be done on the charge. The electrical potential energy of the charge increases by the same amount.

$$\Delta E_p = Eq\Delta x$$

If the electric field is uniform (E is constant):

$$\Delta E_p = Eqd$$

where d is the total distance moved in the opposite direction to the electric field.

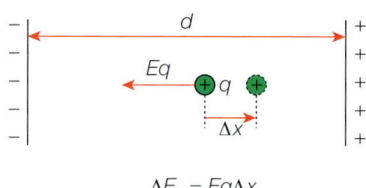

$$\Delta E_p = Eq\Delta x$$

▲ **Figure 5.8** Electric field and potential energy

> ★ **Exam tip**
>
> Notice that the symbol E represents electric field strength, but E_p represents potential energy.

Power

Power is the rate of doing work (or transferring energy).

$$\text{power (W)} = \frac{\text{work done (J)}}{\text{time taken (s)}}$$

Using the equation for work done: $W = Fs$, gives:

$$\text{power} = \frac{Fs}{t} = F\left(\frac{s}{t}\right) = Fv \quad \text{The SI unit of power is the watt.}$$

Worked Example

A 1 kW electric motor is used to lift a load of 300 N with the aid of a simple pulley system, as shown in Figure 5.9. A force of 200 N applied by the motor will just lift the load.

(a) Explain why the force needed to lift the load is less than 300 N.

(b) Calculate the efficiency of the pulley system.

(c) The motor lifts the load 8.0 m in 5.0 s. What is the overall efficiency of the motor and pulley system? State any assumptions you make.

Answer

(a) To lift the load 1.0 m the motor has to pull the rope a distance of 2.0 m. The work done by the motor is $200 \times 2.0 = 400$ J; the work done on the load is only $300 \times 1.0 = 300$ J.

(b) For every 1.0 m the load is raised:

$$\text{efficiency} = \frac{\text{useful energy output}}{\text{total energy input}} \times 100\% = \frac{300}{400} \times 100 = 75\%$$

(c) output power = useful energy out per second $= \dfrac{300 \times 8.0}{5.0} = 480$ W

input power = 1×10^3 W

$$\text{overall efficiency} = \frac{\text{useful power output}}{\text{total power input}} = \frac{480}{1 \times 10^3} \times 100\% = 48\%$$

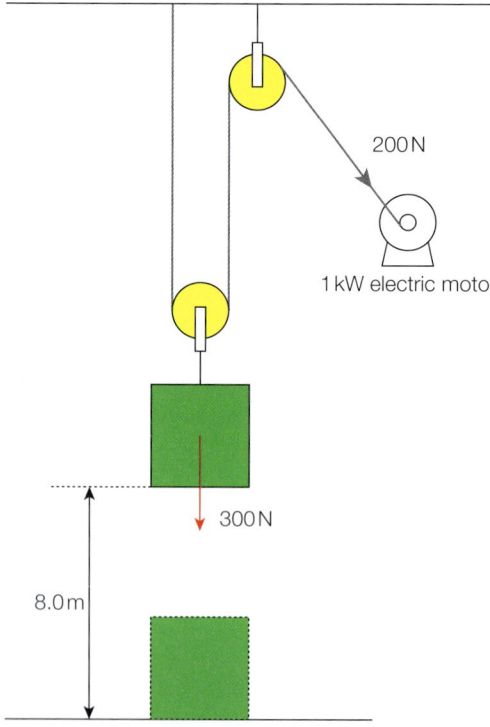

▲ **Figure 5.9** Pulley system

> The load also gains kinetic energy since it is moving, but when the motor stops, this energy is quickly 'lost' so is not 'useful' energy.

↑ Raise your grade

(a) Explain what is meant by *work done*.

work done = force × distance moved ✗

> Work done = force × distance moved **in the direction of the force**.

[1]

(b) Distinguish between *gravitational potential energy and elastic potential energy*.

Gravitational P.E. is the stored energy a mass has due to its position   ✗

> Answer should include reference to **gravitational field** e.g add '…its position in a gravitational field'.

Elastic P.E. is energy stored because something has been stretched   ✔

[2]

(c) The picture shows part of a rollercoaster ride.

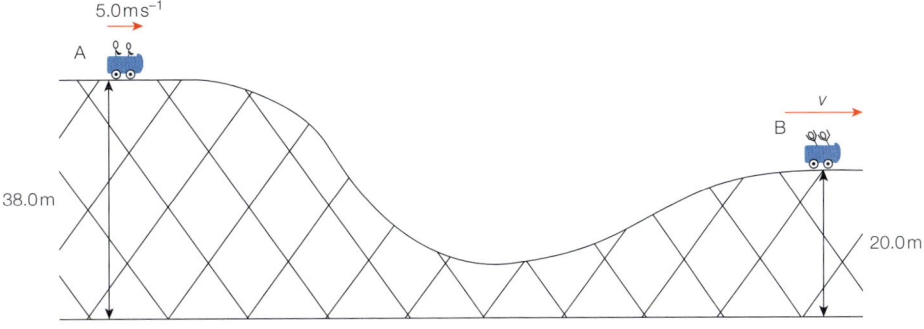

The carriage and passengers, of total mass 500 kg, is moving horizontally at A with a speed of 5.0 m s⁻¹. Air resistance is negligible and other friction forces can be ignored.

Show that the change in gravitational potential energy of the carriage and passengers, as the carriage moves from A to B, is 88 kJ.

$\Delta E_p = mg\Delta h = 500 \times 9.81 \times (38.0 - 20.0) = 88$ kJ ✔

> Method.

[1]

(d) Calculate:

(i) the kinetic energy of the carriage and passengers at B.

> The kinetic energy the carriage already had at A should be added to this value.

[2]

Potential energy lost = kinetic energy gained = 88 kJ ✗ ✔

> Uses principle of conservation of energy.

kinetic energy = 88 kJ

(Correct value is 95 kJ.)   [2]

(ii) the speed *v* of the carriage and passengers at B.

$\frac{1}{2}mv^2 = 88 \times 10^3 \rightarrow v = \sqrt{\dfrac{2 \times 88 \times 10^3}{500}} = 18.8$ m s⁻¹ ✔✔

speed = 19 m s⁻¹

> Correct method used.

> Correct calculation allowing for e.c.f.

(e) In fact, air resistance and other frictional forces are significant. The speed of the carriage at B is 13 m s⁻¹. The length of the track from A to B is 30 m. Calculate the average frictional force acting on the carriage and passengers as it moves from A to B.

$Fs = \Delta E_p - \Delta E_k = 88 \times 10^3 - \frac{1}{2} \times 500 \times (13^2 - 5^2) = 5.2 \times 10^4$ ✔

> Correct method.

$F = \dfrac{5.2 \times 10^4}{30} = 1.7$ kN ✔

> Correct calculation.

average frictional force = 1.7×10^3 N   [2]

Exam-style questions

1 A uniform square paving stone, of dimensions 40.0 cm × 40.0 cm × 8.0 cm, has a mass of 30 kg and is lying flat on the ground.

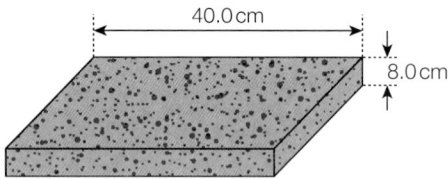

How much work is needed to stand the paving stone on its end?

A 47 J   B 59 J

C 106 J   D 118 J   [1]

2 Starting from rest, a skier skis down a 30° slope of length 25 m. Ignoring frictional forces, what is her speed at the end of the slope?

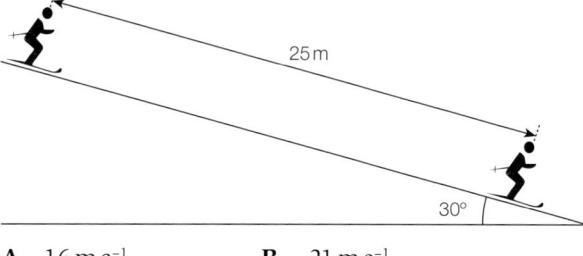

A 16 m s⁻¹   B 21 m s⁻¹

C 24 m s⁻¹   D 31 m s⁻¹   [1]

3 A constant force F is applied to an object which is initially at rest.

Which graph shows the variation of W, the work done by the force, against s, the distance travelled by the object?

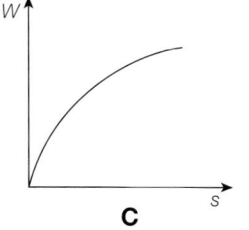 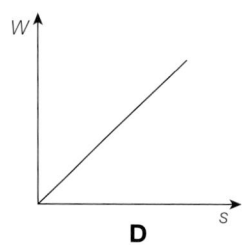

[1]

4 (a) Distinguish between *gravitational potential energy* and *electrical potential energy*. [2]

(b) A tennis ball of mass 60 g is dropped from a height of 4.0 m onto level ground.

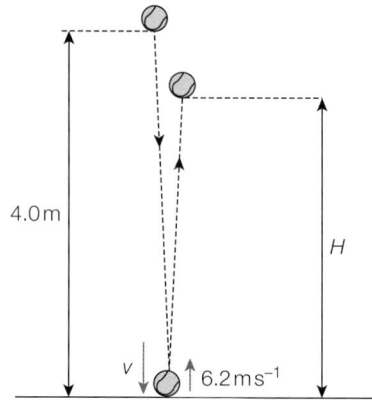

Ignoring air resistance, calculate:

(i) the gravitational potential energy lost by the ball when it reaches the ground

(ii) the speed v of the ball just before it hits the ground. [3]

(c) The ball rebounds with a speed of 6.2 m s⁻¹.

Calculate:
(i) the fraction of energy 'lost' in the collision

(ii) the maximum height H reached by the ball after the first bounce. [3]

5 (a) (i) Define *power*.

(ii) Use your definition to show that:

power = force × velocity [3]

(b) A lorry of mass 3000 kg moves at a speed of 15 m s⁻¹ along a horizontal road. A resistive force of 2.4 kN acts on the lorry.

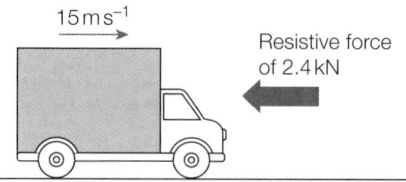

The lorry accelerates at 0.50 m s⁻². Calculate:

(i) the driving force produced by the lorry's engine

(ii) the output power of the engine. [3]

Deformation of solids 6

Force and solid materials

When a pair of forces is applied to a solid material it **deforms**; that is, it changes shape. Forces that stretch a material are called **tensile** forces (see Figure 6.1a); forces which compress a material are called **compressive** forces (see Figure 6.1b).

a Tensile forces   **b** Compressive forces

▲ **Figure 6.1** Forces on a solid material

The deformation is **elastic** if the material returns to its original shape once the forces have been removed. If there is some permanent deformation (e.g., compression or extension) when the forces have been removed, **plastic** deformation has occurred.

> **Key terms**
>
> **Compressive:** forces which compress or squash.
>
> **Deformation:** change of shape.
>
> **Elastic:** returns to original shape when deforming forces are removed.
>
> **Plastic:** some permanent deformation when forces are removed.
>
> **Tensile:** forces which stretch.

Stretching springs

The graph in Figure 6.2 shows the extension (stretch) of a spring as different loads are placed on it.

- **O–A:** The graph is a straight line, through the origin. Doubling the load doubles the stretch (extension) – the spring obeys **Hooke's law:**

$$F = kx$$

  where F is the load applied, x is the extension, and k is called the **spring constant** and is a measure of the stiffness of the spring. The SI units for k are N m^{-1}.

- **A–B:** Point **A** is called the limit of proportionality – this is the point that the spring no longer obeys Hooke's law. Point **B** is called the **elastic limit.** The spring stretches elastically. If the load is removed at any point between O and **B**, the spring will return to its original length.

- **Beyond B:** If the spring is stretched beyond point **B** some plastic deformation occurs. When the load is removed the load–extension graph follows the dotted line. With the load removed completely, there is a permanent extension of the spring.

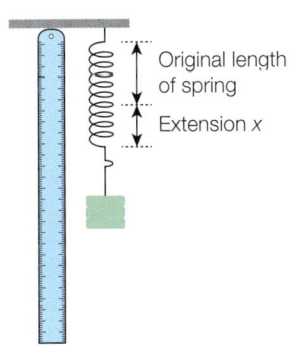 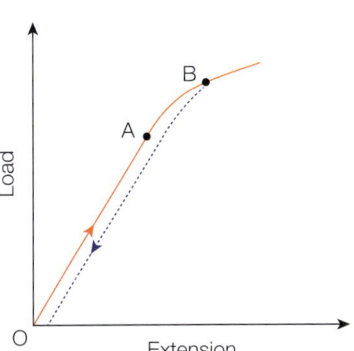

Original length of spring

Extension x

▲ **Figure 6.2** Stretching a spring

> **Key terms**
>
> **Elastic limit:** this is the point where further stretching will deform the spring.
>
> **Hooke's law:** doubling the load on a spring doubles the extension.
>
> **Spring constant:** is a measure of stiffness of the spring.

Energy considerations

The work done in stretching a spring is the area under the load–extension graph (see Figure 6.3). If the spring obeys Hooke's law and is not stretched beyond the elastic limit, the work done on the spring (called the strain energy) is:

Substituting from $F = kx$ gives

$$E_p = \tfrac{1}{2}Fx$$

$$E_p = \tfrac{1}{2}kx^2 \text{ or } E_p = \frac{F^2}{2k}$$

As the spring stretches elastically, all the energy stored is recoverable as mechanical energy when the load is removed from the spring.

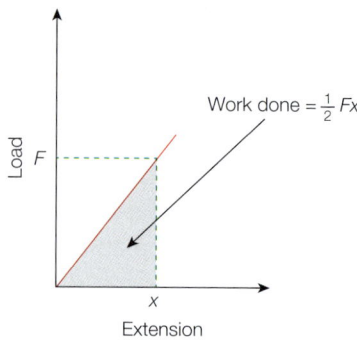

▲ **Figure 6.3** Energy stored in a stretched spring

Worked Example

A spring, with unstretched length 12.0 cm, stretches 4.0 cm when supporting a load of 5.0 N.

(a) Determine the spring constant k of the spring.

(b) Find the energy stored in the spring when it is stretched by 7.0 cm.

State any assumptions you make.

Answer

(a) $k = \dfrac{F}{x} = \dfrac{5.0}{4.0 \times 10^{-2}} = 125\,\text{N m}^{-1}$

(b) $E_p = \dfrac{1}{2}kx^2 = \dfrac{1}{2} \times 125 \times (7.0 \times 10^{-2})^2 = 0.31\,\text{J}$

Assumptions made: the spring obeys Hooke's law, and the elastic limit is not exceeded.

★ **Exam tip**

When using an equation like the equation for the energy stored in a spring, the units used must be **consistent**.

The extension must be in metres (m), the force applied in newtons (N), and the stiffness k in newtons/metre (N m^{-1}), so that the final answer is in joules (J).

Stretching materials

The graph (Figure 6.4) shows how a ductile metal (one that can be drawn into a thin wire, such as copper) stretches when supporting different loads.

- **O–A:** The material obeys Hooke's law. If the load is removed, the material returns to its original length. It has behaved elastically. A is the **Hooke's law limit**.

- **A–B:** The material is now past the Hooke's law limit, but still behaves elastically. If the load is removed, the material again returns to its original length – B is the **elastic limit**.

- **B–C:** The material has been stretched beyond its elastic limit. The material will not return to its original length, but instead return along the dotted line on the graph. There is now some **permanent deformation** of the material.

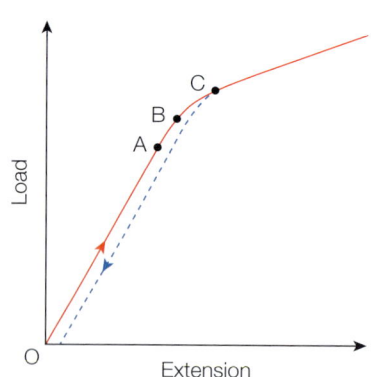

▲ **Figure 6.4** Load against extension graph for ductile metals

If a material is stretched beyond its elastic limit (see Figure 6.4), the work done in stretching the material is the area under the loading curve (solid line). The recoverable mechanical energy is the area under the unloading curve (dotted line).

The difference between the two areas is the energy lost as internal energy (heat) in the material.

Stress, strain, and the Young modulus

When stretching different materials, such as metal wires, **stress** and **strain** are more useful quantities to calculate than just the force applied and the extension, as the stress–strain graph illustrates properties of the material itself rather than one specific length or diameter of the material.

Stress σ

$$\text{stress } \sigma = \frac{\text{force applied}}{\text{cross-sectional area}} = \frac{F}{A}$$

The SI units for stress are $N\,m^{-2}$, or pascal (Pa).

Strain ε

$$\text{strain } \varepsilon = \frac{\text{extension}}{\text{original length}} = \frac{x}{l_0}$$

Strain has no units. It can be expressed as a number or a percentage.

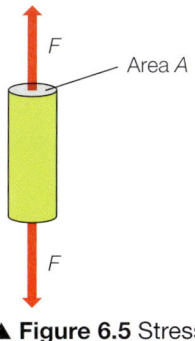

▲ **Figure 6.5** Stress

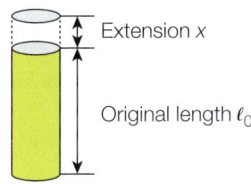

▲ **Figure 6.6** Strain

Worked Example

A steel wire of length 3.0 m and diameter 1.0 mm stretches 1.9 mm when supporting a mass of 10 kg. [Use $g = 9.81\ m\,s^{-2}$.] Calculate:

(a) the stress   (b) the strain.

Answer

(a) $\sigma = \dfrac{F}{A} = \dfrac{10 \times 9.81}{\pi \times (0.5 \times 10^{-3})^2} = 1.23 \times 10^8\ N\,m^{-2}$

(b) $\varepsilon = \dfrac{x}{l_0} = \dfrac{1.9 \times 10^{-3}}{3.0} = 6.3 \times 10^{-4}\ (0.063\,\%)$

> ★ **Exam tip**
>
> When calculating A (the cross-sectional area) don't forget to halve the diameter to find the **radius** when using $A = \pi r^2$.

Young modulus E

The **Young modulus** of a material is a measure of the stiffness of that material. The larger the value of E, the stiffer the material – the greater the stress needed to produce a particular strain.

$$\text{Young modulus} = \frac{\text{stress}}{\text{strain}}$$

$$E = \frac{\sigma}{\varepsilon}$$

The units for the Young modulus are $N\,m^{-2}$ or pascal (Pa).

> ★ **Exam tip**
>
> Be careful not to confuse E for the Young modulus with E for energy.

Worked Example

A nylon rope of length 3.5 m and diameter 5.0 mm stretches 12 mm when supporting a load of 80 kg. Determine the Young modulus of the nylon.

Answer

$$E = \frac{\sigma}{\varepsilon} = \frac{\left(\dfrac{F}{A}\right)}{\left(\dfrac{x}{l_0}\right)} = \frac{\left(\dfrac{80 \times 9.81}{\pi \times (2.5 \times 10^{-3})^2}\right)}{\left(\dfrac{12 \times 10^{-3}}{3.5}\right)} = 1.2 \times 10^{10}\ Pa$$

> **Key terms**
>
> **Strain:** is the force applied divided by the cross-sectional area.
>
> **Stress:** is the extension divided by the original length.
>
> **Young modulus:** E, the stress divided by the strain. It is a measure of the stiffness of the material.

Measuring the Young modulus *E*

Figure 6.7 shows a simple way of measuring *E* using a thin wire.

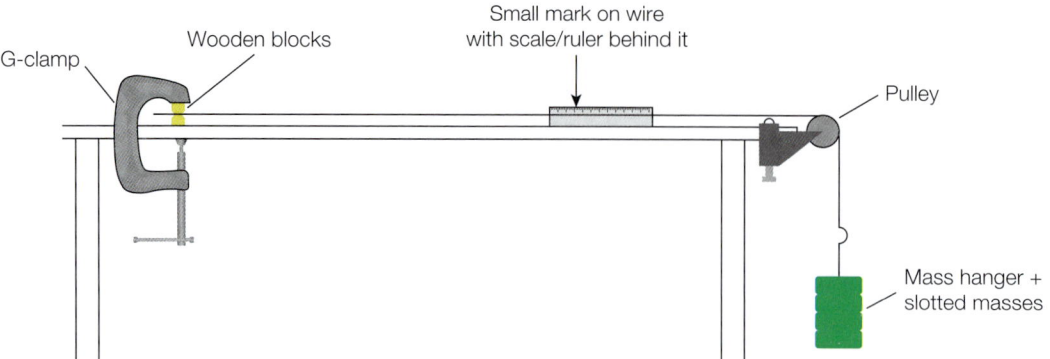

▲ **Figure 6.7** Measuring the Young modulus *E*

This method is useful for testing materials that can be drawn into long, thin strips such as copper or steel wire, nylon or polythene. A mark is made on the wire, and the length of the wire from the wooden blocks recorded. Weights are steadily added, and the extension of the wire recorded for different loads.

The length of the wire can be measured with a metre rule and the diameter with vernier calipers. Stress and strain can then be calculated from the load and extension values, and a graph of stress against strain plotted. The value of *E* can then be found from the graph.

> ### ★ Exam tip
>
> **Making improvements**
>
> How can this experiment be improved? Good suggestions might include:
>
> - using a longer length of wire (this will reduce the percentage uncertainty in the measurement of the length of the wire)
> - measuring the extension with a travelling microscope
> - measuring the diameter of the wire with a micrometer (a micrometer is accurate to ±0.01 mm, whereas vernier calipers are only accurate to ±0.1 mm).

Elastic and plastic deformation

The loading and unloading force–extension graphs for different materials provide useful information about their elastic and plastic properties.

In Figure 6.8a, the metal wire is elastic up to the elastic limit. If the load exceeds the elastic limit value, the material will not return to its original length. The area under the loading curve is the work done in stretching the material (the strain energy). Up to the elastic limit this energy is stored as potential energy and can be recovered as mechanical energy.

In Figure 6.8b, the material is elastic but does not obey Hooke's law. The area under the unloading curve is the energy which is **recoverable** as mechanical energy. The area between the two curves is the energy **lost** as internal energy (heat) in the material. This is why car tyres are hot after a long journey.

In Figure 6.8c, the material behaves plastically. When the load is removed there is only a small reduction in length. The area between the two curves is the energy lost as internal energy (heat) in the material.

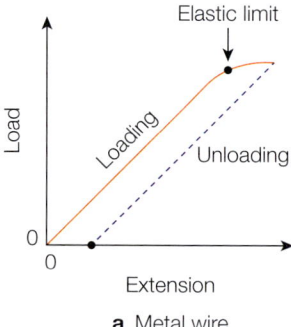

a Metal wire

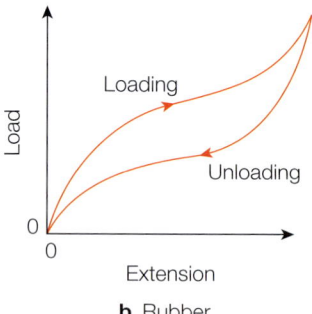

b Rubber

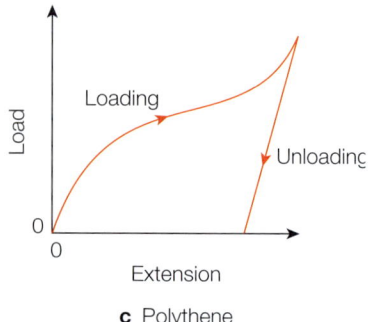

c Polythene

▲ **Figure 6.8** Stretching different materials

↑ Raise your grade

(a) Define, for a metal wire, [2]

    (i) stress

        Stress is the force on the wire divided by the cross-sectional area of the wire. ✔

    (ii) Young modulus

        Young modulus equals $\dfrac{\text{tensile stress}}{\text{tensile strain}}$. ✔

> A correct definition.
>
> A correct definition.

(b) A steel cable of length 0.50 m and diameter 4.0 mm is suspended from the ceiling and supports a chandelier (a light fitting). The mass of the chandelier is 25.0 kg. The Young modulus of steel is 2.1×10^{11} Pa.

Calculate:

(i) the weight of the chandelier.

    $W = mg = 25.0 \times 9.81 = 245\,\text{N}.$ ✔

                weight = ...245.... N [1]

(ii) the extension of the cable caused by the weight of the chandelier.

    $x = \dfrac{FL}{EA} = \dfrac{245 \times 0.50}{2.1 \times 10^{11} \times \pi \times \left(4 \times 10^{-3}\right)^2} = 4.6 \times 10^{-5}$ ✔ ✗ ✗

> Correct method.

> The candidate has forgotten to convert the answer from metres to millimetres and to use the radius of the wire; not the diameter.

                extension = ...4.6×10^{-5}.... mm [3]

(c) The steel cable consists of a large number of very thin strands of steel bound together, rather than a single wire. Suggest a reason for this. [1]

A lot of thin strands are stronger than a single thick cable ✗

> If the combined cross-sectional area of all the thin strands is the same as the cross-sectional area of the thick cable, the strengths of the two cables are exactly the same. A large number of thin cables is much easier to bend (more flexible) than a single thick cable.

Exam-style questions

1 A spring stretches 4 cm when supporting a load of 20 N.

Three identical springs to the one described are connected as shown in the diagram.

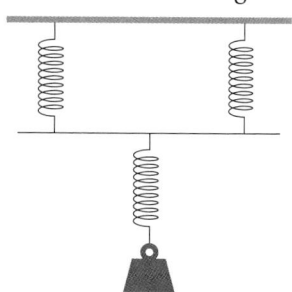

What is the extension of the three springs when supporting a load of 30 N?

A 3 cm B 6 cm

C 9 cm D 12 cm [1]

2 The graph shows the extension of a spring for different loads.

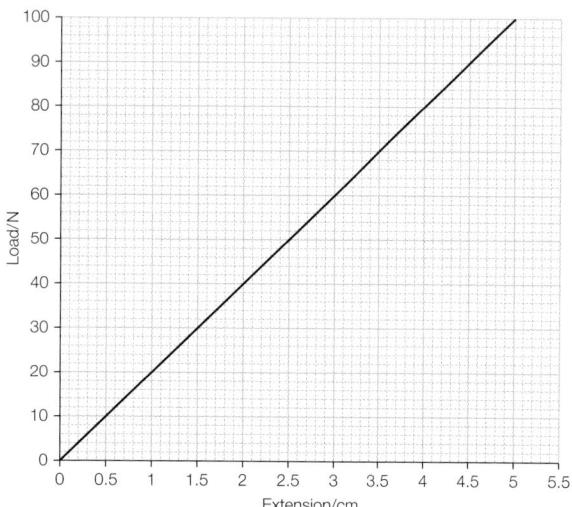

What is the energy stored in the spring when a load of 80 N is applied to the spring?

A 1.6 J B 3.2 J

C 160 J D 320 J [1]

3 What are the SI base units of stress?

A $kg\,m^{-1}\,s^{-2}$ B $kg\,m\,s^{-2}$

C $kg\,m^{-1}\,s^{2}$ D $kg\,m\,s^{2}$ [1]

4 A tensile force is applied to a thin wire causing it to stretch by an amount x. The same force is now applied to a wire of the same material, but with three times the length and twice the diameter. How much will this wire stretch?

A $\dfrac{2x}{3}$ B $\dfrac{3x}{4}$ C $\dfrac{4x}{3}$ D $\dfrac{3x}{2}$ [1]

5 A lift in a building is supported by six steel cables, each 20 m long. Each cable consists of 20 strands of steel wire of diameter 2.0 mm. The Young modulus of steel is 2.1×10^{11} Pa.

A man of mass 80 kg steps into the lift. How far will the lift descend?

A 0.01 mm B 0.05 mm

C 0.10 mm D 0.20 mm [1]

6 A metal wire of length 1.200 m and diameter 0.61 mm stretches 0.43 mm when a load of 1.30 kg is hung from one end.

(a) Calculate:

   (i) the stress on the wire

   (ii) the strain of the wire

   (iii) the Young modulus of the metal. [4]

(b) Justify the number of significant figures you have given for your answer to (a)(iii). [1]

7 (a) Define the Young modulus. [1]

(b) The upper leg bone (femur) in an adult human has a length of 50 cm and a minimum diameter of 2.8 cm.

The Young modulus of bone is 8.5×10^{9} Pa.

Estimate:

   (i) the mass of a man

   (ii) the maximum stress on the bone when the man stands on one leg

   (iii) the compression of the bone when the man stands on one leg. [4]

(c) State, with a reason, whether your answer to (b)(iii) is likely to be an overestimate or an underestimate. [1]

8 You have been asked to investigate the mechanical properties of nylon. You are provided with a reel of nylon thread.

Design a laboratory experiment to determine the mechanical properties of the material. Draw a diagram showing the arrangement of your equipment. In your account you should pay particular attention to:

(a) the procedure to be followed

(b) the measurements to be taken

(c) the control of variables

(d) the analysis of the data

(e) the safety precautions to be taken. [15]

Progressive waves

What are progressive waves?

Progressive waves transfer energy from one point to another.

There are two types of progressive wave.

- **Transverse waves:** Waves on a rope are examples of transverse waves. As a transverse wave passes along the rope, the particles of the rope oscillate in a direction **perpendicular to the direction of energy transfer** (see Figure 7.1).

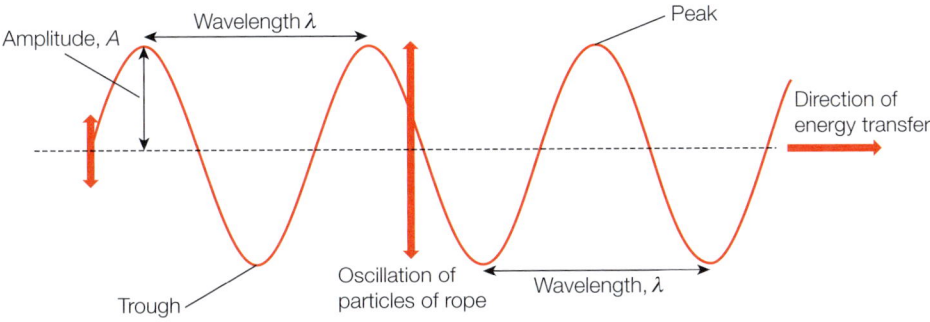

▲ **Figure 7.1** Transverse waves

Water waves, secondary seismic waves and electromagnetic waves are examples of transverse waves.

- **Longitudinal (compression) waves:** Longitudinal, or compression, waves on a slinky (a long spring) can best illustrate the properties of longitudinal waves. The individual rings of the slinky oscillate back and forth **parallel to the direction of energy transfer**.

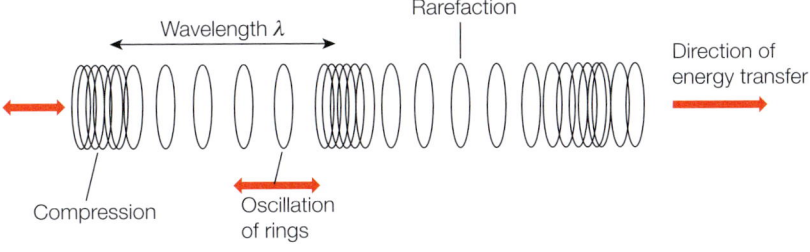

▲ **Figure 7.2** Longitudinal (compression) waves

Sound waves and primary seismic waves are longitudinal waves. As a sound wave travels through air, the air molecules continually move closer together (compression) and then further apart (rarefaction) creating areas of high pressure and low pressure. The greater the amplitude of the sound wave, the greater the pressure difference between the areas of compression and expansion.

Key terms

The key descriptions of transverse waves include:

Amplitude, A (m): The maximum displacement from the equilibrium (rest) position.

Frequency, f (Hz): The number of complete waves passing any point in one second (and the number of complete oscillations of a vibrating particle each second).

Wavelength, λ (m): The distance from one peak to the next (or from one trough to the next).

Key terms

The key descriptions of longitudinal waves include:

Frequency, f (Hz): The number of complete waves passing any point in one second.

Wavelength, λ (m): The distance from the centre of one compression to the centre of the next (or from one rarefaction to the next).

Displacement of a particle

The **displacement** of an individual particle in a transverse or longitudinal wave is shown in Figure 7.3.

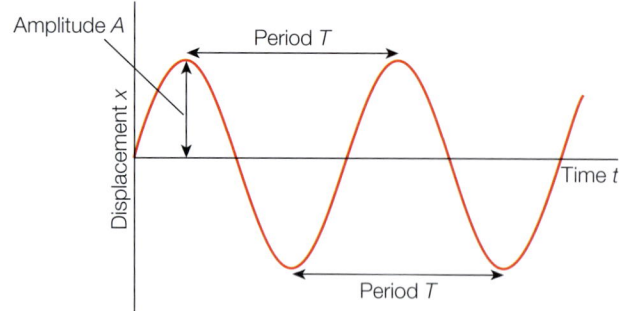

▲ **Figure 7.3** Displacement–time graph for a particle on the wave

The period of a wave

The **period** T of the oscillation is the time taken for one complete oscillation (and the time for one complete wave to pass any given point). It is related to the frequency of the wave by the equation:

$$f = \frac{1}{T}$$

The frequency f is measured in hertz (Hz). $1\,\text{Hz} = 1$ wave/second.

Wave equation

$$\begin{matrix} \textbf{speed of a} \\ \textbf{wave} \end{matrix} = \begin{matrix} \text{length of one wave} \\ \text{(the wavelength)} \end{matrix} \times \begin{matrix} \text{number of waves passing in} \\ \text{one second (the frequency)} \end{matrix}$$

$$v = f\lambda$$

Derivation of wave equation

Starting from the basic speed equation:

$$\text{distance travelled} = \text{speed} \times \text{time}$$

$$s = vt$$

If a wave has a velocity of v and travels for one period of oscillation (T), then the distance it has travelled will be one wavelength (λ) as a complete wave has passed through the starting position.

This means that:

$$\lambda = vT$$

From the relationship between period and frequency:

$$f = \frac{1}{T} \text{ so } T = \frac{1}{f}$$

Therefore:
$$\lambda = v\frac{1}{f}$$

Finally:
$$v = \lambda f$$

Phase difference

Particles at different points along a wave are out of step with each other – there is a **phase difference** between them.

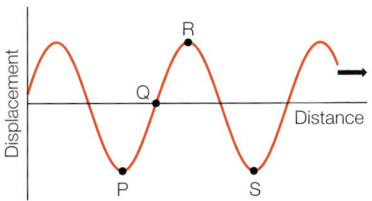

▲ **Figure 7.4** Phase difference

In Figure 7.4 points P and R are exactly half an oscillation out of step with each other – when P starts to move up, R starts to move down (they are in **antiphase**). Points P and Q are one quarter of an oscillation out of step with each other (when P is at its maximum displacement, Q is at the equilibrium position). Points P and S are exactly one cycle out of phase with each other so are in phase (both are moving up or down at exactly the same time). Phase difference can be expressed in degrees or radians where 360°, or 2π radians, represents one complete cycle (see Table 7.1).

▼ **Table 7.1** Phase difference

	Phase difference/cycles	Phase difference/°	Phase difference/radians
P → Q	$\frac{1}{4}$	90	$\frac{\pi}{2}$
P → R	$\frac{1}{2}$	180	π
P → S	1	360 (0)	2π (0)

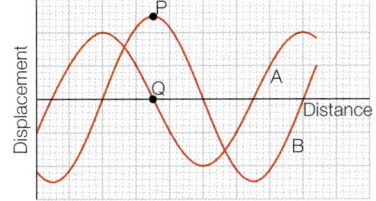

▲ **Figure 7.5** Phase difference

Phase difference can also be used to describe how two waves compare with each other.

In Figure 7.5 the two waves A and B are out of phase with each other. Point P on wave B has reached its maximum displacement and will start to move downwards. The corresponding point Q on wave A is at the equilibrium position and moving upwards; that is, wave A is **lagging** behind wave B by quarter of a cycle. Another way of saying this is wave B **leads** wave A by $\frac{\pi}{2}$ radians or 90°.

Electromagnetic waves

The family of waves which includes visible light is called the **electromagnetic spectrum** (see Table 7.2). All electromagnetic waves:

- are transverse waves
- travel at the speed of light $(3.0 \times 10^8 \, \text{m s}^{-1})$ in a vacuum.

> ★ **Exam tip**
>
> Try to memorise Table 7.2. You may be asked to recall the order of magnitude of the main parts of the spectrum and, from these, the corresponding frequencies can be calculated using $c = f\lambda$, where c is the speed of light $(3.0 \times 10^8 \, \text{m s}^{-1})$.

▼ **Table 7.2** Electromagnetic spectrum

Electromagnetic wave	gamma rays	X-rays	ultraviolet	visible	infrared	microwaves	radio waves
Typical wavelengths/m	10^{-12}	10^{-10}	10^{-8}	4×10^{-7}–7×10^{-7} (400–700 nm)	10^{-5}	10^{-2}	10^{-1}–10^5

Visible light

Visible light is the only part of the electromagnetic spectrum detected by the human eye. Its wavelength ranges from 400 nm (violet) to 700 nm (red) in a vacuum or air.

A cathode-ray oscilloscope (c.r.o.) is useful for investigating waveforms. The two key controls on the instrument panel, as shown in Figure 7.6, are:

- **y-gain:** this states the number of volts per division (volts/div.) in the y direction. A division is usually 1 cm (1 square) on the screen. The smaller the number of volts/div., the more sensitive the scale.

- **time-base:** this indicates how quickly the electron beam moves across the screen, and is usually calibrated in seconds/division (s/div.). The larger the value of the time-base, the slower the dot moves across the screen.

Figure 7.6 shows an example of an oscilloscope display:

The y-gain is set to 20 mV/div. so the amplitude of the voltage signal is 60 mV. (The peak-to-peak voltage is 120 mV.)

- The time-base is set to 10 ms/div. Three complete cycles occur in ten divisions.

- Period T of the signal $= 10 \times \dfrac{10}{3} = 33.3$ ms

- Frequency $f = \dfrac{1}{T} = \dfrac{1}{33 \times 10^{-3}} = 30$ Hz

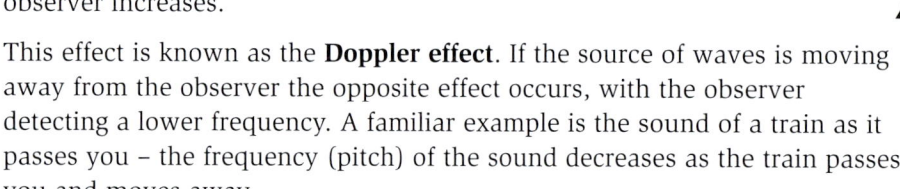

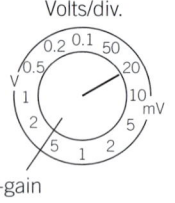
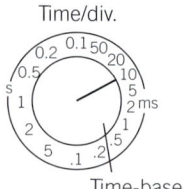

Volts/div.   Time/div.

y-gain   Time-base

▲ **Figure 7.6** Oscilloscope controls

Doppler effect

When a source of waves travels towards a stationary observer, the wavelength of the waves decreases and the frequency detected by the observer increases.

This effect is known as the **Doppler effect**. If the source of waves is moving away from the observer the opposite effect occurs, with the observer detecting a lower frequency. A familiar example is the sound of a train as it passes you – the frequency (pitch) of the sound decreases as the train passes you and moves away.

For a source of sound waves moving towards a stationary observer:

$$f_o = f_s \frac{v}{v - v_s}$$

where f_s is the frequency of sound of the source, v_s the velocity of the source, v the velocity of sound, and f_o the frequency detected by the observer. For a source of sound waves travelling away from a stationary observer:

$$f_o = f_s \frac{v}{v + v_s}$$

> **★ Exam tip**
>
> Numerical questions on the Doppler effect will only be concerned with sound waves, travelling towards or away from a stationary observer.

Worked Example

A train is travelling towards a station at a speed of $35\,\mathrm{m\,s^{-1}}$. An observer standing on a platform in the station hears the train emitting sound of frequency 800 Hz. What frequency will the observer hear when the train has passed through the station? [Speed of sound in air $= 330\,\mathrm{m\,s^{-1}}$.]

Answer

For the train travelling towards the observer: $800 = f_s \left(\dfrac{330}{330 - 35} \right)$ (eqn 1)

For the train travelling away from the observer: $f_o = f_s \left(\dfrac{330}{330 + 35} \right)$ (eqn 2)

Combining eqn 1 and eqn 2: $f_o = \left(\dfrac{330}{330 + 35} \right) \times \left(\dfrac{330 - 35}{330} \right) \times 800 = 650\,\mathrm{Hz}$

Doppler shift

The Doppler effect occurs with many types of wave including light waves and microwaves. The wavelengths of light detected from distant stars are longer than the characteristic wavelengths expected of the line spectra of gases such as hydrogen and helium, suggesting that other stars and galaxies are moving away from the Earth. This is often referred to as the **Doppler shift** or red shift (as the wavelengths are longer, moving towards the red end of the visible spectrum).

> **💡 Remember**
>
> $$f_o = f_s \frac{v}{v \pm v_s}$$
>
> – for sound waves moving towards a stationary observer
>
> + for waves moving away from a stationary observer.

Intensity of a wave

The intensity of a wave is a measure of its power. The intensity I of a wave is proportional to the square of the amplitude A of the wave:

$$I \propto A^2$$

Halving the amplitude of a wave reduces its intensity by a factor of four.

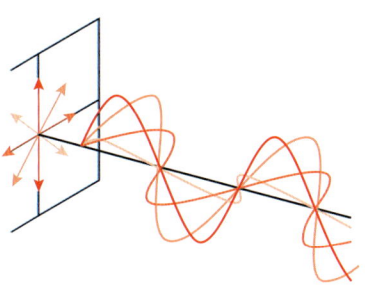

▲ **Figure 7.7a** Unpolarised transverse wave – oscillations are in all perpendicular planes

Polarisation

Most transverse waves are **unpolarised**; the oscillations in the wave happen in every direction perpendicular to the direction in which the wave travels, as shown in Figure 7.7a. Transverse waves can be **polarised** so that the oscillations are limited to one specific plane perpendicular to the direction of propagation, as shown in Figure 7.7b.

Longitudinal waves cannot become polarised as their oscillations can only be in the same direction as they travel.

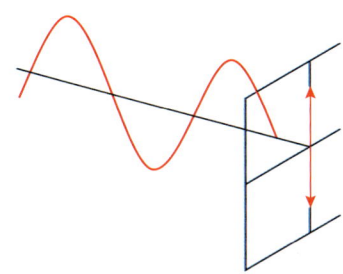

▲ **Figure 7.7b** Polarised transverse wave – oscillations are only in one plane

Polarising electromagnetic waves

Electromagnetic waves are usually unpolarised when emitted by a source, but they can become polarised when they interact with matter. The method of polarisation depends on the rays:

- Visible light will become polarised when it passes through a Polaroid filter. The arrangement of the molecules in the filter causes the waves to be polarised depending on the orientation, as shown in Figure 7.8.

- Microwaves are polarised when they pass through a metal grille. As before, the direction of polarisation will depend on the orientation of the grille, as shown in Figure 7.9.

- Visible light is also polarised by reflection from surfaces like glass or water. The plane of polarisation is perpendicular to the surface of the boundary, as shown in Figure 7.10.

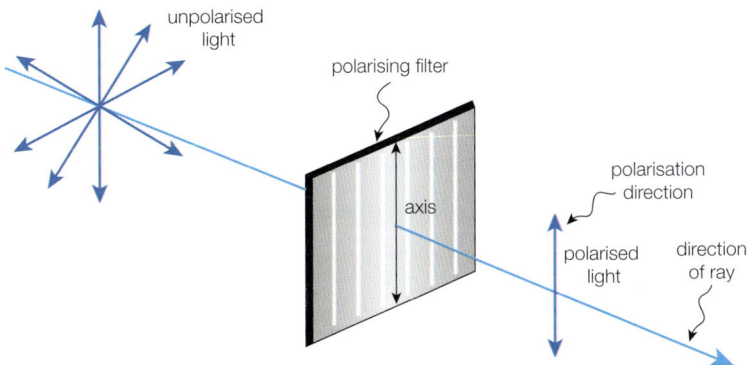

▲ **Figure 7.8** A Polaroid filter polarises visible light

Malus' law

Polarising a previously unpolarised wave will reduce its intensity as some energy is absorbed as it passes through the polariser. If the wave was completely unpolarised, then the intensity is reduced to half of the incident intensity.

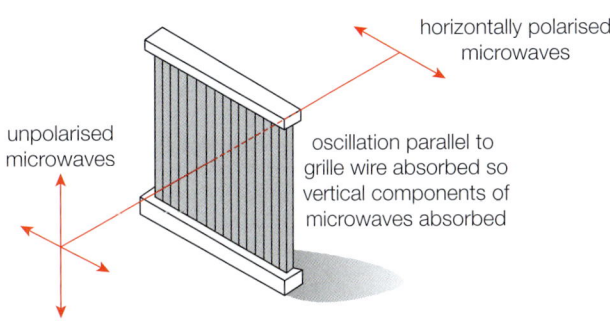

▲ **Figure 7.9** A metal grille polarises microwaves

When an already polarised wave passes through a polariser its intensity will be reduced further. The reduction in intensity depends on the relative angle of the polarised light and the polariser.

This intensity of the wave which passes through the polariser is given by Malus' law:

$$I = I_0 \cos^2 \theta$$

where I_0 is the intensity of the wave when it reaches the polariser, I is the intensity which passes through the polariser, and θ is the angle between the plane of polarisation of the wave and the polariser.

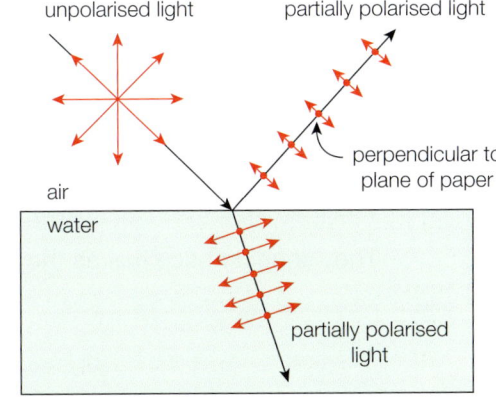

▲ **Figure 7.10** Visible light is polarised by reflection from water

Worked Example

By what fraction will the intensity of a polarised microwave be reduced when it passes through a polarising grille which is at an angle of 60° with respect to the polarisation of the wave?

$$\frac{I}{I_0} = \cos^2\theta, \qquad \frac{I}{I_0} = \cos^2 60, \qquad \frac{I}{I_0} = \frac{1}{4}$$

The wave intensity will be ¼ of the incident value.

Key terms

Displacement of a wave: the distance of a particle of the wave from its equilibrium position at any particular time.

Doppler effect: the effect of relative motion between a source of waves and an observer on the observed frequency, causing it to differ from the emitted frequency.

Doppler shift: the difference between the observed frequency and the emitted frequency of the waves from a source due to relative motion between the source and the observer.

Electromagnetic spectrum: the family of transverse waves from gamma rays to radio waves, which includes visible light.

Longitudinal (compression) waves: the direction of vibration is parallel to the direction of travel of the waves, e.g. a slinky.

Phase difference: the fraction of a cycle between the vibrations of two vibrating particles, measured in radians or degrees.

Polarised waves: transverse waves that vibrate in one plane only.

Speed of a wave: equals the wavelength multiplied by the frequency.

Transverse waves: the direction of vibration is perpendicular to the direction of travel of the waves, e.g. waves on a rope.

↑ Raise your grade

A train is travelling towards a station at a speed of 35 m s⁻¹. An observer standing on a platform in the station hears the train emitting sound of frequency 800 Hz. [Speed of sound in air = 330 m s⁻¹.]

(a) What frequency will the observer hear as the train approaches the station? [2]

$$f_0 = f_s \frac{v}{v - v_s} = 800 \times \frac{330}{330 - 35} = 895 \text{ Hz} \quad ✔✔$$

A clear calculation with the correct answer.

(b) What frequency will the observer hear when the train has passed through the station?

$$f_0 = f_s \frac{v}{v - v_s} = 800 \times \frac{330}{330 - 35} = 895 \text{ Hz} \quad ✗✗$$

The train is now moving away from the observer so the student should be using $f_0 = f_s \dfrac{v}{v + v_s}$.

They should notice this as they should expect the two answers to be different.

Exam-style questions

1 Transverse waves on a rope are travelling at a speed of $10.0\,\mathrm{m\,s^{-1}}$. The frequency of the waves is $5.0\,\mathrm{Hz}$.

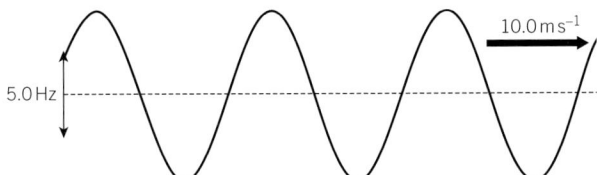

What is the phase difference between two points on the rope a distance $1.00\,\mathrm{m}$ apart?

A $\dfrac{\pi}{2}$ B π C $\dfrac{3\pi}{2}$ D 2π [1]

2 Gamma rays, ultraviolet waves and microwaves are all electromagnetic waves. Which option lists these waves in order of increasing frequency?

A gamma rays, ultraviolet, microwaves

B gamma rays, microwaves, ultraviolet

C ultraviolet, microwaves, gamma rays

D microwaves, ultraviolet, gamma rays   [1]

3 An organ pipe of length $50.0\,\mathrm{cm}$ is closed at one end. The speed of sound in air is $330\,\mathrm{m\,s^{-1}}$. What are the two lowest frequencies that can be produced by the pipe?

A $165\,\mathrm{Hz}$ and $330\,\mathrm{Hz}$ B $165\,\mathrm{Hz}$ and $495\,\mathrm{Hz}$

C $330\,\mathrm{Hz}$ and $660\,\mathrm{Hz}$ D $660\,\mathrm{Hz}$ and $880\,\mathrm{Hz}$ [1]

4 A racing car approaches a stationary observer with a constant speed u. If the speed of sound in air is v, what is the change in frequency heard by the observer as the car passes him?

A $\dfrac{fuv}{\left(v^2 - u^2\right)}$ B $\dfrac{2fuv}{\left(v^2 - u^2\right)}$

C $\dfrac{fuv}{\left(v^2 + u^2\right)}$ D $\dfrac{2fuv}{\left(v^2 + u^2\right)}$ [1]

5 A special loudspeaker, with a power output of $1.0 \times 10^{-4}\,\mathrm{W}$, emits sound energy equally in all directions.

(a) Calculate the *intensity* of sound (the sound energy per second per m²) at the following distances from the speaker:

(i) $3.0\,\mathrm{m}$ (ii) $9.0\,\mathrm{m}$ [2]

[Surface area of a sphere of radius r is $4\pi r^2$.]

(b) Compare the amplitude of vibration of air molecules at the two distances in (a).   [2]

6 (a) Define, for a transverse wave:

(i) the amplitude A (ii) the wavelength λ. [2]

(b) A student is investigating waves using a ripple tank. The drawing shows the waves at one particular moment.

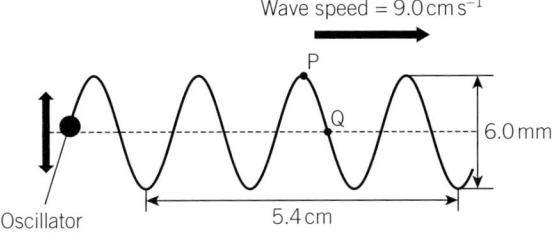

Determine:

(i) the amplitude of the waves

(ii) the wavelength of the waves

(iii) the frequency of the oscillator

(iv) the phase difference between points P and Q.   [4]

(c) Sketch a graph of the displacement of point P against time, for a period of $0.5\,\mathrm{s}$. Include appropriate scales.   [3]

7 (a) Describe the *Doppler effect*.   [2]

(b) An ambulance has a siren which emits a note of frequency $700\,\mathrm{Hz}$. It is travelling towards a stationary observer at a speed of $30\,\mathrm{m\,s^{-1}}$.

Determine:

(i) the frequency heard by the observer

(ii) the change in frequency heard by the observer once the ambulance has gone past.   [3]

8 A beam of polarised light passes through a polaroid and becomes polarised. It then passes through a second polaroid filter which is rotating at a constant rate.

(a) Describe what happens to the intensity of the light leaving the second polaroid for one complete rotation of the filter.   [5]

(b) The intensity of the incident beam reaching the second polaroid is $2.0\,\mathrm{Wm^{-2}}$. Calculate the intensity of the beam after it leaves the filter if the angle between the polarised wave and the polaroid is $20°$.   [2]

8 Superposition

Superposition

Principle of superposition

When two waves meet and overlap, the total displacement at any point is the vector sum of the individual displacements at that point – this is called the **principle of superposition**.

Interference

When two or more waves combine to produce a new wave, they can interfere constructively or destructively.

- **Constructive interference**: the two waves are **in phase**. They superpose ('add up') to produce a wave that has a larger amplitude than the original waves (see Figure 8.1a).

- **Destructive interference**: two waves are 180° (π radians) out of phase. They superpose so that the amplitude of the resultant wave is smaller. If the two waves have the same amplitude they cancel out completely (see Figure 8.1b).

Stationary waves

Stationary waves are produced when interference takes place between two progressive waves of equal frequency and amplitude travelling in opposite directions with the same speed along the same line. For example, if a rope is tied to a post at one end and made to oscillate at a particular frequency at the other end, the wave reflected by the post will overlap with the outward wave to produce a stationary wave.

Imagine two waves of equal frequency and amplitude, travelling towards each other at the same speed.

- At some point the two waves will be in phase and interfere constructively to form a wave with an amplitude which is twice the amplitude of one of the waves, as shown in Figure 8.2a.

- A quarter of a time period later, one wave will have moved a quarter of a wavelength to the right; the other wave will have moved the same distance to the left. The waves are now exactly half a wavelength out of step (in **antiphase**) and will cancel out completely, as shown in Figure 8.2b.

- A further time $T/4$ later the two waves are again in phase and will superpose constructively (Figure 8.2c).

> 💡 **Remember**
>
> The distance between two nodes (or two antinodes) is $\frac{\lambda}{2}$, and the distance between a node and an antinode is $\frac{\lambda}{4}$.

a Constructive interference

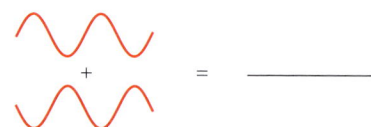

b Destructive interference

▲ **Figure 8.1** Interference

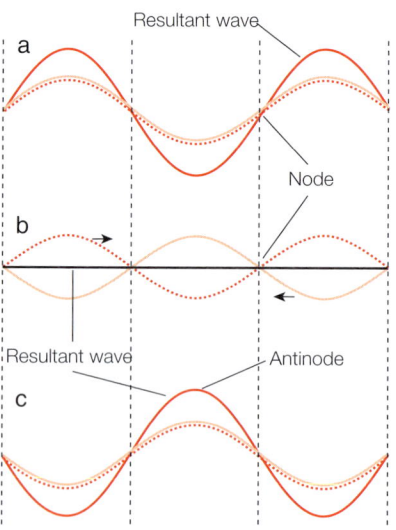

▲ **Figure 8.2** How stationary waves are formed

Examples of stationary waves

Stretched strings

When waves are produced on a stretched string, (for example, by plucking a guitar string) several different stationary waves can be formed.

- **Fundamental frequency (first harmonic)**: The simplest stationary wave on a stretched string is shown in Figure 8.3. This is the **fundamental mode of vibration** (also called the **first harmonic**).
  Nodes are points where the amplitude of vibration is zero.

  Antinodes are points where the amplitude of vibration is a maximum.

The second- and higher-order stationary waves occur at higher frequencies of vibration.

- **Second order (second harmonic)**: Points P and Q are in antiphase (π or 180° out of phase) – P is about to move up as Q is about to move down (see Figure 8.4).

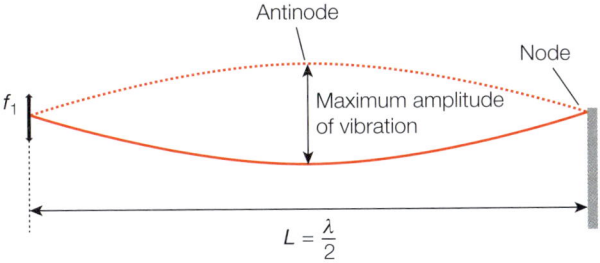

▲ **Figure 8.3** Fundamental mode of vibration

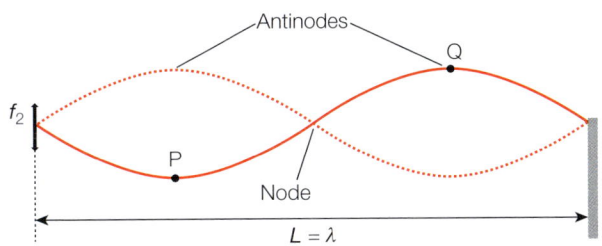

▲ **Figure 8.4** Second-order stationary wave

Air columns

Stationary sound waves can be produced in pipes and other columns of air. A sound wave travelling down a pipe can interfere with a sound wave reflected back from the end of the pipe to form a stationary longitudinal wave. Just as with stretched strings, there are several modes of vibration.

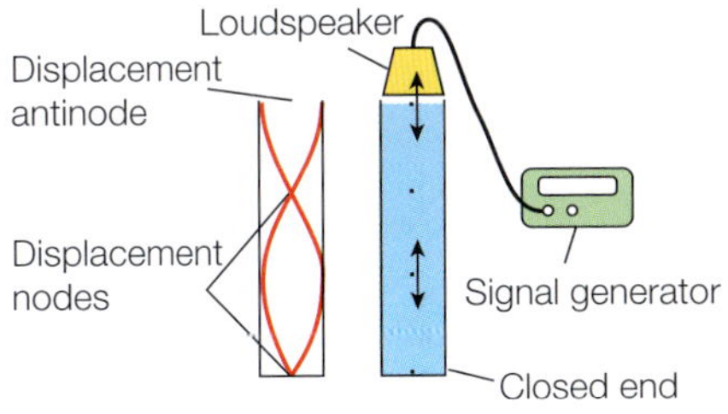

Closed pipe

The experiment in Figure 8.5 shows how stationary waves can be produced in an air column. As the frequency of the signal generator is gradually increased a number of louder sounds are heard at specific frequencies. Figure 8.5 shows how the second-order stationary sound wave is formed.

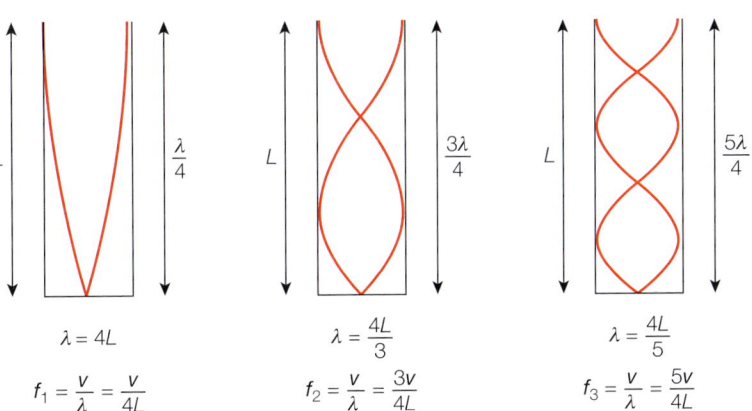

▲ **Figure 8.5** Stationary longitudinal wave

The air molecules cannot vibrate freely at the closed end making this a **displacement node**. The air molecules have no restrictions on their movement at the open end – this is a **displacement antinode**. The displacement nodes are pressure antinodes – the air is being constantly compressed and expanded at these points. The first three stationary waves in a pipe closed at one end are shown in Figure 8.6.

$$\lambda = 4L$$
$$f_1 = \frac{v}{\lambda} = \frac{v}{4L}$$

$$\lambda = \frac{4L}{3}$$
$$f_2 = \frac{v}{\lambda} = \frac{3v}{4L}$$

$$\lambda = \frac{4L}{5}$$
$$f_3 = \frac{v}{\lambda} = \frac{5v}{4L}$$

▲ **Figure 8.6** Stationary waves in a pipe closed at one end

Key term

Antinodes: fixed points in a stationary wave pattern where the amplitude is a maximum.

Open pipe

Stationary waves can also be produced in pipes open at both ends. In this case, both ends are displacement antinodes as the air molecules are free to vibrate with the maximum amplitude (see Figure 8.7).

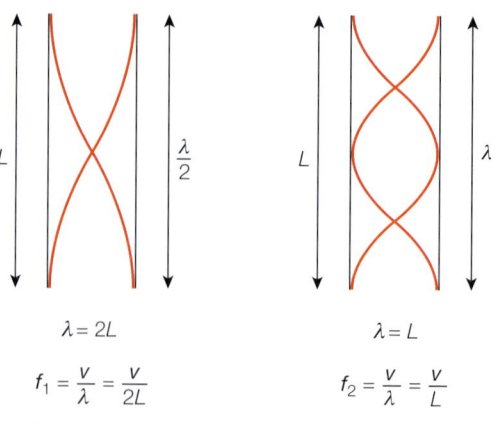
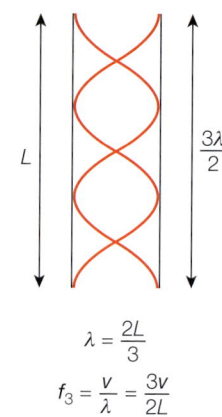

$$\lambda = 2L$$

$$f_1 = \frac{v}{\lambda} = \frac{v}{2L}$$

$$\lambda = L$$

$$f_2 = \frac{v}{\lambda} = \frac{v}{L}$$

$$\lambda = \frac{2L}{3}$$

$$f_3 = \frac{v}{\lambda} = \frac{3v}{2L}$$

▲ **Figure 8.7** Stationary waves in open pipes

Worked Example

A tuning fork is sounded above a cylinder of water, as shown in Figure 8.8. A tap is opened and the level of the water gradually falls. The sound becomes louder when the level of the water falls to certain levels.

Two successive loud sounds occur when the water level is 69.0 cm and 35.8 cm above the bench.

Determine the frequency of the tuning fork.
[Speed of sound in air = 340 m s^{-1}.]

Answer

The difference in the two levels of water must be the distance between two successive nodes $\left(\frac{\lambda}{2}\right)$, as shown in Figure 8.9:

$$\frac{\lambda}{2} = 69.0 - 35.8 = 33.2 \text{ cm} \quad \text{so} \quad \lambda = 66.4 \times 10^{-2} \text{ m}$$

$$f = \frac{v}{\lambda} = \frac{340}{66.4 \times 10^{-2}} = 512 \text{ Hz}$$

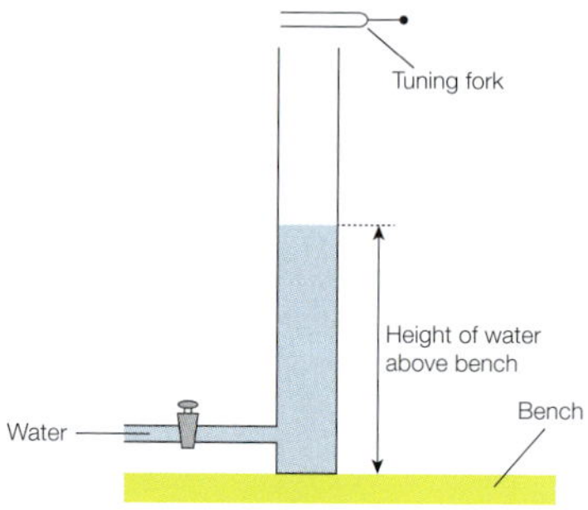

▲ **Figure 8.8** Investigating resonant frequencies

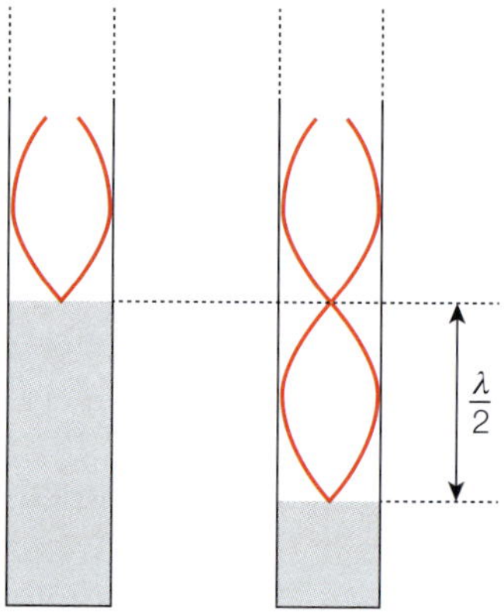

▲ **Figure 8.9**

Microwaves

Microwaves can also be used to demonstrate the properties of stationary waves. A microwave transmitter emits microwaves towards a metal plate which reflects the microwaves back towards the transmitter, as shown in

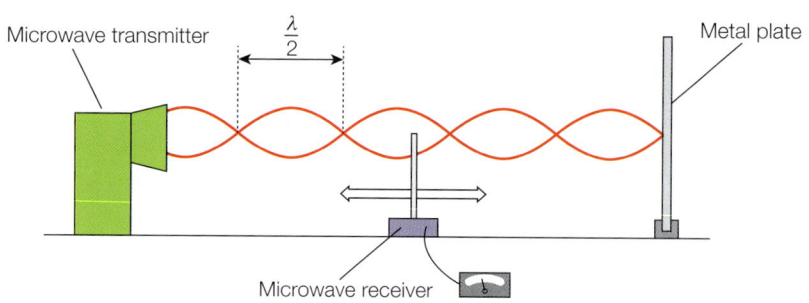

Microwave transmitter

$\frac{\lambda}{2}$

Metal plate

Microwave receiver

▲ **Figure 8.10** Microwave stationary waves

The transmitted and reflected waves superpose to form a stationary wave. As the microwave receiver is moved along a line directly between the microwave transmitter and the metal plate, it detects successive, equally spaced, strong, and weak signals. The distance between successive minima is the distance between two adjacent nodes – half a wavelength.

The nodes on a stationary wave have zero energy; the antinodes have maximum energy. As the positions of the nodes and antinodes do not change, no energy is transferred by a stationary wave.

> **Key terms**
>
> **Fundamental mode of vibration (also called first harmonic):** the simplest stationary wave.
>
> **Nodes:** points where the amplitude of vibration is zero (no displacement).
>
> **Stationary waves:** wave pattern with nodes and antinodes formed when two or more progressive waves of the same frequency and amplitude pass through each other.

Diffraction

When waves pass through a gap, or past a partial obstruction, they bend and spread out beyond the geometric 'shadow' region of the gap or obstruction, as shown in Figure 8.11. This effect is called **diffraction** and the waves are said to be 'diffracted'.

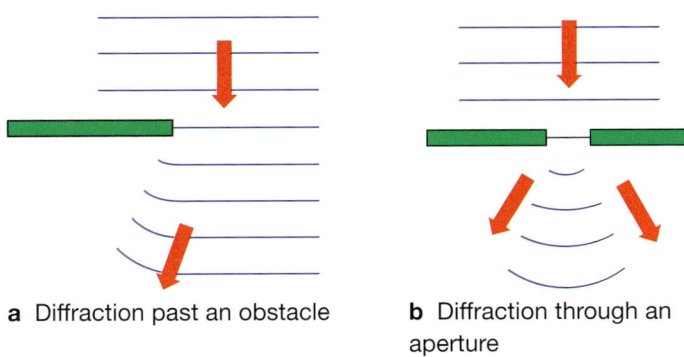

a Diffraction past an obstacle   **b** Diffraction through an aperture

▲ **Figure 8.11** Diffraction

> **Key terms**
>
> **Diffraction:** the spreading of waves when they pass through a gap or round an obstacle.
> **Principle of superposition:** the effect of two waves adding together when they meet.

The amount of spreading of a wave as it passes through an aperture depends on the size of the aperture. Most diffraction occurs when the aperture is a similar size to the wavelength of the waves, as shown in Figure 8.12a.

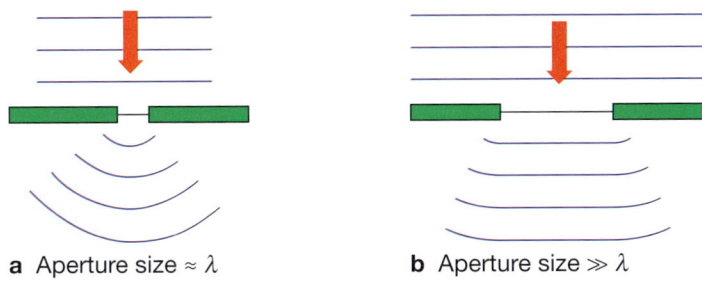

a Aperture size ≈ λ **b** Aperture size ≫ λ

▲ **Figure 8.12** Diffraction and aperture size

Two-source interference

If two sources of waves are in close proximity to each other (for example, two wave 'dippers' in a ripple tank, or two loudspeakers) the two sets of waves overlap. If the two sets of waves have the same frequency and similar amplitudes, and the two sources are **coherent** (have a fixed phase difference between them), they can produce an interference pattern, with points of constructive interference and points of destructive interference.

Figure 8.13 shows successive wavefronts produced by two vibrating dippers in a ripple tank (the lines represent the peaks, or crests, of waves). At points such as P, two wave peaks are interfering constructively and producing a larger amplitude; at points such as Q a wave peak from one source is overlapping with a wave trough from the other, cancelling out (destructive interference). At points such as R, two troughs are meeting and interfering constructively to produce a deeper trough.

If two loudspeakers are connected to the same signal generator and placed a short distance apart, a similar interference pattern to the one described in the ripple tank can be observed (see Figure 8.14).

A person walking from P to Q hears a series of loud and quiet sounds. Where the sound is loud, the **path difference** (the difference in the distance travelled by a sound wave from one loudspeaker compared to the other) must be a whole number of wavelengths, and so constructive interference occurs.

Where there is a quiet sound, the path difference must be an odd number of half-wavelengths so that the sound waves interfere destructively. The sound waves may not cancel out completely because the sound wave from one speaker will have travelled further than the other, so will have a smaller amplitude.

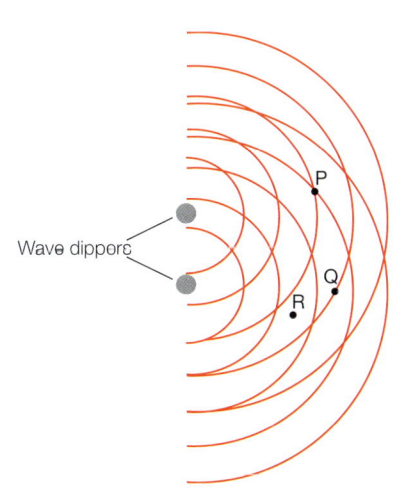

▲ **Figure 8.13** Interference patterns in a ripple tank

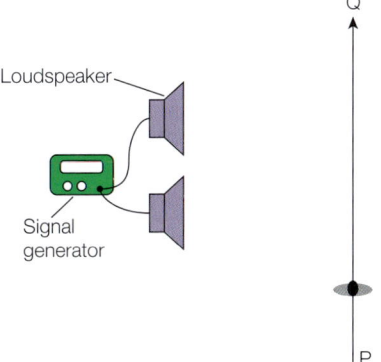

▲ **Figure 8.14** Interference patterns with sound waves

Young's double-slits experiment

Young's double-slits experiment provides evidence for the wave-like nature of light by producing interference fringes on a screen. Figure 8.15 shows how the experiment is set up.

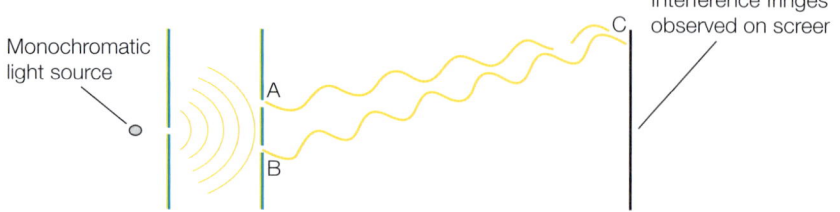

▲ **Figure 8.15** Young's double-slits experiment

Light from a monochromatic light source (e.g., a laser) passes through a single slit and diffracts (spreads out). It then passes through two narrow slits A and B. There is a fixed phase difference between the light emerging from slits A and B as the waves are part of the same wavefront; that is, the light from A and B is **coherent**.

Interference fringes (areas of light and dark) can be observed on a screen some distance away. Consider the case when the light from slits A and B emerges in phase. In Figure 8.15, the distance BC is slightly longer than the distance AC – there is a **path difference** between the light waves from slit A and slit B. If this distance is an odd number of half-wavelengths, the two waves will be out of phase (in antiphase) when they reach C and will interfere destructively – a dark fringe will be observed. If the path difference is a whole number of wavelengths the two waves will be in phase at C and interfere constructively, producing a bright fringe.

From Figure 8.16 the two waves emerging from slits A and B can be considered parallel as the distance D to the screen is much greater than the slit separation a. The path difference BP is $a\sin\theta$. The first-order bright fringe occurs when the path difference is λ, the second-order when the path difference is 2λ, and so on. The nth bright fringe occurs when the path difference is $n\lambda$. For a bright fringe:

$$n\lambda = a\sin\theta$$

Analysing Young's slits experiment

From Figure 8.16, $x_n = D\tan\theta$, but since θ is small, $\tan\theta \approx \sin\theta$, so:

$$x_n = D\sin\theta = D\frac{n\lambda}{a}$$

Similarly, the $(n+1)$th bright fringe is given by $x_{n+1} = D\frac{(n+1)\lambda}{a}$.

The distance between adjacent bright fringes is x, where:

$$x = (x_{n+1} - x_n) = \frac{D\lambda}{a} \quad \text{so} \quad \lambda = \frac{ax}{D}$$

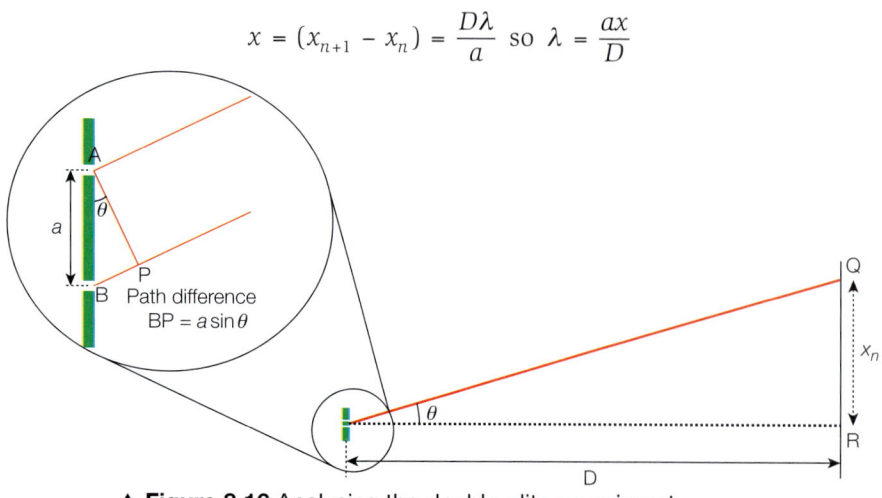

▲ **Figure 8.16** Analysing the double-slits experiment

Key terms

Coherent: when two sources of waves emit waves with a constant phase difference.

Constructive interference: when two waves are in phase.

Destructive interference: when two waves are out of phase.

Path difference: the difference in distances from two coherent sources to an interference fringe.

Maths Skills

For small angles:

$\tan\theta \approx \sin\theta \approx \theta$ where θ is measured in radians.

The wavelengths of visible light are small ($\approx 5 \times 10^{-7}$ m) so the fringes are very close together. To increase the fringe separation:

- Make the slit separation a as small as possible.
- Make the distance D between the slits and the screen as large as possible.
- Use light with as long a wavelength as possible (e.g., red light rather than blue light).

Carrying out the experiment in a darkened room also makes the fringes easier to detect.

Worked Example

Light of wavelength 589 nm is incident on a pair of slits, forming an interference pattern on a screen 1.40 m away. The bright fringes on the screen are 0.20 cm apart. Determine the separation of the two slits.

Answer

$$a = \frac{\lambda D}{x} = \frac{589 \times 10^{-9} \times 1.40}{0.20 \times 10^{-2}} = 4.12 \times 10^{-4} \text{ m } (0.412 \text{ mm})$$

Diffraction gratings

The double-slit in Young's experiment can be replaced by multiple slits. Increasing the number of slits has the effect of making much sharper and clearer **maxima** (bright lines or 'fringes'). A diffraction grating consists of many parallel slits extremely close together, ruled on a transparent plate. The light passing through each slit is diffracted, and constructive interference occurs only at very specific angles, the light waves cancelling each other out in all other directions.

For the light waves from two adjacent slits to be in phase and add up constructively, the path difference BP must be a whole number of wavelengths.
From Figure 8.17, if the distance between adjacent slits is d:

$$d \sin \theta = n\lambda$$

where n is an integer. There is also a **zero-order** maximum in the same direction as the incident beam (the waves coming from every slit of the grating have all travelled the same distance, so they are all in phase).

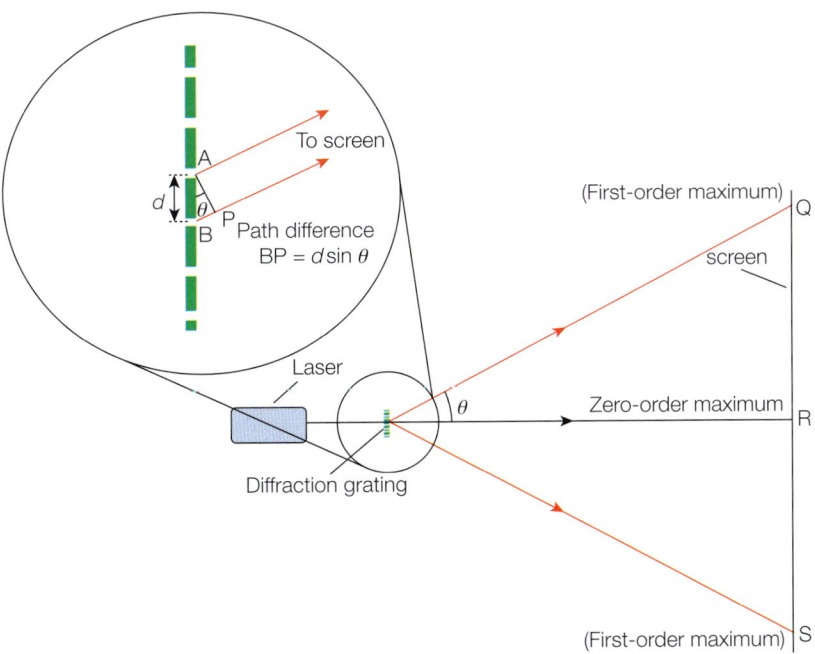

▲ **Figure 8.17** Analysing the diffraction grating experiment

↑ Raise your grade

(a) In relation to light waves, explain what is meant by the terms:

(i) monochromatic

A single colour ✗

> A more precise, scientific definition is needed. Monochromatic light waves are waves with a single wavelength (e.g., the light from a laser).

(ii) constructive interference. [3]

When two waves meet and make a bigger light wave ✔ ✗

> The right idea, but a better answer would be 'When two waves overlap, and are **in phase**, so that their **amplitudes** add up to make a light wave with a larger amplitude'.

(b) A student attempts to measure the wavelength of light using Young's slits, as shown below.

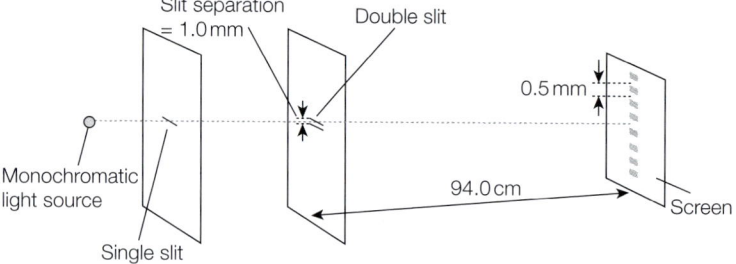

Slit separation = 1.0 mm · Double slit · 0.5 mm · Monochromatic light source · 94.0 cm · Screen · Single slit

(i) The light is diffracted by the single slit. Explain what is meant by *diffraction*.

The light bends as it passes through the slit ✔ ✗

> correct, but the candidate should have added '... and **spreads out**, beyond the geometric 'shadow' of the slit' for the second mark

(ii) Explain why the light from the double slit is *coherent*.

The light waves from the two slits are in phase. ✗

> The light entering each of the two slits is from the same wavefront so there must be a fixed phase difference between the light waves emerging from them (the light waves don't need to be in phase to be coherent).

(iii) Use the student's results to calculate the wavelength of the light used.

$$\lambda = \frac{ax}{D} = \frac{1.0 \times 10^{-3} \times 0.5 \times 10^{-3}}{94.0 \times 10^{-2}} = 5.32 \times 10^{-7} \, \text{m} \; ✔$$

> Correct calculation.

> Correct equation/method.

wavelength = 5.32 × 10⁻⁷ m ✔ [5]

The student wants to make the fringes further apart so that they are easier to see and measure accurately. Suggest two changes the student could make to the experiment to achieve this.

1 Move the screen further away from the slits ✔ Correct suggestion.

2 Move the two slits further apart ✗ [2]

> The opposite is true. From the two-slits equation $x = \frac{\lambda D}{a}$, so the fringe spacing x increases if a, the distance between the slits, is decreased.

Exam-style questions

1 Two waves superpose as shown below.

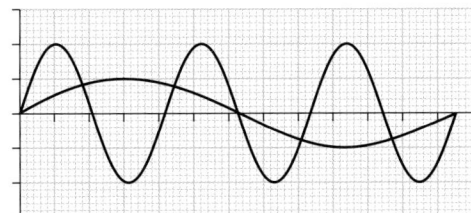

Which diagram shows the resultant of the two waves? [1]

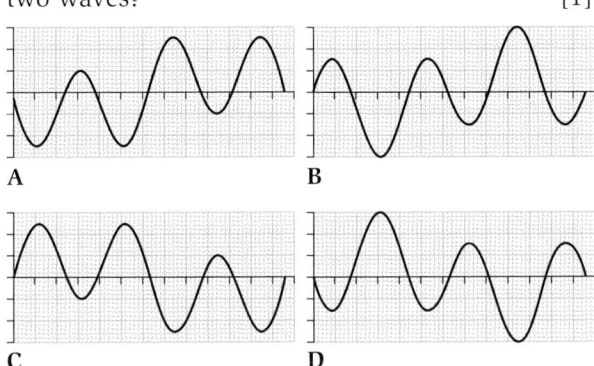

A   B

C   D

2 A string is stretched between points X and Y. One end of the string is vibrated, setting up a stationary wave, as shown below.

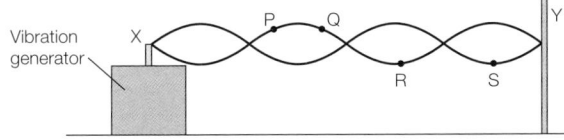

Which statement is correct?

A The string is oscillating at its fundamental frequency.

B The distance RS is one wavelength.

C Points P and Q are in phase.

D Points R is a node. [1]

3 An organ pipe of length 0.500 m is closed at one end. What are the two lowest resonant frequencies the pipe can produce?

[The speed of sound is $340 \, \text{m s}^{-1}$.]

A 170 Hz   340 Hz

B 170 Hz   510 Hz

C 227 Hz   340 Hz

D 227 Hz   510 Hz [1]

4 Which statement correctly describes the meaning of *diffraction*?

A When waves meet in phase and their amplitudes add up.

B When waves meet out of phase and cancel out.

C When waves pass an obstacle and bend, entering the geometric shadow of the object.

D When waves from two sources have a fixed phase difference. [1]

5 A diffraction grating has 400 lines mm^{-1}. When monochromatic light passes through the grating, the third-order maxima subtend an angle of 40°, as shown below.

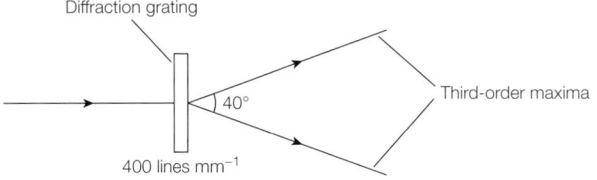

What is the wavelength of the light?

A 285 nm   **B** 410 nm   **C** 540 nm   **D** 820 nm [1]

6 White light passes through a diffraction grating and a series of visible spectra are observed on a screen some distance away, together with a white light maximum at a point directly in line with the diffraction grating and the light source (zero order). A series of visible spectra are seen either side of the zero order.

(a) Describe what is meant by *diffraction*. [2]

(b) Explain why:

(i) the central zero-order maximum is white

(ii) the first-order maximum for green light is in a different position from the first-order maximum for red light. [2]

(c) Blue light of wavelength 460 nm produces a third-order maximum at an angle of 17.7°. A wavelength of red light produces a second-order maximum at the same angle.

Calculate:

(i) the number of lines per millimetre of the diffraction grating

(ii) the wavelength of the red light. [3]

9 Electricity

Current and charge

An **electric current** is any flow of electrically-charged particles. In metal conductors the charge carriers are electrons; in liquids the charge carriers are ions.

The amount of **charge** on a charged particle is **quantised** – it is always a multiple of the electronic charge $e = 1.6 \times 10^{-19}$ C, where C stands for **coulomb**, the unit of electric charge.

The **current** I is the amount of charge flowing past a point each second. If a charge Q flows in a time t, the current I is given by $I = \dfrac{Q}{t}$

Key terms

Charge: measured in coulombs.

Electric current: any flow of electrically-charged particles.

Quantised: a discrete set of values.

Worked Example

A current of 60 mA flows in a wire for 1 minute.

Calculate:

(a) how much charge passes

(b) how many electrons pass. [$e = 1.6 \times 10^{-19}$ C.]

Answer

(a) $Q = It = 60 \times 10^{-3} \times 60 = 3.6$ C

(b) number of electrons = $\dfrac{3.6}{1.6 \times 10^{-19}}$

$= 2.3 \times 10^{19}$ electrons (2 s.f.)

When an electric current passes through a metal wire the charge is carried by electrons which 'drift' from one end of the wire to the other. Figure 9.1 shows a wire of cross-sectional area A carrying a current I. The **number density** n of the material is the number of free charge carriers per cubic metre. It is sometimes also called the charge density.

In time t, the charge carriers travel a distance $L = vt$, where v is the drift velocity, and so the number of charge carriers passing any given point in time t is nAL, where n is the number density of charge carriers and A is the cross-sectional area of the wire.

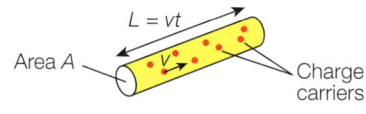

▲ **Figure 9.1** Charge carriers in a conductor

If the charge on each charge carrier is q, the current I is:

$$I = \frac{nALq}{t} = nAqv \qquad \left(\text{as } v = \frac{L}{t} \right)$$

Remember

$I = nAqv$

You need to be able to derive this equation.

Worked Example

Estimate the drift velocity of electrons in a copper wire of diameter 0.56 mm when a current of 60 mA passes through it. Assume the charge density of copper is 10^{29} m^{-3} and use $e = 1.6 \times 10^{-19}$ C.

Answer

$$v = \frac{I}{nAq} = \frac{60 \times 10^{-3}}{10^{29} \times \pi \times \left(0.28 \times 10^{-3}\right)^2 \times 1.6 \times 10^{-19}}$$

$$= 1.5 \times 10^{-5} \text{ m s}^{-1} \ (0.015 \text{ mm s}^{-1})$$

Remember

Conventional current I is shown flowing from positive to negative; in reality it is electrons which flow the other way – from the negative terminal to the positive terminal.

Potential difference and the volt

In the circuit shown in Figure 9.2, the electrons effectively 'pick up' energy as they pass through the cell and 'deposit' it at the lamp. The amount of energy delivered to the lamp per **coulomb** of charge passing through the lamp is called the **potential difference** (p.d.), or voltage V, across the bulb:

$$V = \frac{W}{Q}$$

where W is the work done when a charge Q passes.

In Figure 9.3 a **voltmeter** connected across the lamp measures how much energy per coulomb is lost, in volts (V). 1 volt is 1 joule per coulomb. A voltmeter is connected in parallel with ('across') a component because it is comparing the energy per coulomb on either side of the component.

In Figure 9.3, if the current flowing through a component is I and the p.d. across the component is V, the **power** P dissipated in the component is:

$$P = \frac{\text{energy per}}{\text{second}} = \frac{\text{no of coulombs}}{\text{per second}} \times \frac{\text{energy lost per}}{\text{coulomb}}$$

or $\qquad P = IV$

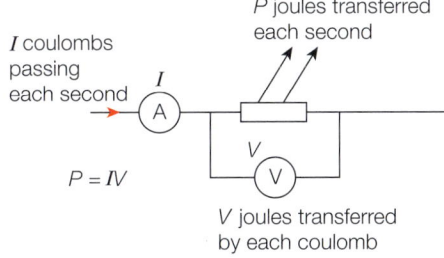

$P = IV$

I coulombs passing each second

P joules transferred each second

V joules transferred by each coulomb

▲ **Figure 9.3** Energy transfer

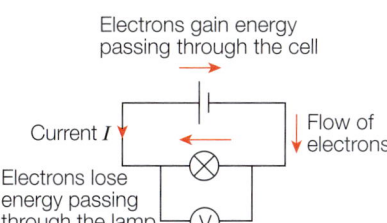

Electrons gain energy passing through the cell

Current *I*

Flow of electrons

Electrons lose energy passing through the lamp

▲ **Figure 9.2** Electron flow

Worked Example

An electric kettle is labelled '230 V, 3 kW'.

(a) Calculate the electric current in the kettle when in use.

(b) Determine the energy transferred to the kettle if it is switched on for 5 minutes.

Answer

(a) $I = \frac{P}{V} = \frac{3000}{230} = 13.0$ A

(b) energy = power × time = $3000 \times (5 \times 60) = 9 \times 10^5$ J   (900 kJ)

Resistance and the ohm

The **resistance** of an electrical component is a measure of how difficult it is to pass a current through the component (see Figure 9.4).

Resistance is defined by the equation:

$$R = \frac{V}{I}$$

If 1 A flows in a component when the potential difference across it is 1 V, then the resistance of the component is 1 ohm (Ω).

Combining the equations $P = IV$ and $V = IR$ gives:

$$P = I^2R$$

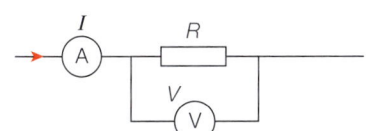

▲ **Figure 9.4** Electrical resistance

I–V characteristics

The circuit shown in Figure 9.5 can be used to investigate how the electrical resistance of a component changes. By adjusting the variable resistor, the current can be measured for different values of p.d. across the resistor.

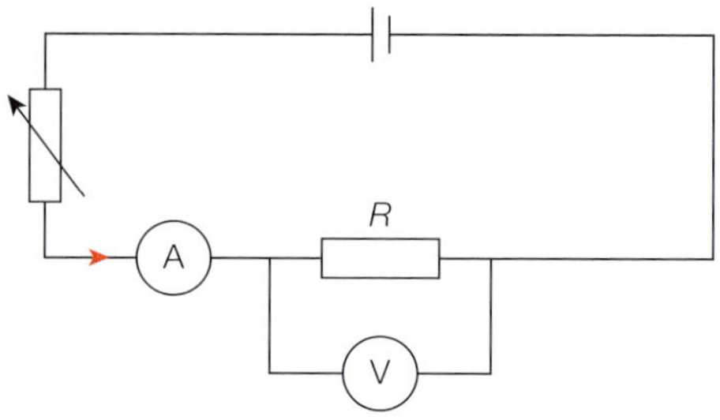

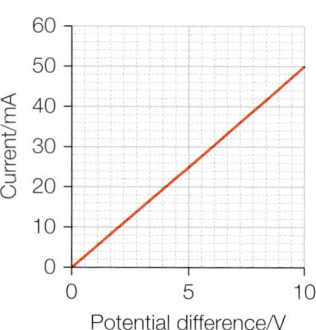

▲ **Figure 9.5** Measuring resistance

I–V characteristic of an ohmic resistor

For some resistors, including metal wires at constant temperature, the current passing through the resistor is directly proportional to the applied voltage, as shown in Figure 9.6.

$$R = \frac{V}{I} = \text{constant}$$

This is known as **Ohm's law**. A resistor that obeys Ohm's law is an **ohmic resistor**. Ohm's law is more of a 'rule' than a law, which only some resistors obey, and then only under certain conditions; for example, constant temperature.

Other electrical components generally do not obey Ohm's law.

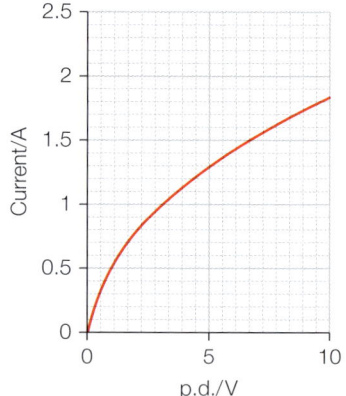

▲ **Figure 9.6** *I–V* characteristic of an ohmic resistor

> ★ **Exam tip**
>
> When describing Ohm's law, it is not enough to write:
>
> $$R = \frac{V}{I}$$
>
> This is just the definition of resistance; for Ohm's law to be obeyed the resistance must be **constant**.
>
> To calculate the resistance, read off a value of *V* and the corresponding value of *I* on the graph and then calculate *V/I*.

I–V characteristic of a filament lamp

For a filament lamp doubling the voltage from 5 V to 10 V increases the current, but doesn't double it, as shown in Figure 9.7. Since *R = V/I* the resistance of the lamp must be **increasing**. This is because as more current passes through the lamp, the filament (usually made of tungsten metal) gets hotter, and so the metal atoms vibrate much more vigorously, making it more difficult for electrons to pass through.

I–V characteristic of a semiconductor diode

When a semiconductor diode is **reverse-biased** (connected in its reverse direction – see Figure 9.8a) it has a very large resistance – almost no current flows as shown in Figure 9.9.

If the diode is **forward-biased** (connected in its forward direction – see Figure 9.8b.) a small p.d (~0.6 V) causes the diode to conduct with almost no resistance as shown in Figure 9.9.

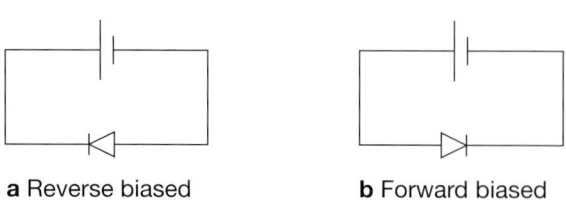

a Reverse biased   **b** Forward biased

▲ **Figure 9.8** Forward- and reverse-biased diodes

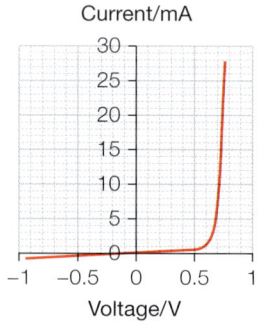

▲ **Figure 9.7** *I–V* characteristic of a filament lamp

▲ **Figure 9.9** *I–V* characteristic of a semiconductor diode

Resistivity

For a conductor such as a metal wire or a carbon resistor (see Figure 9.10), the resistance of the conductor depends on its dimensions and on the material it is made from. Doubling the length of a wire doubles the resistance. Doubling the diameter of a wire increases the cross-sectional area of the wire by a factor of four which **decreases** the resistance by the same factor (it is as if four of the original wires have been connected in parallel).

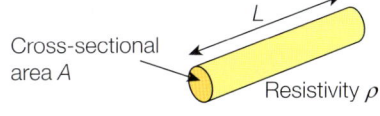

▲ Figure 9.10 Resistivity

For a conductor of length L and cross-sectional area A, we can write:

$$R = \frac{\rho L}{A}$$

where ρ is the **resistivity** of the material. ρ is measured in Ωm. Good conductors such as copper have very low resistivities (about $10^{-8}\,\Omega$m). Good insulators have very high resistivities (about 10^{12}–$10^{16}\,\Omega$m). Semiconductors have resistivities between 10^{-3} and $10^{8}\,\Omega$m.

> **Remember**
>
> Resistance, $R = \dfrac{\rho L}{A}$
>
> The units of resistivity ρ are Ωm.

Worked Example

Constantan wire is an alloy of copper and nickel with a resistivity of $49.0 \times 10^{-5}\,\Omega$mm. What is the resistance of a constantan wire of length 50.0 cm and diameter 0.46 mm?

Answer

$$R = \frac{\rho L}{A} = \frac{4.90 \times 10^{-7} \times 0.500}{\pi \times \left(0.23 \times 10^{-3}\right)^2} = 1.47\,\Omega$$

> ★ **Exam tip**
>
> Check that you've halved the diameter to find the radius when using πr^2 to find the cross-sectional area of the wire, and don't forget to square r!

Sensing devices

Light-dependent resistor (LDR)

The resistance of some semiconductor materials is altered by the amount of light falling on them (Figure 9.11). The light energy is absorbed by the material, releasing electrons in the material to become 'conduction electrons'. A typical **light-dependent resistor** (LDR) has a resistance of 10 MΩ in complete darkness but this falls to a few hundred ohms in bright light.

LDRs are used in a range of electronic devices to operate light-sensitive switches including automatic street lights and camera light-meters. Its symbol is shown in Figure 9.12.

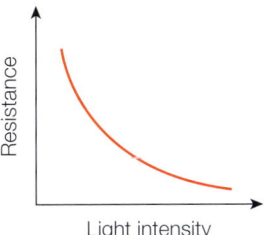

▲ Figure 9.11 Resistance of an LDR against light intensity

▲ Figure 9.12 Symbol for an LDR

Negative temperature coefficient thermistor

The resistance of thermistors changes with temperature. A **negative temperature coefficient** thermistor has a smaller resistance at higher temperatures, as shown in Figure 9.13 – the rise in temperature increases the number of electrons that are 'free' to conduct.

The change of resistance can be very large for a small increase or decrease in temperature. Thermistors are used in thermostats, fire alarms and digital thermometers. They monitor the oil and water temperature in cars and prevent high currents flowing through computers and electric motors when first switched on. Its symbol is shown in Figure 9.14.

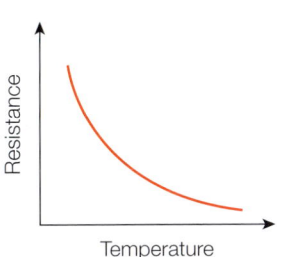

▲ Figure 9.13 Resistance of a thermistor against temperature

▲ Figure 9.14 Symbol for a thermistor

↑ **Raise your grade**

(a) Define the *ohm*. [1]

<u>The ohm is the S.I. unit of resistance</u> ✗

> The statement is true but is not a definition of the ohm.
>
> A better answer would be '1 ohm is 1 volt per ampere'.

(b) A three-way light bulb has two filaments and power outputs of 60 W, 120 W, and 180 W. It is connected to a 120 V mains supply as shown.

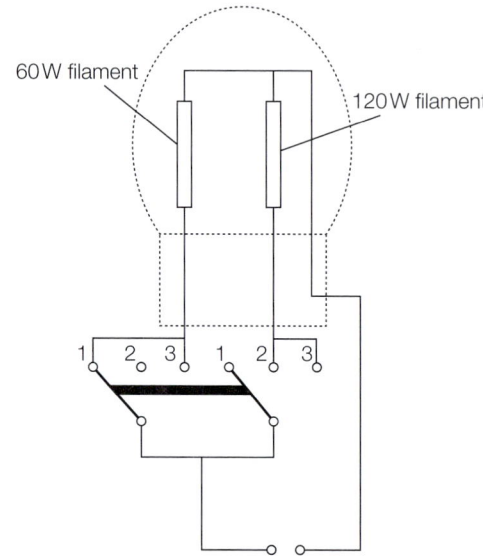

(i) Show that, when the switch is in position 1, the resistance of the 60 W filament is 240 Ω. [1]

$$P = \frac{V^2}{R} \Rightarrow R = \frac{V^2}{P} = \frac{120^2}{60} = 240\,\Omega \checkmark$$

> Method and correct substitution.

(ii) The 60 W filament is a tungsten wire of length 580 mm and diameter 0.046 mm. Calculate the resistivity of tungsten. ✓ Substitution and calculation. [4]

$$\rho = \frac{RA}{L} = \frac{240 \times \pi \times (0.023 \times 10^{-3})^2}{580 \times 10^{-3}} = 6.9 \times 10^{-7}\,\Omega\,m^{-1}$$ ✓

> The units of resistivity are Ω m, not Ω m⁻¹.

Method

resistivity of tungsten = $6.9 \times 10^{-7}\,\Omega\,m^{-1}$ ✗ [4]

(c) The switch is moved to position 2. Calculate the resistance of the 120 W filament. [2]

$$P = \frac{V^2}{R} \Rightarrow R = \frac{V^2}{P} = \frac{120^2}{120} = 120\,\Omega \checkmark$$

> Substitution and calculation.

Method.

resistance of the 120 W filament = 120 Ω

(d) The switch is moved to position 3. Calculate the total current drawn from the 120 V supply. [2]

$$\text{Total resistance} = 240 + 120 = 360\,\Omega \text{ ✗}$$

$$\text{Current } I = \frac{V}{R} = \frac{120}{360} = 0.33\,A \text{ ✗}$$

> The two filaments are connected in parallel, not in series. The combined resistance is 80 Ω, and the total current is $\frac{120}{80} = 1.5\,A$

current drawn from the supply = 0.33 A

Exam-style questions

1 An electric toaster is labelled 230 V, 1 kW. It is switched on for 3 minutes. How many electrons pass through the toaster in this time?

A 8.2×10^{17} **B** 1.6×10^{19}

C 4.9×10^{21} **D** 8.0×10^{21} [1]

2 An electric lamp has a resistance of 720 Ω when switched on. If a charge of 25 C passes through the lamp in one minute, what is the power of the lamp?

A 30 W   **B** 75 W   **C** 125 W   **D** 300 W   [1]

3 A thin slice of germanium of thickness 0.2 mm and width 2.5 mm, carries a current of 30 μA.

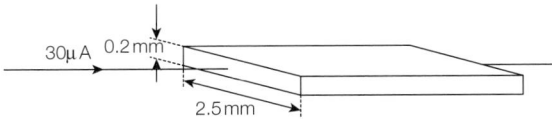

The density of charge carriers in germanium is 6.0×10^{20} m^{-3}. Assuming all the charge carriers are electrons, what is their drift velocity?

A 0.063 m s^{-1} **B** 0.63 m s^{-1}

C 6.3 m s^{-1} **D** 0.063 m s^{-1} [1]

4 A student is investigating how the resistance of a metal wire varies with the diameter of the wire.

She measures the resistance R of wires made of the same material and the same length, but different diameters d.

Which graph should she plot to obtain a straight line?

A R against d^2 **B** R against $\sqrt{d}$

C R against $\dfrac{1}{d^2}$ **D** R against $\dfrac{1}{\sqrt{d}}$ [1]

5 A wire resistor has a resistance of R ohms. What is the resistance of a wire of the same material, but three times as long, with twice the diameter?

A $\dfrac{2}{3}R$ **B** $\dfrac{3}{4}R$ **C** $\dfrac{4}{3}R$ **D** $\dfrac{3}{2}R$ [1]

6 A 50 W heater is made from a metal wire of resistivity 4.9×10^{-7} Ω m and diameter 0.56 mm. The heater is connected to a 12 V car battery.

What is the length of the wire?

A 0.12 m   **B** 0.48 m   **C** 1.4 m   **D** 5.8 m   [1]

7 A power cable consists of six strands of copper wire, each of diameter 1.2 mm. The resistivity of copper is 1.7×10^{-8} Ω m.

What is the resistance of 1.0 km of the cable?

A 0.10 m   **B** 0.42 m   **C** 2.5 m   **D** 15 m   [1]

8 A wire of resistance 5.0 Ω is to be made by extruding 1 cm³ of copper wire.

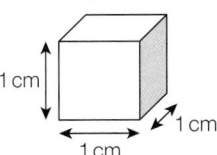

The resistivity of copper is 1.7×10^{-8} Ω m.

Calculate:

(a) the length of the wire   [2]

(b) the diameter of the wire.   [2]

9 The graph below shows how the resistance of a thermistor changes with temperature.

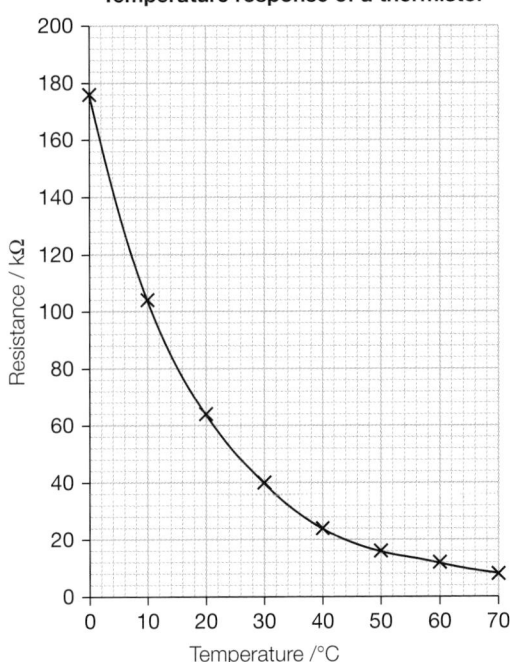

Temperature response of a thermistor

The thermistor is connected in series with a 6.0 V battery.

(a) Calculate the current in the circuit when the temperature of the thermistor is 30°C.   [2]

(b) Describe how the changes in resistance for the thermistor could have been investigated to produce this graph.   [4]

Knowledge check

You should be able to:

- describe changes in currents and potential difference in electrical circuits
- describe energy gains and losses when charges pass through a voltage (p.d.)
- analyse circuits in terms of resistance, potential difference and current using $V = IR$.

Practical circuits

Electromotive force (e.m.f.), potential difference (p.d.), and internal resistance

- **Electromotive force**: The **electromotive force** (e.m.f.) E of a power supply, such as a cell or a laboratory power pack, is the energy given to each coulomb of charge as it passes through the supply. The name is slightly misleading as it is not a force at all!

- **Internal resistance**: Cells or power supplies have an **internal resistance** r due to their physical composition; charges do not move totally freely though them.

- **External potential difference**: As charge moves around a circuit it transfers energy to other forms (e.g., heat and light). The energy transferred by each coulomb as it passes through a component (such as a lamp or resistor) is the **potential difference** (p.d.) across the component.

In the circuit shown in Figure 10.1:

$$E = Ir + IR$$

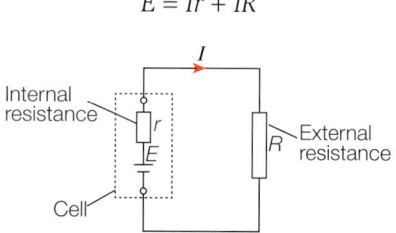

▲ **Figure 10.1** Electromotive force, internal resistance, and terminal potential difference

> **Remember**
>
> $$\text{voltage} = \frac{\text{energy}}{\text{charge}}$$
>
> 1 volt = 1 joule coulomb^{-1}
> $1\,V = 1\,J\,C^{-1}$

The precise meanings of terms such as e.m.f. and terminal p.d. are often confused. Table 10.1 gives exact descriptions of what each term means.

▼ **Table 10.1** Electromotive force, terminal potential difference, and 'lost volts'

Symbol	Name	Meaning
E	e.m.f.	energy supplied to each coulomb as it passes through the cell
Ir	the 'lost' volts	energy lost by each coulomb in passing through the cell
IR	external p.d.	potential difference across the external resistor
$E - Ir$	terminal p.d.	potential difference between the terminals of the cell or power supply

> **Key terms**
>
> **Electromotive force (e.m.f.) E:** the amount of electrical energy per unit charge produced inside a source of electrical energy.
>
> **External potential difference:** as charge moves around a circuit it transfers energy to other forms (e.g., heat and light).
>
> **Internal resistance:** some energy is needed to 'push' the charge through the cell as the cell itself has some electrical resistance. The resistance of the cell or power supply is called the **internal resistance** r.

Kirchhoff's laws

Kirchhoff's first law (conservation of charge)

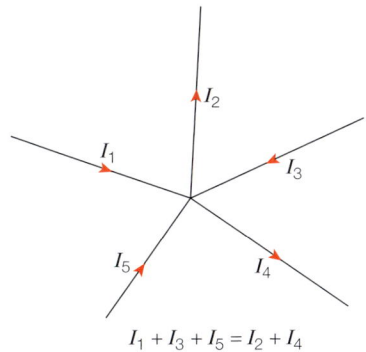
▲ **Figure 10.3** Kirchhoff's first law

$$I_1 + I_3 + I_5 = I_2 + I_4$$

> **Remember**
>
> **Kirchhoff's first law:** The total current flowing into a junction is equal to the total current flowing out of the junction.
>
> It is effectively a statement of the conservation of electric charge (see Figure 10.3).

Worked Example

Determine the readings on ammeters A_3 and A_4 in Figure 10.4.

Answer

At junction P, the current divides with $38 - 12 = 26\,\text{mA}$ heading towards Q.

At junction Q, the current divides:

reading on ammeter $A_3 = 26 - 15 = 11\,\text{mA}$

reading on ammeter $A_5 = 15 + 11 + 12 = 38\,\text{mA}$.

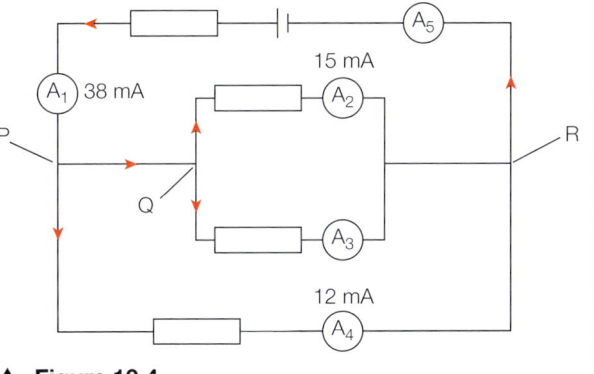
▲ **Figure 10.4**

Kirchhoff's second law (conservation of energy)

> **Remember**
>
> **Kirchhoff's second law:** For any complete loop in a circuit, the sum of the e.m.f.s round the loop is equal to the sum of the potential drops around the loop.
>
> It is a statement of the conservation of energy in electrical circuits (see Figure 10.5).

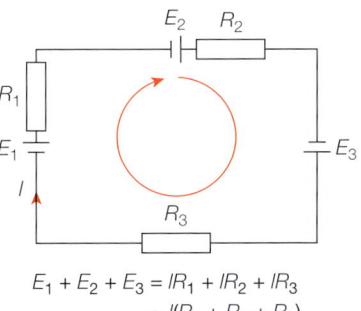

$$E_1 + E_2 + E_3 = IR_1 + IR_2 + IR_3$$
$$= I(R_1 + R_2 + R_3)$$

▲ **Figure 10.5** Kirchhoff's second law

A useful way of thinking about Kirchhoff's second law is to imagine a single coulomb of charge going around the circuit. It gains energy passing through a source of e.m.f. and loses energy in passing through resistors. By the time a coulomb completes a circuit, it will have lost as much energy as it has gained.

Worked Example

In the circuit shown in Figure 10.6, calculate the currents I_1 and I_2.

Answer

For the left-hand loop: $V = IR$

(eqn 1) $7 = 3I_1 + 2(I_1 + I_2)$

For the right-hand loop: $V = IR$

(eqn 2) $5 = I_2 + 2(I_1 + I_2)$

Re-arranging eqn 1 and eqn 2:

$7 = 5I_1 + 2I_2$ and $5 = 3I_2 + 2I_1$

Solving these equations gives $I_1 = I_2 = 1\,\text{A}$.

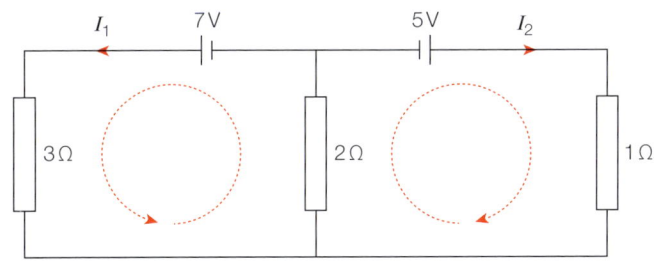
▲ **Figure 10.6**

> **Key terms**
>
> **Kirchhoff's first law:** the total current flowing into a junction is equal to the total current flowing out of the junction.
>
> **Kirchhoff's second law:** for any complete loop in a circuit, the sum of the e.m.f.s round the loop is equal to the sum of the potential drops around the loop.

Resistors in series

The current passing through several resistors connected in series must be the same (see Figure 10.7).

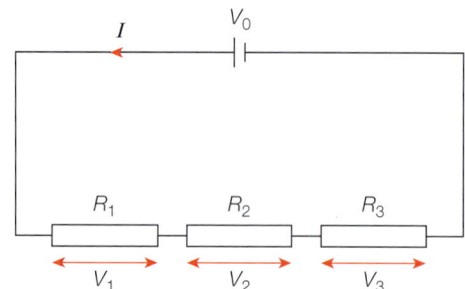

▲ **Figure 10.7** Resistors in series

From Kirchhoff's second law:

$$V_0 = V_1 + V_2 + V_3 = IR_1 + IR_2 + IR_3$$
$$V_0 = I(R_1 + R_2 + R_3)$$

The single resistor of resistance R that is equivalent to the three resistors in parallel is:

$$R = R_1 + R_2 + R_3$$

Resistors in parallel

The potential difference across a number of resistors connected in parallel is the same across each resistor (see Figure 10.8).

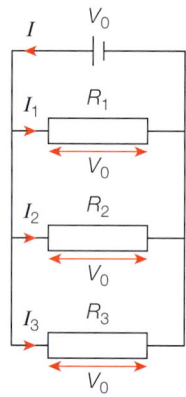

▲ **Figure 10.8** Resistors in parallel

From Kirchhoff's first law:

$$I = I_1 + I_2 + I_3 = \frac{V_0}{R_1} + \frac{V_0}{R_2} + \frac{V_0}{R_3}$$

$$I = V_0 \left(\frac{1}{R_1} + \frac{1}{R_2} + \frac{1}{R_3} \right)$$

The single resistor of resistance R that could replace the three resistors in parallel is given by:

$$\frac{1}{R} = \left(\frac{1}{R_1} + \frac{1}{R_2} + \frac{1}{R_3} \right)$$

Resistors connected in parallel have a smaller combined resistance.

> **★ Exam tip**
>
> For resistors in series:
>
> $$R = R_1 + R_2 + R_3 + \dots$$
>
> This formula is provided in Exam Papers 1, 2, and 4, but you need to be able to derive it.

> **★ Exam tip**
>
> For resistors in parallel:
>
> $$\frac{1}{R} = \left(\frac{1}{R_1} + \frac{1}{R_2} + \frac{1}{R_3} \dots \right)$$
>
> When using this equation, don't forget to turn your answer 'upside-down' to find R.
>
> This formula is provided in Exam Papers 1, 2, and 4, but you need to be able to derive it.

Worked Examples

1 Six equal resistors, each of resistance R, are connected as shown in Figure 10.9.

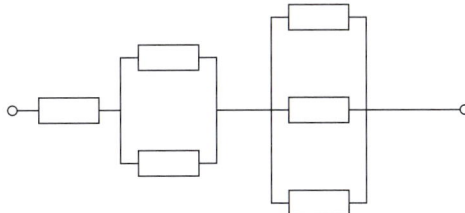

▲ **Figure 10.9**

Determine the total resistance of this combination of resistors.

Answer

The three resistors in parallel combine to give a total resistance of $\dfrac{R}{3}$.

The two resistors in parallel combine to give a resistance of $\dfrac{R}{2}$.

The total resistance is:

$$R + \frac{R}{2} + \frac{R}{3} = \frac{11R}{6}$$

2 Calculate the resistance of the resistor that must be placed in parallel with a resistor of resistance $15\,\Omega$ to have a combined resistance of $6\,\Omega$.

Answer

$$\frac{1}{6} = \frac{1}{15} + \frac{1}{R} \quad \text{so} \quad \frac{1}{R} = \frac{1}{6} - \frac{1}{15} = \frac{5-2}{30} = \frac{3}{30} = \frac{1}{10}$$

$$R = 10\,\Omega$$

Potential dividers

Two or more resistors can be connected in series to form a **potential divider** circuit. As the name suggests, a potential divider is a way of sharing, or dividing up, a potential difference between the different resistors. Figure 10.10 shows a simple potential divider circuit.

$$V_{\text{S}} = V_1 + V_2 = I(R_1 + R_2) \quad \text{so} \quad I = \frac{V_{\text{S}}}{R_1 + R_2}$$

Hence

$$V_1 = IR_1 = \left(\frac{R_1}{R_1 + R_2}\right) V_{\text{S}} \qquad \text{similarly} \qquad V_2 = IR_2 = \left(\frac{R_2}{R_1 + R_2}\right) V_{\text{S}}$$

If the resistors have the same resistance, half the supply voltage V_{S} is across each of the two resistors. If one resistor is twice as big as the other, the p.d. across the larger resistor is twice as much as the p.d. across the smaller resistor.

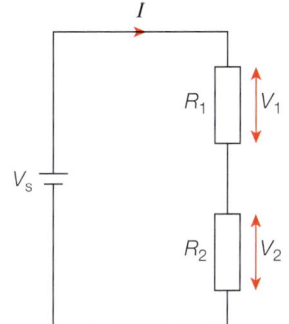

▲ **Figure 10.10** Potential divider

Key terms

Potential difference (p.d.): the energy transferred by each coulomb as it passes through a component (such as a lamp or resistor).

Potential divider (also known as a potentiometer): two or more resistors in series connected to a source of fixed potential difference.

Worked Examples

1   In the circuit shown in Figure 10.11, calculate the potential difference across the 5 kΩ resistor.

Answer

The 'fraction' of the supply p.d. across the 5 kΩ resistor is:
$\left(\dfrac{5}{5+15}\right) = \dfrac{1}{4}$, so the p.d. across the 5 kΩ resistor $= \dfrac{1}{4} \times 12 = 3\,\text{V}$

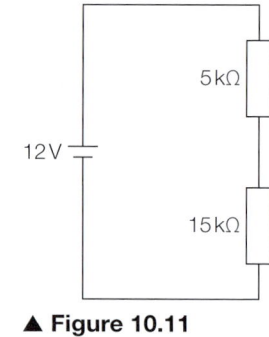

▲ **Figure 10.11**

2   The resistance of a negative temperature coefficient thermistor decreases as its temperature increases. When its temperature is 20 °C its resistance is 90 kΩ; when its temperature is 50 °C its resistance is 30 kΩ.

For the circuit shown in Figure 10.12:

(a)   Determine the output p.d. V_{out} when the temperature is:

(i)   20 °C   (ii)   50 °C.

(b)   State how you would alter the circuit so that the output p.d. decreased as the temperature increased.

Answer

(a)   (i) $V_{\text{out}} = \left(\dfrac{10}{10+90}\right) \times 10 = 1.0\,\text{V}$ (ii) $V_{\text{out}} = \left(\dfrac{10}{10+30}\right) \times 10 = 2.5\,\text{V}$

(b)   Swapping the thermistor and the fixed resistor around would mean that V_{out} would decrease as the temperature increased.

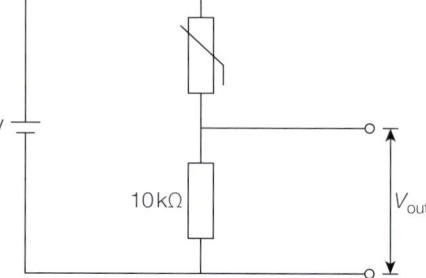

▲ **Figure 10.12**

A continuously variable potential divider (Figure 10.13) can be made by replacing the fixed resistors with a variable resistor (e.g., a length of resistance wire or a rheostat).

The output p.d. V_0 can have any value from 0 V to V_s, by moving the sliding connector P up or down.

$$V_O = \left(\dfrac{x}{l}\right)V_s$$

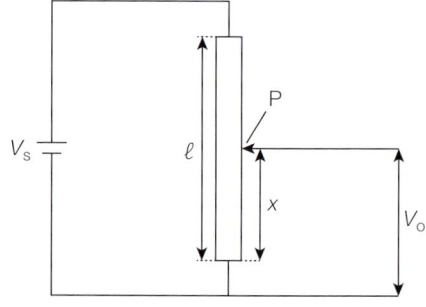

▲ **Figure 10.13** Continuously variable potential divider

Potentiometers

A potential divider is also known as a **potentiometer**, particularly when used to compare potential differences. Figure 10.14 is an example of a simple potentiometer circuit used to measure the e.m.f. E of a **dry cell** (the correct name for a torch 'battery').

The e.m.f. of the **driver cell** is V_0 and is known to a high degree of accuracy. The slider is moved up or down until the reading on the centre-zero galvanometer (a sensitive ammeter) reads zero – the potentiometer is then balanced. The e.m.f. of the dry cell can be found from:

$$E = \left(\dfrac{x}{l}\right)V_0$$

The advantages of this method of measuring potential differences are:

* it is a *null* method (the galvanometer only has to read '0' accurately)

* no current is drawn from the unknown p.d. so there are no 'lost' volts across the internal resistance of the dry cell

* the accuracy can be increased by using a longer resistance wire.

A disadvantage of this method is the apparatus is bulky and slow to use compared to a digital voltmeter.

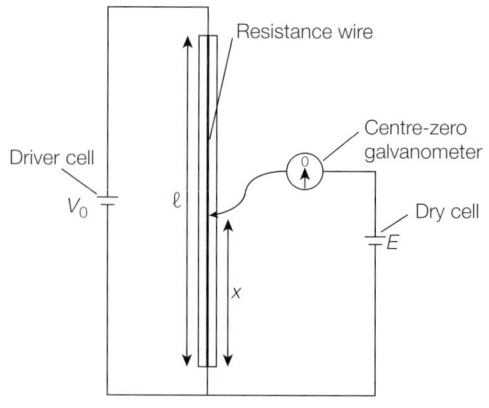

▲ **Figure 10.14** Potentiometer circuit

Worked Example

A potentiometer is connected to a dry cell of unknown e.m.f. E as shown in Figure 10.15. The balance length (when the galvanometer reads zero) is 54.7 cm.

When the dry cell is replaced with a reference cell of e.m.f. 1.02 V the balance length changes to 37.4 cm. Calculate the e.m.f. of the dry cell.

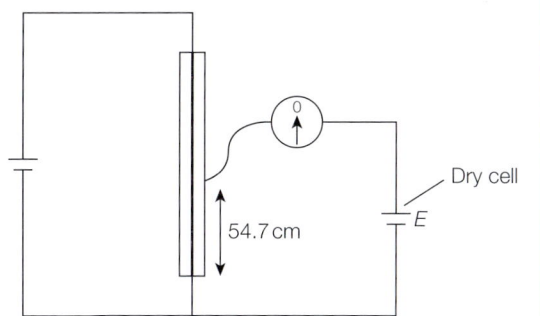

▲ **Figure 10.15**

Answer

For the dry cell:
$$E = \left(\frac{x}{l}\right)V_0, \quad E = \frac{54.7}{l}V_0 \quad \text{(eqn 1)}$$

For the reference cell: $102 = \left(\frac{37.4}{l}\right)V_0$ so $\frac{V_0}{l} = \frac{1.02}{37.4}$ (eqn 2)

Substituting eqn 2 into eqn 1:
$$E = 54.7 \times \frac{1.02}{37.4} \quad \text{so} \quad E = 1.49 \ V$$

Electronic sensors

Circuits designed to detect changes in the environment, such as temperature or light intensity, depend on producing an output p.d. that changes with the variable being monitored. The sensing device is a resistor which changes resistance with changes in the variable, and is part of a potential divider circuit.

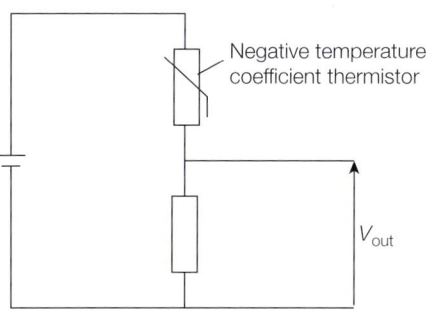

▲ **Figure 10.16** Temperature sensor

Temperature sensor

The resistance of a negative temperature coefficient thermistor decreases as the temperature increases. In the potential-divider circuit shown in Figure 10.16, a temperature increase will reduce the resistance of a thermistor and increase the output p.d. V_{out}. A decrease can be achieved by either swapping the fixed resistor and the thermistor, or using a positive temperature coefficient thermistor.

Light sensor

A light sensor (Figure 10.17) can be made in a similar way, replacing the thermistor with a light-dependent resistor (LDR). As the incident light intensity increases, the resistance of the LDR decreases and the p.d. across the LDR decreases. The output V_{out} increases.

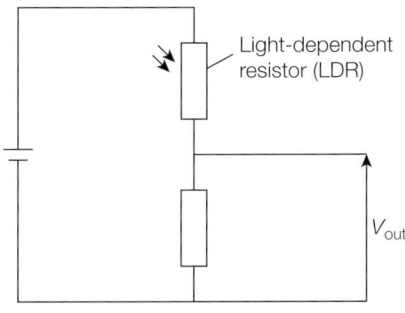

▲ **Figure 10.17** Light sensor

⬆ Raise your grade

(a) In relation to a cell, explain the meaning of:

(i) electromotive force (e.m.f.) [2]

The force from the cell pushing the current ✗

> e.m.f. is **not** a force. The e.m.f. is the electrical energy per unit charge produced inside the cell.

(ii) terminal potential difference (p.d.).

The voltage from the cell ✗

> As the name suggests, it is the potential difference between the two terminals of the cell.

(b) State Kirchhoff's second law. [2]

The sum of the e.m.f.s is equal to the sum of the potential drops ✓ ✗

> For the second mark, the candidate should have added 'around any complete loop of a circuit'.

(c) A dry cell, of e.m.f. E and internal resistance r is connected to an external resistor of resistance R. The current I is recorded for two different values of R, as shown [3]

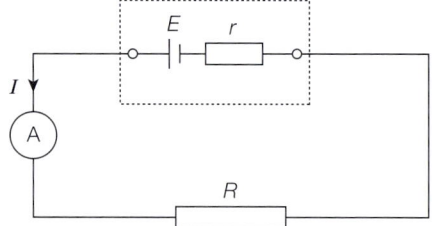

R/Ω	I/A
10	0.5
18	0.3

(i) Show that the internal resistance r is 2 Ω.

Using Kirchhoff's 2ⁿᵈ law: $E = 0.5(r + 10) = 0.3(r + 18)$ ✓   Method.

$$0.5r + 5 = 0.3r + 5.4 \ ✓ \qquad \text{Calculation.}$$

$$r = 2\Omega$$

(ii) Calculate the e.m.f. E.

$E = 0.5(r + 10) = 0.5 \times 12 = 6.0V$ ✓   Calculation.

E = ...6.0... V [3]

(d) A different cell, of e.m.f. 9.0 V and internal resistance 0.5 Ω, is connected to a 4 Ω resistor. Calculate: [4]

(i) the terminal p.d.

current $i = \dfrac{9.0}{4.5} = 2.0A$ ✓ Method. ✓ Calculation.

terminal p.d. = e.m.f. – 'lost volts' = $9.0 – (2.0 \times 0.5) = 8.0$ V

terminal p.d. = ...8.0... V

(ii) the energy per second dissipated in the 4 Ω resistor.

$P = IR = 2.0 \times 4 = 8$ W

> ✗ ✗ A very careless mistake at this level, which has cost two easy marks: $P = I^2R = 16$ W.

energy dissipated per second = ...8... W

Exam-style questions

1 The diagram shows different currents entering or leaving a circuit junction.

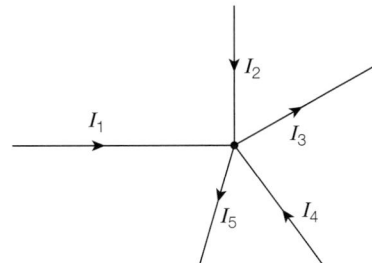

Which statement is correct?

A $I_1 + I_2 = I_3 + I_4 + I_5$

B $I_1 + I_2 = I_3 + I_4 - I_5$

C $I_1 + I_2 = I_3 - I_4 - I_5$

D $I_1 + I_2 = I_3 - I_4 + I_5$ [1]

2 For the circuit shown, which statement is correct?

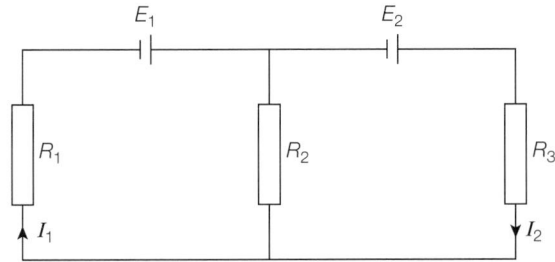

A $E_1 = (I_1 - I_2)R_2 + I_1R_1$

B $E_1 = (I_1 + I_2)R_2 + I_1R_1$

C $E_2 = I_2R_3 + (I_1 + I_2)R_2$

D $E_2 = I_2R_3 - (I_1 + I_2)R_2$ [1]

3 A cell, of e.m.f. 24 V and internal resistance 4 Ω, is connected to two resistors, each of resistance 8 Ω, as shown.

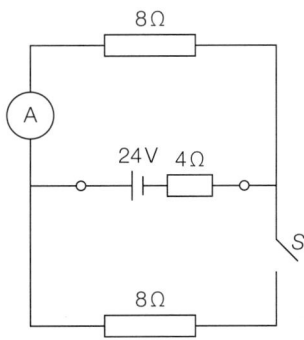

What is the **change** in the ammeter reading when switch S is closed?

A 0.5 A   **B** 1.0 A   **C** 1.5 A   **D** 2.0 A   [1]

4 A battery, of e.m.f. E and internal resistance r, is connected to a high-resistance voltmeter and a variable resistor of resistance R as shown.

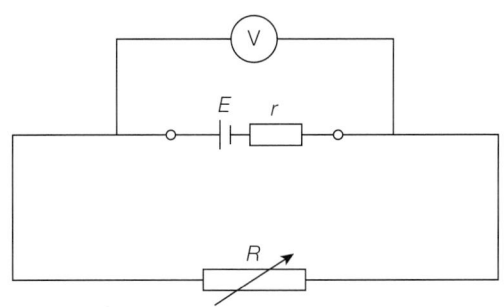

When $R = 1\,\Omega$ the voltmeter reading was 3 V.

When $R = 3\,\Omega$ the voltmeter reading was 6 V.

Which line in the table gives the correct values of E and r?

	E/V	r/Ω
A	6.0	1.5
B	6.0	3.0
C	12.0	1.5
D	12.0	3.0

[1]

5 A potentiometer circuit is used to measure the output p.d. of a thermocouple using a uniform resistance wire of length 99.6 cm and resistance 12 Ω, as shown.

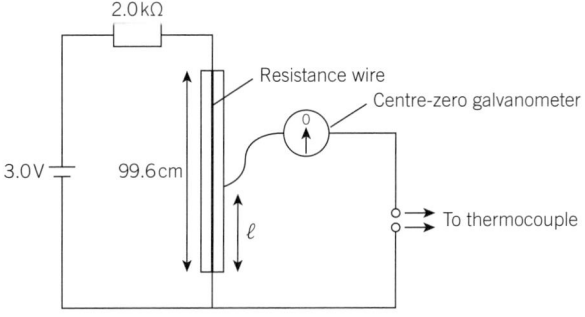

The output p.d. of the thermocouple is known to be between 0 V and 15 mV.

(a) Explain why the 2.0 kΩ resistor is needed.   [1]

(b) The centre-zero galvanometer reads zero when $\ell = 68.3$ cm.

Calculate:

(i) the current in the 2.0 kΩ resistor

(ii) the thermocouple p.d.   [4]

11 Particle physics

Atoms, nuclei, and radiation

Alpha-particle scattering – developing the nuclear model of the atom

At the beginning of the 20th century it was known that atoms contained both positively and negatively-charged parts, but not how these were organised inside the atom.

An experiment carried out by Geiger and Marsden was key to beginning to reveal the structure of the atom.

A thin piece of gold foil was bombarded by a (narrow) beam of alpha particles in an evacuated chamber (see Figure 11.1). Geiger and Marsden counted the number of alpha particles deflected at different angles by the gold foil by observing small flashes of light as the alpha particles hit a zinc sulphide screen. They found that:

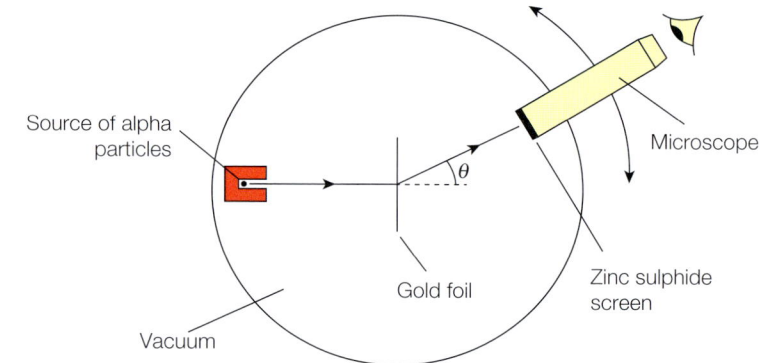

▲ **Figure 11.1** Geiger and Marsden's experiment

- most of the alpha particles passed straight through the gold foil undeflected, or only deflected by a small angle

- a few were deflected by a large angle, sometimes greater than 90°.

Rutherford showed that these observations could be explained by the atom being mostly empty space, with most of the mass of the atom concentrated in a (positively) charged nucleus (see Figure 11.2).

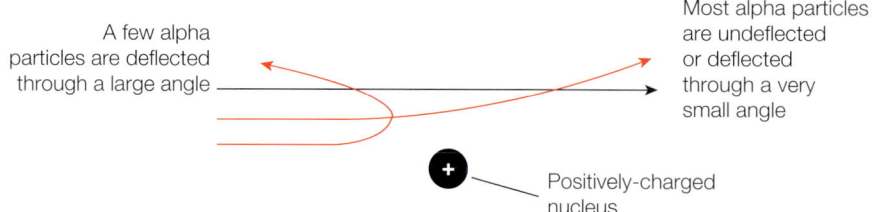

▲ **Figure 11.2** Alpha particle scattering

Later experiments showed that there were two types of particle in the nucleus: positively-charged protons and uncharged neutrons with electrons 'in orbit' around the nucleus. There were an equal number of protons and electrons, with the charge on each proton being $+e$.

Nucleon number, proton number and isotopes

The nuclear model of the atom consists of a nucleus formed of **protons** and **neutrons**, surrounded by **electrons**, as shown in Figure 11.3.

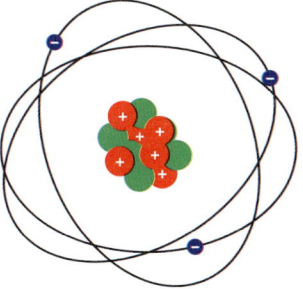

> **Key terms**
>
> **Nucleon number** A: the total number of particles in the nucleus (neutrons + protons).
>
> **Proton number** Z: the number of protons in the nucleus (equal to the number of electrons orbiting the nucleus if the atom is neutral).

▲ **Figure 11.3** Nuclear model of the atom

The number of neutrons in the nucleus is $A - Z$. The proton number Z is different for each element (Figure 11.4). For example, if Z is 3, the element is lithium.

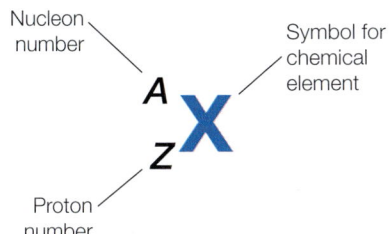

▲ **Figure 11.4** Proton number and nucleon number

Isotopes are different versions of the same element. Their nuclei have the same number of protons but different numbers of neutrons. For example, $^{12}_{6}C$ and $^{13}_{6}C$ are different isotopes of carbon with exactly the same chemical properties – they both have six protons in their nuclei, but $^{12}_{6}C$ has six neutrons and $^{13}_{6}C$ has seven neutrons.

Worked Example

$^{60}_{27}Co$ is an isotope of cobalt. State how many:

(a) protons   **(b)** neutrons   **(c)** nucleons   **(d)** electrons there are in an atom of this isotope.

Answer

(a) There are 27 protons (the proton number).   **(b)** There are $60 - 27 = 33$ neutrons.

(c) The total number of particles in the nucleus is 60.   **(d)** The number of electrons is the same as the proton number, 27.

In any nuclear process both the nucleon number and the proton number are conserved; for example:

$$^{9}_{4}Be + ^{4}_{2}He \rightarrow ^{12}_{6}C + ^{1}_{0}n$$

The total number of nucleons remains constant (13) as does the total number of protons (6). The nuclei of some isotopes are unstable – at some point they will 'decay' by emitting an alpha particle, beta particle or gamma ray. Some of the mass of the nucleus is converted to energy of the emitted radiation (e.g. the kinetic energy of the alpha particle). This process is called **radioactive decay**.

α, β, and γ decay

Alpha (α) decay

An **alpha particle** consists of two protons and two neutrons (identical to a $^{4}_{2}H$ nucleus). The proton number of a nuclide emitting an alpha particle will decrease by two and the nucleon number will decrease by four.

For example, radium-226 decays into radon-222 by emitting an alpha particle:

$$^{226}_{88}Ra \rightarrow ^{222}_{86}Rn + ^{4}_{2}He + energy$$

The mass of the radon-222 nucleus added to the mass of the helium-4 nucleus is slightly less than the mass of the radium-226 nucleus. The mass lost (called the **mass defect**) has been 'converted' into energy (the α-particle has kinetic energy, for example), but the total **mass–energy** is the same.

> 💡 **Remember**
>
> Both nucleon number and proton number are conserved in nuclear reactions, as is mass–energy.

Matter and antimatter

For every known type of particle there is a corresponding **antiparticle** with the same mass and opposite electric charge. Some of the other properties of antimatter are also opposite to their corresponding matter particles, which is why neutrons are not the same as antineutrons. Examples of matter and antimatter parties are shown in Table 11.1. Some antimatter particles are referred to by their own unique name, like the positron, while most simply have 'anti' as a prefix.

If a particle collides with its antiparticle, they annihilate each other, producing photons. Equally, a photon of sufficient energy can create a particle and its corresponding antiparticle in a process called 'pair production'.

▼ Table 11.1

Matter particle	Antimatter particle
proton p^+	antiproton $\bar{p}$
electron e^-	antielectron e^+ (positron)
neutron n	antineutron $\bar{n}$

Beta (β) decay

There are two types of **beta decay**:

- **β⁻-decay** occurs when a neutron in the nucleus changes into a proton and an electron. The electron is emitted as a fast-moving β⁻-particle. A very light, electrically-neutral antiparticle, called an electron antineutrino, is also emitted.

  For example, carbon-14 decays or 'transmutes' into nitrogen-14 by β⁻-decay:

$$^{14}_{6}\text{C} \rightarrow \, ^{14}_{7}\text{N} + \, ^{0}_{-1}e + \bar{v} + \text{energy}$$

> v is the symbol for an electron neutrino, a neutral particle with almost no mass. $\bar{v}$ is the symbol for an electron antineutrino.

- **β⁺-decay** occurs when a proton decays into a neutron and a positron $\left(^{0}_{+1}e\right)$ and an electron neutrino (v) are emitted. The positron is emitted as a fast-moving β⁺-particle. The electron neutrino is a very light, electrically neutral particle.

  A positron is the antimatter equivalent of an electron – if a positron and an electron were to meet they would annihilate each other, becoming electromagnetic energy.

  For example, carbon-10 decays into boron-10 by β⁺-decay:

$$^{10}_{6}\text{C} \rightarrow \, ^{10}_{5}\text{B} + \, ^{0}_{+1}e + v + \text{energy}$$

💡 **Remember**

The general equation for β⁻-decay is $n \rightarrow p + e^- + \bar{v}$.

The general equation for β⁺-decay is $p \rightarrow n + e^+ + \bar{v}$.

Gamma (γ) decay

Gamma decay occurs when a γ-ray is emitted (a high-frequency electromagnetic wave). There is no change in either the proton number or the nucleon number. Gamma decay can occur alongside alpha or beta decay, when an unstable nucleus adjusts to a more stable energy level.

Properties of alpha, beta, and gamma radiations

The Table 11.2 lists some important properties of α-, β-, and γ-radiations.

▼ **Table 11.2** Properties of α-, β-, and γ-radiations

	Alpha (α)	Beta (β)	Gamma (γ)
Nature	2 protons + 2 neutrons ($^{2}_{4}$He nucleus)	fast-moving electron (or positron for β⁺-decay)	high-frequency electromagnetic wave
Charge	+2e	−e (or +e for β⁺-decay)	no charge
Range in air	a few cm	~ 1 m	unlimited
Stopped by …	a few sheets of paper	several millimetres of aluminium	several centimetres of lead

Alpha particles and beta particles are deflected in opposite directions in a uniform electric field, but gamma rays are undeflected as they have no charge (see Figure 11.5). The force on an alpha particle is twice as large as the force on an electron moving at the same speed, but the deflection of the alpha particle is much smaller due to its much greater mass ($m_\alpha \approx 7000\,m_e$.)

When alpha particles and beta particles enter a magnetic field they are deflected in opposite directions; gamma rays are undeflected, as shown in Figure 11.6.

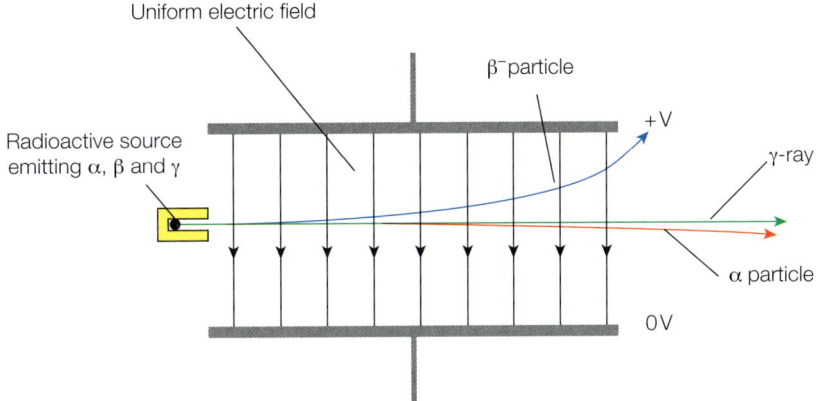

▲ **Figure 11.5** Deflections of α, β-, and γ radiation in an electric field

Fundamental particles

Electrons, neutrons, and protons were once thought to be **fundamental particles** (i.e., they did not consist of combinations of other particles). It was later discovered that, although electrons are still believed to be fundamental particles, protons and neutrons consist of combinations of smaller particles. These particles were given the name **quarks**.

The **standard model of particle physics** asserts that there are 12 fundamental particles, which can be divided into two groups, according to their properties, as shown in Table 11.3.

Quarks: there are six types of quark. Protons and neutrons are made up of different combinations of quarks.

Leptons: there are six types of lepton. An electron is one example of a lepton. All leptons have very small masses (lepton means light in Greek).

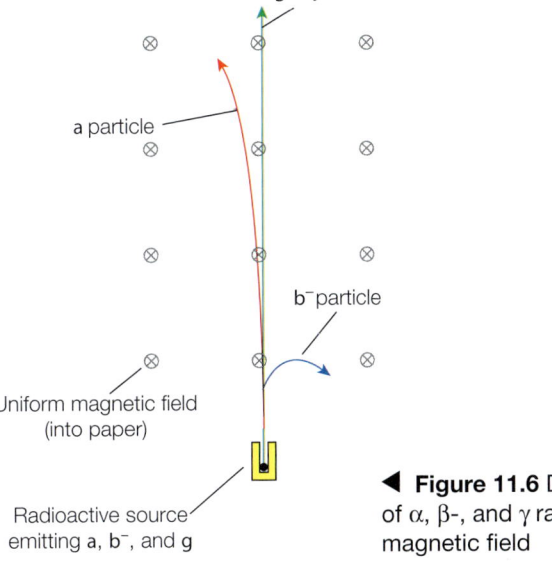

◀ **Figure 11.6** Deflections of α, β-, and γ radiation in a magnetic field

▼ **Table 11.3** Quarks and leptons

				Charge/e
Quarks	up, u	charm, c	top, t	$+\dfrac{2}{3}$
	down, d	strange, s	bottom, b	$-\dfrac{1}{3}$
Leptons	electron, e	muon, μ	tau, τ	-1
	electron-neutrino, v_e	muon-neutrino, v_μ	tau-neutrino, v_τ	0

Quarks occur in groups of two or three, never separately. The top quark is the heaviest with a mass approximately 200 times the mass of a proton. As well as the 12 fundamental particles, there are 12 equivalent antiparticles.

A proton consists of two up quarks and one down quark (uud), held together by the **strong nuclear force**. A neutron consists of one up quark and two down quarks (udd). Particles that consist of combinations of quarks and antiquarks are called **hadrons** (hadrons are defined as particles held together by the strong nuclear force). **Baryons** are particles consisting of three quarks (Figure 11.7). They include protons and neutrons. **Mesons** (Figure 11.8) are particles consisting of one quark and one antiquark. Antibaryons consist of three antiquarks (Figure 11.9).

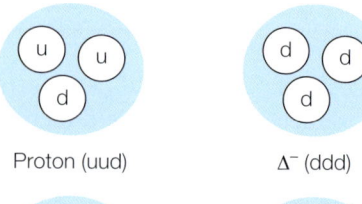

Proton (uud) Δ⁻ (ddd)

Neutron (udd) Δ⁺⁺ (uuu)

▲ **Figure 11.7** Some baryons

Worked Example

One type of hadron consists of two down quarks and one strange quark. State the charge on this hadron.

Answer

Both the down quark and the strange quark have a charge $-\dfrac{1}{3}e$, so the total charge must be $-e$.

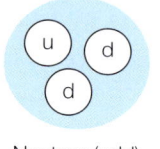 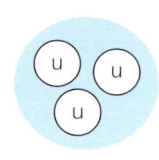

p⁻ meson (ūd) K⁰ meson (ds̄)

▲ **Figure 11.8** Some mesons

Quarks and beta decay

In β⁻-decay, a down quark changes into an up quark in one of the neutrons in a nucleus, making it a proton, and in doing so emits an electron (the β⁻-particle) and an electron antineutrino. In β⁺-decay, one of the protons in a nucleus changes into a neutron by one of the up quarks changing into a down quark, emitting a positron (the β⁺-particle) and an electron neutrino in the process.

The force (or interaction) responsible for beta decay, causing a neutron to change into a proton (or a proton into a neutron), is the **weak nuclear force** (or **weak interaction**).

There are four fundamental forces that control the interactions between fundamental particles, as shown in Table 11.4.

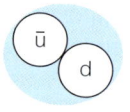 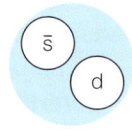

Mesons
consist of a quark and an antiquark e.g., us̄

Hadrons

Baryons
consist of 3 quarks, e.g., uud for a proton

Antibaryons
consist of 3 antiquarks, e.g., ūūd̄ for an antiproton

▲ **Figure 11.9** Hadrons

▼ **Table 11.4** Fundamental forces

Force	Range	Acts on
Gravity	no limit	all objects
Electromagnetic	no limit	charged objects
Strong nuclear force	10^{-15} m	quarks and antiquarks
Weak nuclear force	10^{-18} m	fundamental particles

Exam-style questions

1   Which statement is supported by the scattering of alpha particles by gold foil? [1]

   A   The atom contains electrons in orbit around the nucleus.

   B   The atom is mostly empty space.

   C   The nucleus is held together by strong forces.

   D   The nucleus of an atom contains positively charged particles and electrically neutral particles.

2   Which statement about the nuclei of two different isotopes of a substance is correct? [1]

   A   The two nuclei contain equal numbers of neutrons.

   B   The two nuclei contain equal numbers of nucleons.

   C   The two nuclei contain equal numbers of protons and neutrons.

   D   The two nuclei contain equal numbers of protons.

3   ^{218}Po, an isotope of polonium, decays into an isotope of lead by emitting an alpha particle. The lead decays into an isotope of bismuth by emitting a β^- particle:

   $$^{218}_{84}\text{Po} \rightarrow ^{X}_{82}\text{Pb} \rightarrow ^{214}_{Y}\text{Bi}$$

   What are the values of X and Y? [1]

	X	Y
A	214	81
B	214	83
C	218	81
D	218	83

4   How many nucleons are there in a nucleus of an atom of $^{23}_{11}$Na? [1]

   A 11   **B** 12   **C** 23   **D** 34

5   The isotope niobium-86 ($^{86}_{41}$Nb) decays by β^+-decay into an isotope of zirconium (Zr). Which equation describes the decay? [1]

   A $^{86}_{41}\text{Nb} \rightarrow ^{86}_{40}\text{Zr} + ^{0}_{-1}\text{e} + v$

   B $^{86}_{41}\text{Nb} \rightarrow ^{86}_{40}\text{Zr} + ^{0}_{+1}\text{e} + v$

   C $^{86}_{41}\text{Nb} \rightarrow ^{86}_{40}\text{Zr} + ^{0}_{-1}\text{e} + \bar{v}$

   D $^{86}_{41}\text{Nb} \rightarrow ^{86}_{40}\text{Zr} + ^{0}_{+1}\text{e} + \bar{v}$

6   The Δ^{++} hadron has a charge $+2e$ and consists of three quarks. If two of the quarks are up quarks what is the third quark? [1]

   A d   **B** $\bar{\text{d}}$ **C** u   **D** $\bar{\text{u}}$

7   Which one of the particles listed is not a fundamental particle? [1]

   A   an electron   **B**   a neutrino

   C   a quark   **D**   a proton

8   Which of the following results from the Geiger Marsden alpha particle scattering experiment indicates that most of the volume of an atom is empty space? [1]

   A   A small number of the alpha particles were deflected through large angles.

   B   Some of the alpha particles were deflected by very small angles.

   C   Most of the alpha particles passed through the atom undeflected.

   D   Some of the alpha particles were deflected by angles greater than 90°

9   A matter particle and its corresponding antimatter particle share some properties but differ in others.

   (a) State an example of a matter particle and its corresponding antiparticle. [1]

   (b) State one similarity between the particle and antiparticle you gave in (a). [1]

   (c) State one difference between the particle and antiparticle you gave in (a). [1]

   (d) Describe what would happen if the particle and antiparticle collided with each other. [2]

12 Motion in a circle

Kinematics of uniform circular motion

Radians and angular displacement

The **angular displacement** θ of an object moving in a circular path is usually measured in **radians** (Figure 12.1).

For a complete circle ($360°$), $l = 2\pi r$, so $\theta = 2\pi$ radians:

$$1 \text{ radian} = \frac{360}{2\pi} = 57.3°$$

Angular speed

The **angular speed** ω of an object is the rate of change of angular displacement with time:

$$\omega = \frac{\Delta\theta}{\Delta t}$$

where $\Delta\theta$ is the angle 'swept' by the radius in time Δt and is measured in radians per second (rad s^{-1} or just s^{-1}). The linear speed v is related to the angular speed ω by the equation:

$$v = r\omega$$

Linear speed is represented as being tangent to the circular motion but its direction is changing at each instant, so we do not call it velocity. For an object moving with constant angular speed ω around a circle of radius r (see Figure 12.2), the time for one complete revolution is T, where:

$$T = \frac{2\pi r}{v} = \frac{2\pi}{\omega}$$

This is usually stated as:

$$\omega = \frac{2\pi}{T}$$

> **Remember**
>
> θ (in redian) $= \dfrac{\text{arc length}}{\text{radius}} = \dfrac{l}{r}$
>
> Angles can be measured in degrees (°) or radians (rad). $360°$ is equivalent to 2π radian

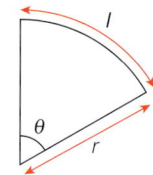

▲ **Figure 12.1** Radian measure

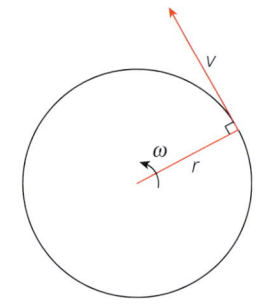

▲ **Figure 12.2** Angular speed

Worked Example

The approximate distance from the Earth to the Moon is 3.84×10^8 m; the Moon takes 27.3 days to orbit the Earth.

Calculate:
(a) the angular speed ω of the Moon around the Earth

(b) the average speed v of the Moon.

Answer

(a) $\omega = \dfrac{2\pi}{T} = \dfrac{2\pi}{27.3 \times 24 \times 60 \times 60} = 2.66 \times 10^{-6} \text{ rad s}^{-1}$

(b) $v = r\omega = 3.84 \times 10^8 \times 2.66 \times 10^{-6} = 1.02 \text{ km s}^{-1}$

> **Remember**
>
> $v = r\omega$
>
> $\omega = \dfrac{2\pi}{T}$

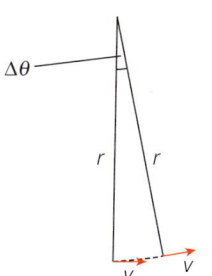

Centripetal acceleration and centripetal force

Centripetal acceleration

Velocity is a **vector** quantity – it has both magnitude and direction. Since an object moving in a circle is constantly changing direction, its velocity is constantly changing. It is **accelerating**.

From Figure 12.3, for an object moving a small angular displacement $\Delta\theta$ in a small time Δt, the change in velocity Δv is:

$$\Delta v = 2v\sin\left(\frac{\Delta\theta}{2}\right) = v\Delta\theta \qquad \text{(as } \Delta\theta \text{ is very small)}$$

So the **centripetal acceleration** a is:

$$a = \frac{\Delta v}{\Delta t} = \frac{v\Delta\theta}{\Delta t} = v\omega$$

Substituting $v = r\omega$, the centripetal acceleration a can also be written:

$$a = r\omega^2 = \frac{v^2}{r}$$

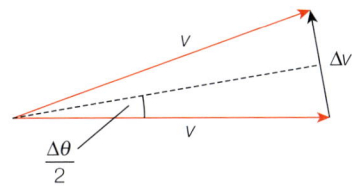

In time Δt

▲ **Figure 12.3** Centripetal acceleration

> **Maths Skills**
>
> For small angles:
>
> $\sin\theta = \theta$
>
> where the angle θ is measured in radians.

Worked Example

A wind turbine (see Figure 12.4) rotates at 14 revolutions per minute (r.p.m.). The diameter of the turbine is 70 m.

(a) Calculate the angular speed of the turbine in rad s^{-1}

(b) Determine the centripetal acceleration of a point:
    (i) at the end of one of the turbine blades
    (ii) at the midpoint of one of the turbine blades.

Answer

(a) $\omega = 14 \text{ r.p.m.} = \dfrac{14 \times 2\pi}{60} = 1.47 \text{ rad s}^{-1}$

(b) (i) $a = r\omega^2 = 35 \times 1.47^2 = 75.6 \text{ m s}^{-2}$

    (ii) $a = r\omega^2 = 17.5 \times 1.47^2 = 37.8 \text{ m s}^{-2}$

> **Remember**
>
> $$a = r\omega^2 = \frac{v^2}{r}$$

Centripetal force

From $F = ma$, a force is needed to accelerate an object. If an object is moving along a curved path it must be accelerating because its velocity (a vector) is changing even if its speed is constant.

The force needed to keep an object of mass m moving in a circle of radius r with constant speed v is:

$$F = ma = mr\omega^2 = \frac{mv^2}{r}$$

The direction of the acceleration is towards the centre of the circle so the force must act towards the centre of the circle (Figure 12.5).

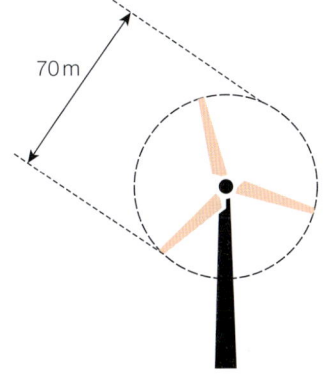

70 m

▲ **Figure 12.4** Wind turbine

> **Remember**
>
> $$F = mr\omega^2 = \frac{mv^2}{r}$$

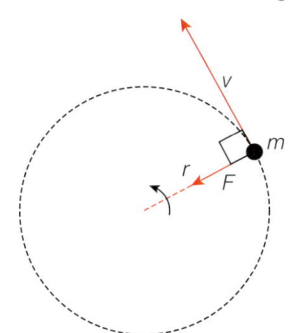

▲ **Figure 12.5** Centripetal force

Worked Example

1 The international space station ISS (Figure 12.6) orbits the Earth at a height of 400 km above the Earth's surface, taking 92 minutes to complete one orbit. The radius of the Earth is 6370 km.

   (a) Calculate:

   (i) the angular speed of the space station

   (ii) the centripetal acceleration of the space station.

   (b) The mass of the space station is 4.2×10^5 kg.

   (i) Calculate the centripetal force acting on the space station.

   (ii) What provides this force?

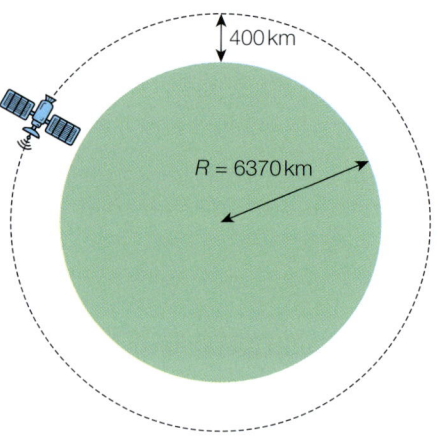

▲ **Figure 12.6** International Space Station orbiting the Earth

Answer

   (a) (i) $\omega = \dfrac{2\pi}{T} = \dfrac{2\pi}{92 \times 60} = 1.14 \times 10^{-3}$ rad s^{-1}

   (ii) $a = r\omega^2 = (6.37 \times 10^6 + 4 \times 10^5) \times (1.14 \times 10^{-3})^2 = 8.80$ m s^{-2}

   (b) (i) $F = ma = 4.2 \times 10^5 \times 8.80 = 3.70 \times 10^6$ N

   (ii) The gravitational pull of the Earth on the space station keeps it in orbit around the Earth.

2 A ball of mass m connected to a string of length l is whirled in a vertical circle at a constant speed v, as shown in Figure 12.7.

   (a) Explain why the ball is accelerating even though it is travelling at constant speed.

   (b) Calculate the tension in the string:

   (i) when the ball is at its lowest point

   (ii) when the ball is at its highest point.

   (c) Determine the minimum velocity needed for the ball to continue to travel in a circular path.

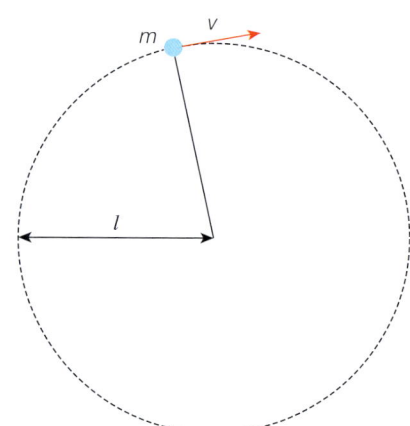

▲ **Figure 12.7** Vertical rotation of a ball

Answer

   (a) Acceleration is the rate of change of velocity – a vector quantity, which has both magnitude and direction. As the direction of the ball is changing continuously, the velocity is also changing. It is accelerating.

   (b) (i) At the lowest point (see Figure 12.8):

   $$T - mg = \frac{mv^2}{r}$$

   $$T = mg + \frac{mv^2}{r}$$

   (ii) At the highest point (see Figure 12.9):

   $$T + mg = \frac{mv^2}{r}$$

   $$T = \frac{mv^2}{r} - mg$$

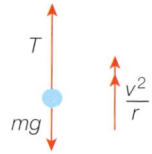

▲ **Figure 12.8**

   (c) From (b) (ii), the tension in the string cannot be less than zero (otherwise the string will be slack and the ball will be in free fall).

   $$T = \frac{mv^2}{r} - mg > 0 \quad \text{so} \quad v^2 > gr$$

   $$v > \sqrt{(gr)}$$

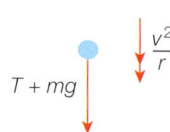

▲ **Figure 12.9**

↑ Raise your grade

A conical pendulum consists of a small ball of mass 50 g suspended on a string and rotated in a circle of radius 0.40 m at a constant speed of 90 revolutions per minute, as shown. [The value of g is 9.81 m s^{-2}.]

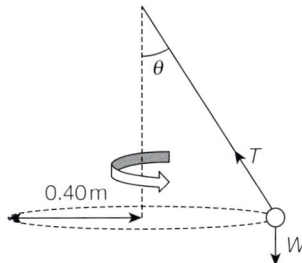

(a) (i) Show that the angular velocity ω of the sphere is 9.4 rad s^{-1}. [1]

$$\omega = 90 \text{ r.p.m.} = \frac{90 \times 2\pi}{60} = 9.4 \text{ rad s}^{-1} \checkmark$$

Correct method and calculation.

(ii) Hence show that the centripetal acceleration of the ball is 35 m s^{-2}. [1]

Correct method and calculation.

$$a = r\omega^2 = 0.40 \times 9.4^2 = 35.3 \text{ m s}^{-2} \checkmark$$

(b) (i) Calculate the centripetal force acting on the ball. [2]

$$F = ma = (50 \times 10^{-3}) \times 35.3 = 1.77 \checkmark \times$$

Correct method and calculation, but units omitted (should be N), so loses second mark.

centripetal force = 1.77

(ii) State the direction of this force. [1]

Outwards from the centre of the circle ✗

Incorrect statement – 'centripetal' means 'towards the centre'.

(c) Show that the angle θ is approximately 75°. [4]

$\rightarrow$: $\quad T\sin\theta = mr\omega^2 = 50 \times 10^{-3} \times 0.40 \times 9.4^2$

$\quad\quad\quad = 1.77 \checkmark$

$F = ma$ applied in horizontal direction correctly.

$\uparrow$: $\quad T\cos\theta = W = mg = 50 \times 10^{-3} \times 9.81 \checkmark$

$\quad\quad\quad = 0.491$

Forces in vertical direction resolved correctly.

$$\tan\theta = \frac{T\sin\theta}{T\cos\theta} = \frac{1.77}{0.491} = 3.60 \checkmark$$

Knowledge of the relationship between sin, cos and tan is essential here.

$\quad\quad\quad 1.30 \times$

The student has used their calculator in radian mode and this gives a final answer in radians. The question asked for an answer in degrees so this final mark is lost.

Exam-style questions

[The value of g is 9.81 m s^{-2}.]

1 (a) Express the following angles in radians:

    (i) 60°   **(ii)** 250°   **(iii)** 95°   [3]

  (b) Express the following radian measures in degrees:

    (i) $\dfrac{\pi}{6}$ **(ii)** $\dfrac{3\pi}{4}$ **(iii)** $\dfrac{9\pi}{7}$ [3]

2 An electric fan rotates at 800 r.p.m. The distance from the tip of a blade of the fan to the centre is 30.0 cm.

  Calculate:

  (a) the angular speed of the fan in rad s^{-1} [2]

  (b) the linear speed of the tip of a blade.   [2]

3 A simple pendulum consists of a small ball of mass 80.0 g tied to a thin string of length 0.60 m.

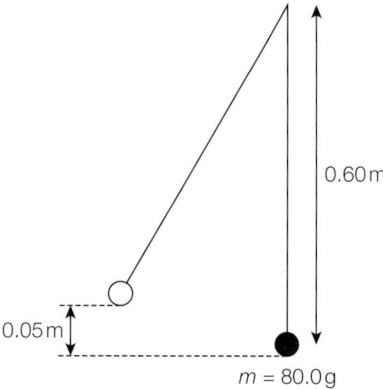

0.60 m

0.05 m

$m = 80.0\,g$

  The ball is pulled to one side so that it is raised 0.05 m above its lowest position and released.

  Calculate:

  (a) the speed of the ball when it reaches its lowest position   [2]

  (b) the maximum tension in the string.   [2]

4 A ball of mass m is connected to a string of length l and whirled around to form a conical pendulum moving with angular velocity ω in a horizontal circle of radius r.

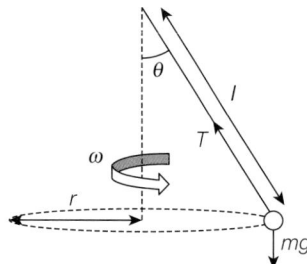

θ

l

T

ω

r

mg

  (a) State an expression, in terms of m, r and ω, for:

    (i) the centripetal acceleration of the ball

    (ii) the centripetal force acting on the ball.   [2]

(b) Hence show that:

  (i) $T = ml\omega^2$

  (ii) $\tan\theta = \dfrac{r\omega^2}{g}$ [2]

5 A ball of mass 0.20 kg is connected to a light, inextensible string and rotates in a vertical circle of radius 0.40 m. The speed of the ball at its lowest point is 6.0 m s^{-1}.

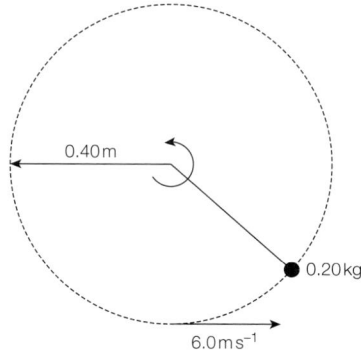

0.40 m

0.20 kg

6.0 m s^{-1}

  (a) Use the principle of conservation of energy to show that the speed of the ball at its highest point is 4.5 m s^{-1}.   [2]

  (b) Hence calculate the tension in the string:

    (i) at its lowest point

    (ii) at its highest point.   [3]

  (c) Describe and explain what would happen if the speed of the ball at its lowest point was 3.0 m s^{-1}.   [2]

6 A cyclist travels in a circle of radius 50.0 m at a speed of 20 m s^{-1}, as shown below. The combined mass of the motorcyclist and motorcycle is 250 kg.

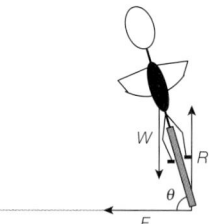

W

R

θ

F

  (a) Calculate:

    (i) the combined weight W of the motorcyclist and the motorcycle

    (ii) the vertical reaction force R.   [2]

  (b) (i) Calculate the centripetal acceleration of the motorcycle and motorcyclist.

    (ii) Hence show that the centripetal force F is 2.0 kN.   [3]

  (c) Explain what provides the force F.   [2]

  (d) Show that the angle θ is 51°. State any assumptions you make.   [2]

Newton's law of gravitation

A gravitational field

A gravitational field is a region of space surrounding a mass in which another mass will experience a force. This force is always attractive; the two masses are pulled together. Gravitational fields permeate all of space, but they are stronger near large masses and grow weaker with increased separation.

Law of gravitation

Any two masses will attract each other. The size of the force of attraction depends on the product of the two point masses. and varies inversely as the square of their distance apart. This can be summarised in **Newton's law of gravitation**:

$$F = \frac{Gm_1m_2}{r^2}$$

where m_1 and m_2 are the masses, and r the distance between the centres of the two masses (Figure 13.1).

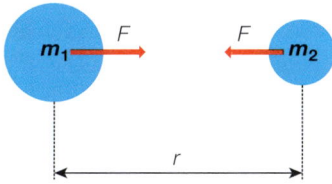

▲ **Figure 13.1** Newton's law of gravitation

G is a constant, called the universal gravitational constant, and is equal to $6.67 \times 10^{-11}\,\text{N}\,\text{m}^2\,\text{kg}^{-2}$.

For objects such as planets and moons, each mass can be treated as if all the mass is acting at the centre of the object (in the same way as a charged sphere can be treated as if all the charge is at the centre of the sphere).

> 💡 **Remember**
>
> $$F = \frac{Gm_1m_2}{r^2}$$

Worked Examples

1 The Earth has a mass of 6.0×10^{24} kg, and the Moon has a mass of 7.4×10^{22} kg. The average distance between the centre of the Earth, and the centre of the Moon is 3.9×10^5 km. Calculate the gravitational force exerted on the Moon by the Earth. This is also the force exerted by the Moon on the Earth.

   Answer

   $$F = \frac{Gm_1m_2}{r^2} = \frac{6.67 \times 10^{-11} \times (6.0 \times 10^{24}) \times (7.4 \times 10^{22})}{(3.9 \times 10^5 \times 10^3)^2} = 1.95 \times 10^{20}\,\text{N}$$

2 Estimate the gravitational attraction between two people standing a metre apart. The gravitational attraction between people is negligible when compared with their attraction to the Earth.

   Answer

   $$g = \frac{Gm_1m_2}{r^2} = \frac{6.67 \times 10^{-11} \times 60 \times 60}{(1.0)^2} = 2.4 \times 10^{-7}\,\text{N}$$

> ★ **Exam tip**
>
> The value of G is provided in Exam Papers 1, 2, and 4.

> 💡 **Remember**
>
> We can estimate the mass of a person to be 60 kg.

Gravitational field strength

The **gravitational field strength** g at a point is the force per unit mass on a small test mass at that point (see Figure 13.2). It is a vector (having both magnitude and direction) and has units of $N\,kg^{-1}$.

From Newton's law, for a mass M the gravitational field a distance r from the centre of the mass is:

$$g = \frac{GM}{r^2}$$

The direction of the force is towards the centre of mass M. Gravitational field lines can be drawn to show the direction of the gravitational field.

Figure 13.3 illustrates that the Earth's gravitational field is **radial** (it varies as $1/r^2$), but on a smaller scale, such as near the Earth's surface, the field strength is constant in both magnitude and direction – a **uniform** field.

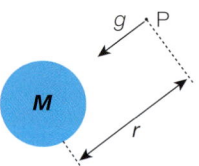

▲ **Figure 13.2** Gravitational field strength g

> 💡 **Remember**
>
> Near the Earth's surface the gravitational field g is constant and has a value of $9.81\,N\,kg^{-1}$.

Worked Example

[GM for Earth is $4.0 \times 10^{14}\,N\,m^2\,kg^{-1}$; radius of the Earth is $6.4 \times 10^3\,km$.]

Calculate the gravitational field strength:

(a) on the Earth's surface

(b) 1.00 km above the Earth's surface

Answer

(a) $g = \dfrac{GM}{r^2} = \dfrac{4.0 \times 10^{14}}{(6.4 \times 10^6)^2} = 9.8\,N\,kg^{-1}$

> These values illustrate why g near the Earth's surface can be considered a constant.

(b) $g = \dfrac{GM}{r^2} = \dfrac{4.0 \times 10^{14}}{(6.4 \times 10^6 \times 1.0 \times 10^3)^2} = 9.8\,N\,kg^{-1}$

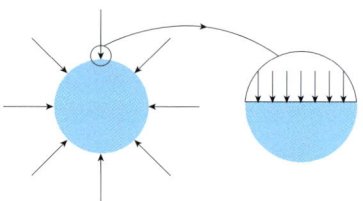

▲ **Figure 13.3** Radial gravitational field

> ⭐ **Exam tip**
>
> Take care to be consistent with units. In this question, GM is in S.I. units, so r must be converted to metres before substituting the value for r into the equation.

Orbital motion

Any object moving round in a circle of radius r at a constant speed v is accelerating because it is constantly changing direction. Acceleration is the rate of change of velocity (a vector), and an object moving in a circle is changing direction. The acceleration is a where

$$a = \frac{v^2}{r}.$$

Using $F = ma$, the force F needed to keep an object of mass m moving in a circle of radius r with constant speed v (see Figure 13.4) is:

$$F = \frac{mv^2}{r}$$

In the case of a satellite orbiting the Earth, or a planet orbiting the Sun, this force is the gravitational force of attraction.

For a mass m orbiting a mass M at a distance r:

$$\frac{GMm}{r^2} = \frac{mv^2}{r} \qquad \text{so} \qquad v = \sqrt{\frac{GM}{r}}$$

The time T for one orbit is:

$$T = \frac{2\pi r}{v} = 2\pi \sqrt{\frac{r^3}{GM}}$$

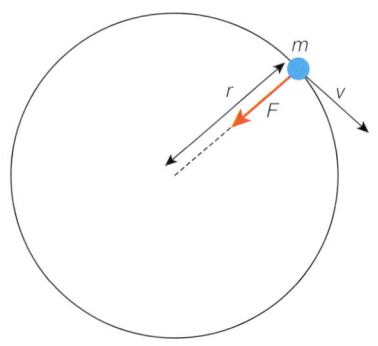

▲ **Figure 13.4** Orbital motion

Worked Example

[GM for Earth = $4.0\times10^{14}\,\mathrm{N\,m^2\,kg^{-1}}$, radius of the Earth = 6.4×10^{3} km]

A satellite is orbiting in a **low Earth** orbit, 1000 km above the Earth's surface. Calculate:

(a) the speed of the satellite

(b) the time taken for one orbit of the Earth.

Answer

(a) $v=\sqrt{\frac{GM}{r}}=\sqrt{\frac{4.0\times10^{14}}{(6.4\times10^{6}+10^{6})}}=7.4\times10^{3}\,\mathrm{m\,s^{-1}}$

(b) $T=\frac{2\pi r}{v}=\frac{2\pi\times(6.4\times10^{6}+10^{6})}{7.4\times10^{3}}=6.28\times10^{3}\,\mathrm{s}$ (about 105 minutes)

Link

For more details about circular motion see Unit 12 *Motion in a circle.*

★ Exam tip

Remember that the radius of the orbit is the radius of the Earth **plus** the height of the orbit above the surface of the Earth.

Geostationary satellites

Satellites in geostationary orbit complete one rotation of the Earth in 24 hours, as shown in Figure 13.5. These satellites are placed directly above the Equator so that, when viewed from Earth, the satellites appear stationary.

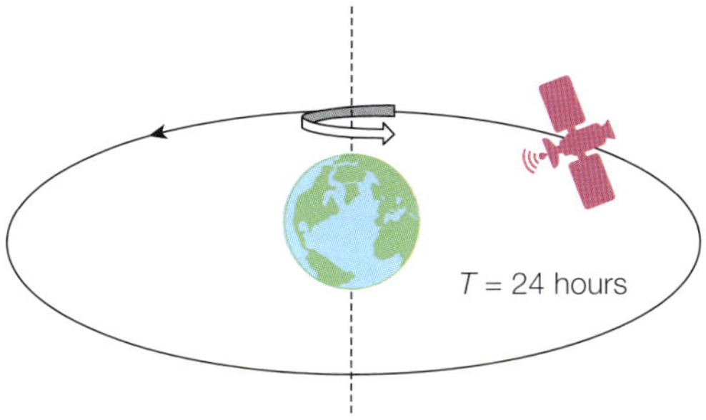

▲ **Figure 13.5** Geostationary orbit

Rearranging $T=2\pi\sqrt{\frac{r^3}{GM}}$

the radius r of a geostationary orbit is:

$$r=\sqrt[3]{\frac{GMT^2}{4\pi^2}}=\sqrt[3]{\frac{4\times10^{14}\times(24\times60\times60)^2}{4\pi^2}}$$

$$r=4.23\times10^{7}\,\mathrm{m}$$

The height h of a geostationary satellite **above the Earth's surface** is:

$$h=4.23\times10^{7}-6.4\times10^{6}=3.59\times10^{7}\,\mathrm{m}$$

Gravitational potential

The **gravitational potential** ϕ at a point P is defined as the work done in bringing a unit mass (1 kg) from infinity to that point (see Figure 13.6). As work would need to be done in taking unit mass **from** the point **to** infinity, gravitational potential is always negative.

Remember

By convention, gravitational potential is zero at infinity, so anywhere else the gravitational potential is less than zero; i.e., negative.

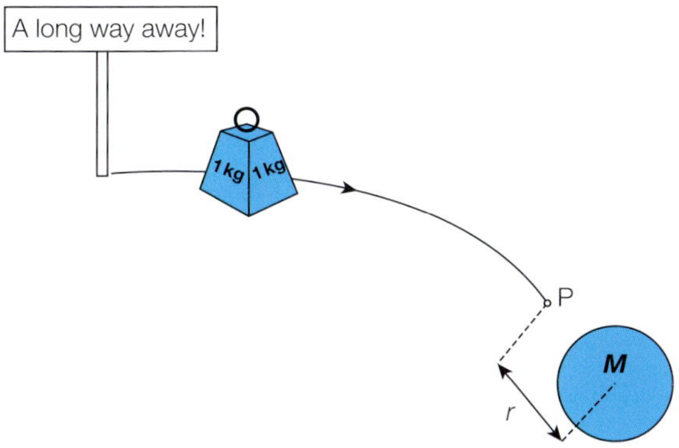

▲ **Figure 13.6** Gravitational potential

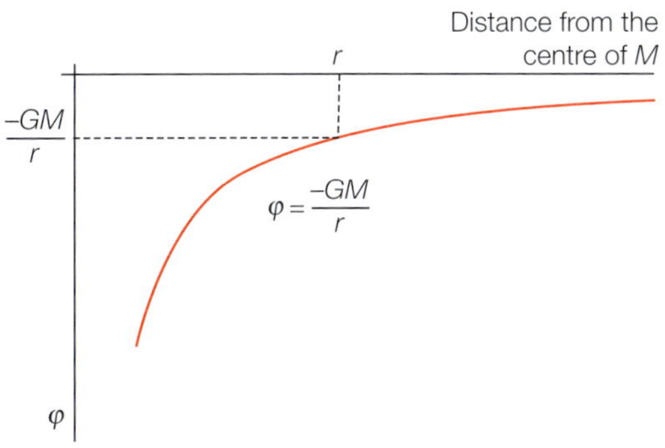

▲ **Figure 13.7** Variation of gravitational potential

For point P, a distance r from a mass M, as shown in Figures 8.6 and 8.7, the gravitational potential is:

$$\phi = -\frac{GM}{r}$$

This means that (GM/r) joules of energy are needed to move 1 kg from P to a long way away from M. The closer point P is to mass M (the smaller the value of r), the greater the value of (GM/r), and so the greater the energy needed to move 1 kg to a point far away from M.

Worked Example

Ignoring air resistance how fast must an object be thrown up in the air to not come down again? (What is its **escape velocity**?)

Answer

On the Earth's surface the gravitational potential is $-\dfrac{GM}{R}$ where R is the radius of the Earth, and M its mass.

An amount of energy equal to $\dfrac{GM}{R}$ is needed for 1 kg to 'escape',

so a mass m would need $\dfrac{GMm}{R}$ joules of energy.

The kinetic energy given to the object must be at least as large as this for the object to escape:

$$\frac{1}{2}mv^2 \geq \frac{GMm}{R}$$

$$v \geq \sqrt{\frac{2GM}{R}}$$

For Earth, $GM = 4.0 \times 10^{14}\,N\,m^2\,kg^{-1}$, $R = 6.4 \times 10^3\,km$, so

$$v_{escape} = \sqrt{\frac{2 \times 4.0 \times 10^{14}}{6.4 \times 10^6}}$$

$$= 1.12 \times 10^4\,m\,s^{-1}$$

The **escape velocity** is the velocity an object needs to completely break free from the gravitational field of a mass and 'escape' to infinity.

The 'escape velocity' from Earth is about 11 km s⁻¹ whereas the escape velocity of the Moon is only 2.4 km s⁻¹.

Gravitational field and gravitational potential difference

The gravitational potential difference between two points is the work done in moving unit mass (1 kg) from one point to the other. If the two points are close enough together, such that the gravitational field strength g does not change significantly, the change in gravitational potential $\Delta\phi$ in moving a small distance Δh is:

$$\Delta\phi = g\Delta h$$

For a mass m moving a distance Δh, the change in gravitational potential energy ΔE_p is $mg\Delta h$ (the equation for gravitational potential energy calculations on, or near, the Earth's surface).

 Link

$$\Delta E_p = mg\Delta h$$

See Unit 6 *Work, energy and power* for examples of the use of this equation.

↑ Raise your grade

The graph shows how the gravitational potential ϕ due to the planet Mars varies with distance r from the centre of the planet. The mean radius of Mars is 3.4×10^6 m.

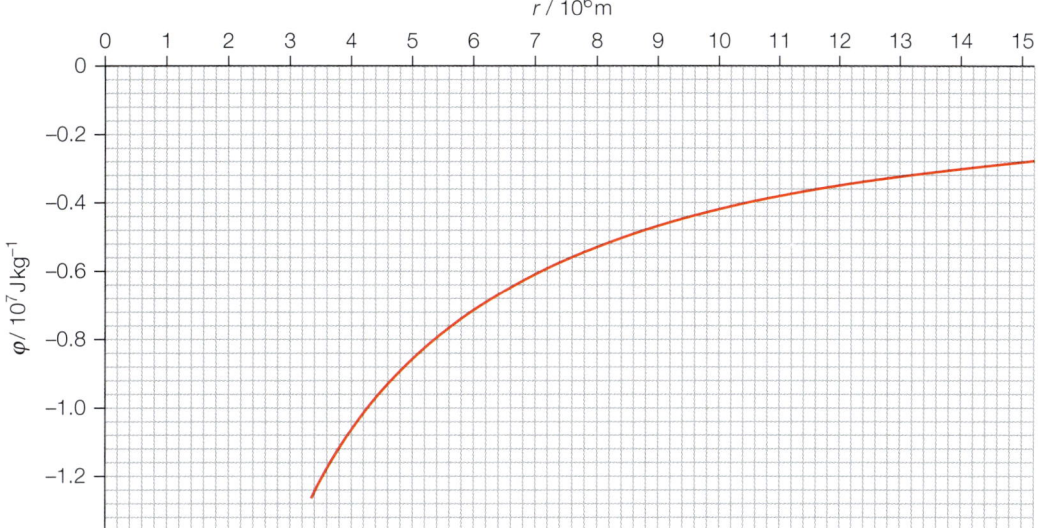

(a) (i) Define *gravitational potential*. [2]

The gravitational potential at a point is the energy
needed to move a mass from infinity to that point. ✔ ✗

> The statement is correct, but incomplete. Gravitational potential is the work done transferring **unit mass** (i.e., 1 kg) from infinity to the point concerned.

(ii) Explain why gravitational potential is negative. [1]

Energy is needed to transfer a mass <u>from</u> a point <u>to</u> infinity – energy is gained moving a
mass the other way. ✔   A good answer.

(b) Use the graph to find the gravitational potential:

(i) on the surface of Mars [1]

1.3×10^7 J kg^{-1} ✔   Good estimate from the graph.

gravitational potential $= 1.3 \times 10^7$ J kg^{-1}

(ii) at a point P, 6.0×10^6 m from the centre of Mars. [1]

7.1×10^6 J kg^{-1} ✔

gravitational potential $= 7.1 \times 10^6$ J kg^{-1}

(d) Calculate the energy needed to move 100 kg from the surface of Mars to the point P. [2]

Energy needed $= 100 \times (1.3 \times 10^7 - 7.1 \times 10^6) = 5.9 \times 10^8$ J ✔ ✔

> Correct method, correct calculation with unit.

energy needed $= 5.9 \times 10^8$ J

Exam-style questions

[Use $G = 6.67 \times 10^{-11}\,\text{N}\,\text{m}^2\,\text{kg}^{-2}$; radius of the Earth $= 6.4 \times 10^3\,\text{km}$]

1   Two average-sized adults are standing approximately 1.0 m apart.

  (a) Estimate the mass of one adult.   [1]

  (b) Hence estimate the gravitational force between the two adults.   [1]

  (c) If both adults are standing on a frictionless surface, and in the absence of air resistance, estimate the size of the acceleration towards each other.   [1]

2   The Moon orbits the Earth once in 27.3 days. Its mean orbital radius is $3.8 \times 10^5\,\text{km}$. Use this information to calculate the mass of the Earth.   [3]

3   Use g on the Earth's surface as $9.81\,\text{N}\,\text{kg}^{-1}$. The Earth's radius is $6.37 \times 10^3\,\text{km}$.

  (a) Calculate the mass of the Earth.   [2]

  (b) Justify the number of significant figures in your answer.   [1]

4   A planet is in a circular orbit around a star of mass M. The radius of the orbit is R, and the time for one orbit is T.

  (a) Show that the velocity v of the planet is:
  $$v = \sqrt{\frac{2GM}{R}}$$ [2]

  (b) Hence show that:
  $$\frac{R^3}{T^2} = \text{a constant}$$ [1]

5   Use g on the surface of the Moon as $1.6\,\text{m}\,\text{s}^{-2}$. The radius of the Moon is $1.74 \times 10^3\,\text{km}$.

  Calculate:

  (a) the mass of the Moon   [1]

  (b) the gravitational potential on the surface of the Moon   [1]

  (c) the *escape velocity* from the Moon's surface.   [2]

6   Europa, one of Jupiter's moons, orbits the planet once every 85 hours at a radius of $6.7 \times 10^5\,\text{km}$.

  Calculate:

  (a) the speed of Europa   [1]

  (b) the mass of Jupiter.   [2]

7   **(a)** Define *gravitational potential*.   [1]

  (b) A spacecraft, of mass $6.0 \times 10^4\,\text{kg}$, is in orbit round the Earth at a height of 200 km.

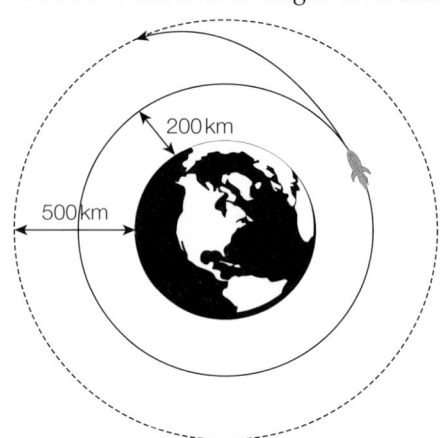

  Calculate:

  (i) the gravitational field strength at this height

  (ii) the gravitational potential at this height.   [3]

  (c) The spacecraft moves to an orbit 500 km above the Earth's surface.

  Calculate:

  (i) the gravitational potential at this new height

  (ii) the increase in gravitational potential energy of the spacecraft.   [3]

8   A weather satellite, of mass 320 kg, is to be placed in geostationary orbit, $3.56 \times 10^7\,\text{m}$ above the Earth's surface.

  (a) **(i)** State the time taken for one orbit.

  (ii) Show that the speed of the satellite is $3.1 \times 10^3\,\text{m}\,\text{s}^{-1}$.

  (iii) Calculate the kinetic energy of the satellite.   [4]

  (b) Calculate the gravitational potential:

  (i) on the Earth's surface

  (ii) at the orbit height of $3.59 \times 10^7\,\text{m}$.   [3]

  (c) Calculate the increase in gravitational potential energy of the satellite in being lifted from the Earth's surface to its orbit height.   [1]

  (d) Suggest one reason why the energy needed to launch the satellite into orbit is much more than the sum of your answers to (a)(iii) and (c).   [1]

Knowledge check

You should be able to:

- describe how temperature can be measured at the surface of a liquid
- describe the changes of state between solids, liquid and gases and the behaviour of particles when materials are in these states.

Heat and temperature

Heat is the movement of energy caused by a temperature difference. If one body (A) is in thermal contact with another body (B) at a lower temperature, then heat energy will flow from A to B (see Figure 14.1).

The extra energy that B gains can either increase the average energy of the molecules of B or cause B to do physical work. For example, if B is a gas, the extra energy can cause the gas to expand (doing external work) or the internal energy of the gas to increase (or both).

If A and B are at the same temperature, no net heat flow takes place. The two bodies are in thermal equilibrium (see Figure 14.2).

If $T_1 > T_2$ heat energy flows from A to B

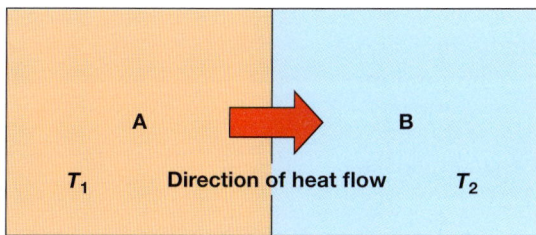

Thermal equilibrium: $T_1 = T_2$ no **net** heat energy flows from A to B

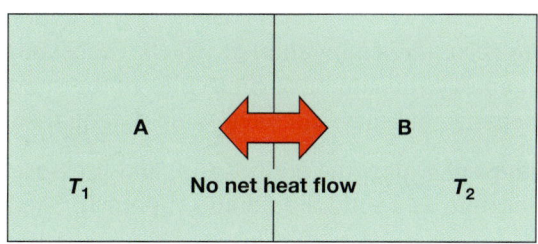

▲ **Figure 14.1** Heat flow when $T_1 > T_2$

▲ **Figure 14.2** Thermal equilibrium when $T_1 = T_2$

Measuring temperature

A wide range of **thermometric properties** can be used to measure temperature, including:

- expansion of a liquid (e.g., mercury or alcohol)
- change in electrical resistance of a metal wire (e.g., platinum)
- change of electrical resistance of a thermistor
- change in the output p.d. of a thermocouple
- change in pressure of a fixed volume of gas.

Some of these physical properties vary linearly with a change in temperature over a large range of temperatures (e.g., the resistance of a metal wire); others do not vary linearly, or do so only over a small range of temperatures (e.g., the resistance of a thermistor).

> **Remember**
>
> A **thermometric property** is one which changes with temperature; e.g., electrical resistance.

Worked Example

A platinum resistance wire has a resistance of $760\,\Omega$ when the temperature is $30\,°C$ and a resistance of $830\,\Omega$ when the temperature is $70\,°C$.

(a) Determine the temperature of the wire when its resistance is $795\,\Omega$.

(b) Calculate the resistance of the wire when the temperature is $55\,°C$

(c) State any assumptions you make in deriving your answers.

Answer

(a) $\dfrac{\text{Change in temperature}}{\text{change in resistance}} = \dfrac{70-30}{830-760} = 0.571\,°C\,\Omega^{-1}$, so when $R=795\,\Omega$,

temperature$=30+(795-760)\times0.571=50\,°C$

(b) $30+(R-760)\times0.571=55\,°C$, $R-760=\dfrac{55-30}{0.571}=43.8$, so $R=804\,\Omega$

(c) The resistance of the wire varies linearly with temperature for the range of temperatures in the question.

Temperature scales

Most temperature scales are thermometric scales in that they rely on the properties of a particular substance to establish the fixed points of the scale. The **Celsius scale**, for example, uses the melting point of ice ($0\,°C$) and the boiling point of water ($100\,°C$) as its fixed points.

The **absolute (thermodynamic) scale** of temperature does not depend on physical properties. Instead the scale relies on two fixed points:

- **absolute zero**: the temperature at which a substance has minimum internal energy (the atoms or molecules of the substance have no random kinetic energy, but may still have some potential energy),

- **triple point of water**: the temperature and pressure at which water exists in equilibrium as a solid, liquid and vapour ($0.01\,°C$ and $611.2\,Pa$).

The size of the unit of temperature on the thermodynamic scale of temperature is chosen to be the same as the size of a degree on the Celsius scale – a $1\,°C$ change in temperature is the same as a change of 1 kelvin ($1\,K$).

> **Remember**
>
> To convert temperature from Celsius scale to absolute scale:
>
> $T/K = T/°C + 273.15$

⬆ Raise your grade

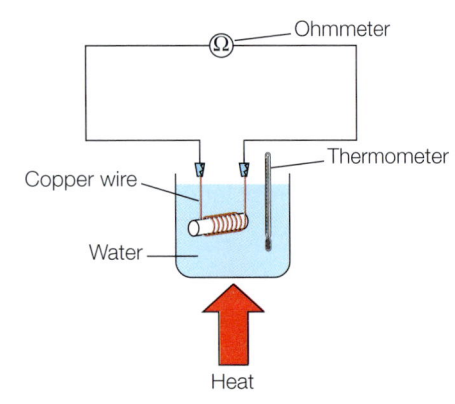

$T/°C$	R/Ω
25	4.4 ± 0.05
38	4.6 ± 0.05
51	4.8 ± 0.05
65	5.0 ± 0.05
69	5.1 ± 0.05
89	5.4 ± 0.05

(a) (i) Plot a graph of R/Ω against $T/°C$. Include error bars for R. [2]

(ii) Draw the straight line of best fit and a worst acceptable straight line on your graph. Both lines should be clearly labelled. [2]

(iii) Determine the gradient and intercept of the line of best fit. [2]

$\text{Gradient} = \dfrac{5.16 - 4.44}{74.0 - 28.0} = 0.0157\ \Omega^{-1}$ ✔

Correct read-offs and substitution into $\dfrac{\Delta y}{\Delta x}$ to find the gradient.

gradient = $0.0157\ \Omega\ °C^{-1}$

Using the point (84.0, 5.32) in $y = mx + c$:

$c = y - mx = 5.32 - 0.0157 \times 84.0 = 4.00\ \Omega$ ✔

Correct read-off and substitution into $y = mx + c$.

intercept = $4.00\ \Omega$

'Worst acceptable' line drawn – 'bottom-right to top-left'.

Points plotted accurately.

Error bars drawn correctly.

'Best fit' line drawn accurately.

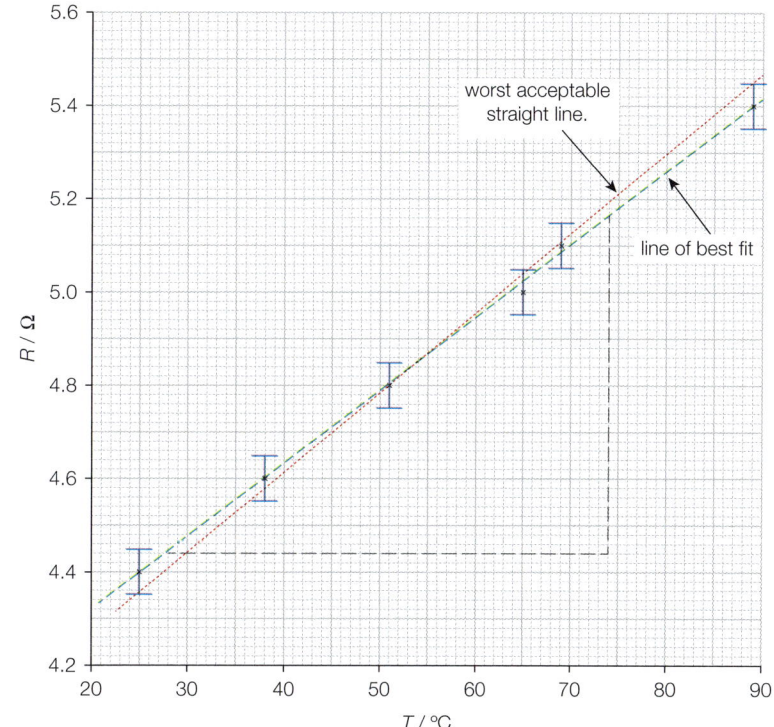

(b) R and T are related by the equation: [2]

$$R = a(1 + bT)\text{ where } a \text{ and } b \text{ are constants.}$$

Use your values from (a)(iii) to determine the values of a and b. Give appropriate units.

$a = y - \text{intercept} = 4.00$
$ab = \text{gradient} = 0.0157$

$b = \dfrac{\text{gradient}}{a} = \dfrac{0.0157}{4.00} = 3.93 \times 10^{-3}$ ✔✗

The candidate has matched a with the y-intercept and ab with the gradient, but has omitted the units for a and b.

The units for a must be the same as the units for R: Ω.

bT must be dimensionless, so b has units of $°C^{-1}$.

$a = \underline{\quad 4.00 \quad}$
$b = \underline{\quad 3.93 \times 10^{-3} \quad}$

Specific heat capacity and specific latent heat

Solids, liquids and gases

Solids

The atoms and molecules in a solid substance are close together, held in place by strong forces of attraction between them (see Figure 14.3a). Heating a solid causes the molecules to vibrate more (the temperature of the solid increases). If enough heat energy is supplied, the molecules can break free of each other, and the substance starts to melt.

Liquids

In liquids, the atoms and molecules are free to slip past each other and move about at random (see Figure 14.3b). The forces of attraction between molecules are smaller, allowing the liquid to flow. Heating a liquid cause the temperature to rise as the molecules gain more kinetic energy. If enough energy is supplied, all the molecules have enough energy to break free and the liquid starts to boil.

Gases

The atoms or molecules that make up a gas are much further apart and moving at high speeds, continually having elastic collisions with the walls of their container and each other (see Figure 14.3c). Heating a gas causes the average molecular speed to increase and the temperature of the gas to rise.

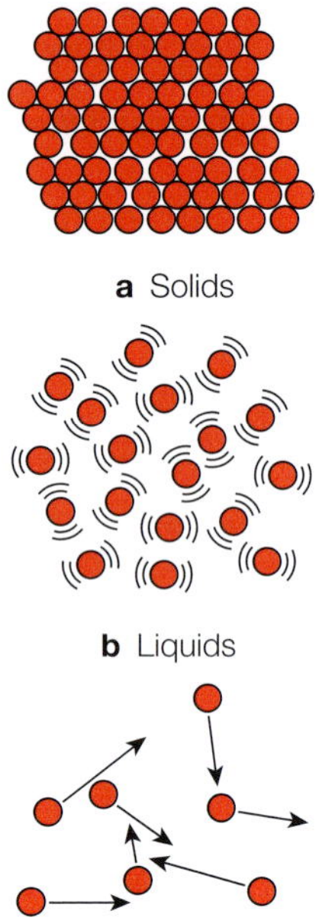

a Solids

b Liquids

c Gases

▲ **Figure 14.3** Solids, liquids, and gases

Melting and boiling

If a solid such as ice is heated at a uniform rate, its temperature rises until it reaches its **melting point**. Once completely melted, its temperature continues to rise until it reaches its **boiling point**. Once boiled, the temperature of the gas continues to rise. Figure 14.4 illustrates this process.

- As a solid is heated from A to B, its temperature rises. The energy needed to raise the temperature of 1 kg of a substance by 1 °C is called the **specific heat capacity** c of the substance.

- From B to C the substance is melting. All the heat energy supplied is being used to weaken or break the bonds between molecules – the temperature of the substance remains constant.

- The energy needed to melt 1 kg of the substance at its melting point is called the **specific latent heat of fusion** L_f. (When a liquid solidifies energy is released.) The amount of energy needed to melt m kg of a substance at its melting point is mL_f.

- From C to D the temperature of the liquid rises until it reaches its boiling point. If the specific heat capacity of the liquid is c, then the heat energy needed to raise the temperature of a mass m of the liquid by an amount $\Delta\theta$ is $mc\Delta\theta$.

- From D to E all the heat energy supplied is being used to break the bonds between molecules completely, and the temperature of the liquid/gas remains constant until this process is complete. The energy needed to change 1 kg of a liquid at its boiling point into gas is called the **specific latent heat of vaporisation** L_v. When a gas condenses back into a liquid this energy is released as heat into the surroundings. The amount of energy needed to vaporise m kg of a substance at its boiling point is mL_v.

- From E to F the molecules gain more kinetic energy and the temperature of the gas rises.

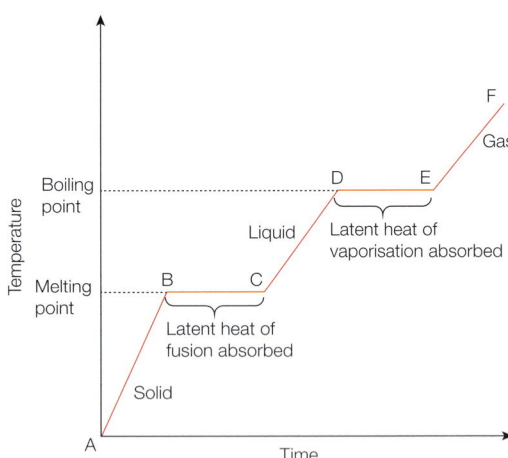

▲ **Figure 14.4** Melting and boiling

Cooling by evaporation

At any one instant the molecules of a liquid have a range of kinetic energies (see Figure 14.5). The more energetic particles which also happen to be at the surface of the liquid may have enough energy to 'escape' completely (become vaporised). If the fastest molecules evaporate, the average energy of the molecules left behind decreases; its temperature falls slightly compared to its surroundings.

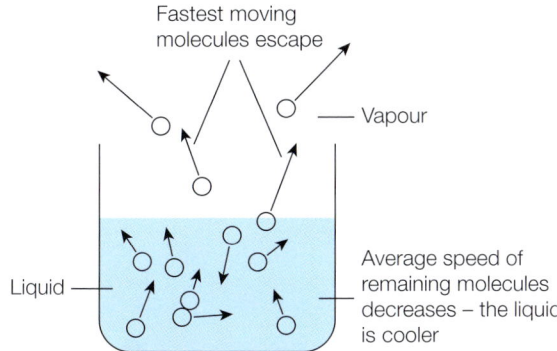

▲ **Figure 14.5** Cooling by evaporation

> ★ **Exam tip**
>
> **Specific** in a definition means 'for 1 kg'. The units of specific heat capacity are J kg^{-1} K^{-1}. The units of specific latent heat are J kg^{-1}.

> 💡 **Remember**
>
> The *specific latent heat* of a substance is the energy needed to change the state of 1 kg of the substance **at constant temperature**.

Finding the specific heat capacity of a metal

A metal of known mass m is placed in an insulated container and heated with an electrical heater which fits into a hole drilled into the metal, as shown in Figure 14.6. A thermometer, fitted into a second drilled hole, measures the temperature change.

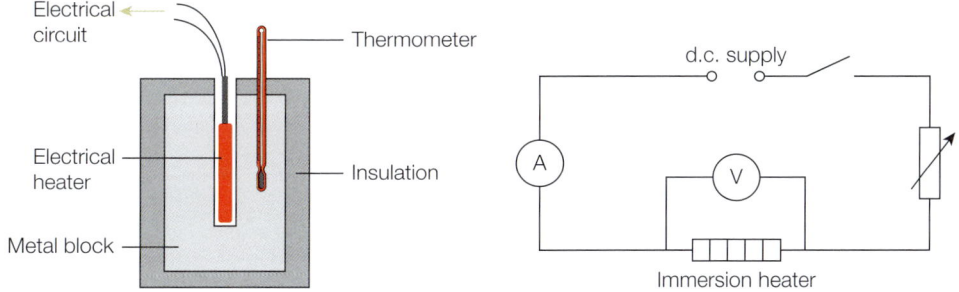

▲ **Figure 14.6** Measuring specific heat capacity

The electrical power of the heater is IV, where I is the current through the heater and V the potential difference across it. A stopwatch records the time t the heater is switched on and the thermometer measures the change in temperature $\Delta\theta$.

If no heat is lost to the surroundings: $mc\Delta\theta = IVt$ so $c = \dfrac{IVt}{m\Delta\theta}$

The value is likely to be an overestimate of the true value as some energy will always be lost to the surroundings, so change in temperature is less.

Worked Example

A cylinder of aluminium, of mass 1.0 kg, is heated by a 50 W electric immersion heater. After 4 minutes the temperature of the cylinder has risen 11 °C. Determine the specific heat capacity of aluminium.

Answer

energy supplied to aluminium $= 50 \times (4 \times 60) = 12\,000$ J

specific heat capacity of aluminium $= \dfrac{12\,000}{1 \times 11} = 1.09 \times 10^{3}$ J kg^{-1} K^{-1}

> The correct value is 920 J kg^{-1} K^{-1}.

Finding the specific latent fusion of a solid

Figure 14.7 shows how the specific latent heat of fusion of ice can be found. Ice at its melting point is heated by an electrical immersion heater. The amount of ice melted in a fixed time can be found by weighing the beaker before and after heating. The heat energy needed to melt an amount m of ice at its melting point is mL_f. The value measured is lower than the actual value as the heat gained from surroundings causes more ice to melt.

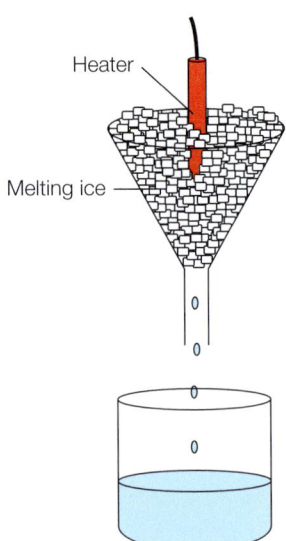

▲ **Figure 14.7** Measuring the specific latent heat of fusion

Worked Example

Ice at its melting point is heated by a 50 W immersion heater. After 10 minutes it is found that 100 g of ice has melted.

(a) Calculate the specific latent heat of ice.

(b) State, with a reason, why your answer is an overestimate or an underestimate of the true value.

Answer

(a) $(100 \times 10^{-3}) \times L_F = 50 \times (10 \times 60)$

$L_F = 3.0 \times 10^5 \, \text{J kg}^{-1}$

The true value is $3.3 \times 10^5 \, \text{J kg}^{-1}$.

(b) This is an underestimate because some of the heat energy to melt the ice comes from the surroundings. This can be compensated for by repeating the experiment with the heater removed or switched off. The amount of water that collects in the beaker can be subtracted from the first value to find a more accurate value for L_F.

Finding the specific latent heat of vaporisation of a liquid

The liquid is first heated to its boiling point, as shown in Figure 14.8.

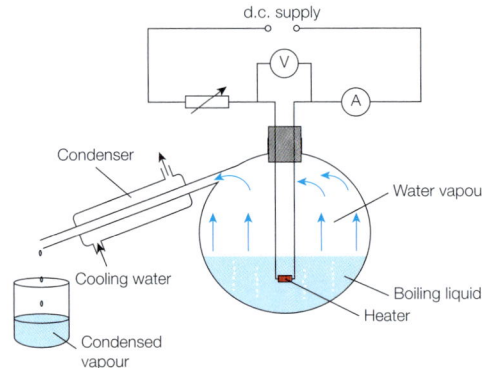

▲ **Figure 14.8** Measuring the specific latent heat of vaporisation

As the liquid boils the vapour passes through a condenser and collects in a beaker. The liquid is boiled for a fixed time t, and the current I in the heater and potential difference V across the heater are recorded. The amount of condensed vapour collected m can be found by weighing the beaker before and after the liquid has boiled. The amount of energy needed to boil m kg at its boiling point is mL_V.

The experiment can be repeated with different values of I and V but boiling the liquid for the same fixed time t. The results can be used to obtain an accurate value for L_V, the specific latent heat of vaporisation of the liquid, by eliminating the heat energy lost to the surroundings during the experiment.

↑ Raise your grade

(a) Define *specific latent heat*.   [3]

> The energy needed to change state ✔✗✗

> A better answer would be 'the energy needed, **per kg**, to change state, **at constant temperature.**' In the case of a liquid, it is the energy needed to change 1 kg from liquid at its boiling point to vapour.

(b) A student carries out an experiment to find the specific latent heat of vaporisation of water.   [1]

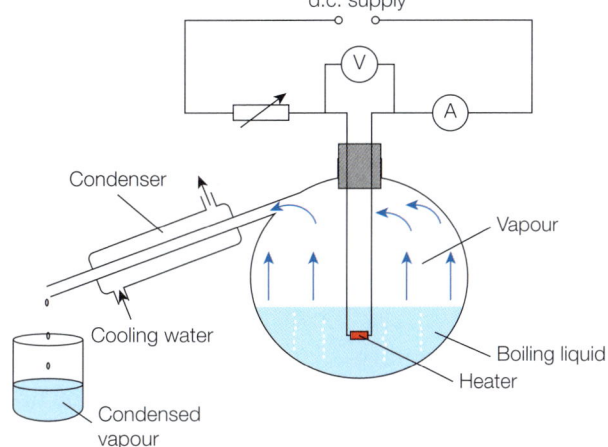

The water is heated up to its boiling point, and then boiled for 6.0 minutes. The mass m of condensed water vapour collected is found by weighing the beaker before and after boiling the water. The experiment is repeated with a different set of values for the current I in the heater and p.d. V across the heater.

The results of the experiment are shown.

I / A	V / V	m / g
3.0	16	47
2.0	12	21

Suggest why the same amount of heat energy H is lost to the surroundings in each experiment.

> A good answer.

> The apparatus is at the same temperature for the same time in both experiments. ✔

(c) Calculate:

(i) the specific latent heat of vaporisation of the liquid L_v [3]

> $3.0 \times 16 \times (6 \times 60) = (47 \times 10^{-3}) \times L_v + H$ (1) ✔
>
> $2.0 \times 12 \times (6 \times 60) = (21 \times 10^{-3}) \times L_v + H$ (2) ✔
>
> (1) − (2): $26 \times 10^{-3} \times L_v = 17280 - 8640$
>
> $L_v = 3.3 \times 10^5$ ✔

> Correct method.
>
> Correct analysis.
>
> Correct final answer.
>
> [3]
>
> $L_v = 3.3 \times 10^5 \, \text{J kg}^{-1}$

(ii) the heat energy H lost to the surroundings.   [1]

> From equation 2:
>
> $H = 2.0 \times 12 \times (6 \times 60) - (21 \times 10^{-3}) \times 3.3 \times 10^5$
>
> $= 1710 \, \text{J}$ ✔

> Correct calculation.
>
> $H = 1710 \, \text{J}$

Exam-style questions

1 **(a)** Convert into kelvin, and to an appropriate number of decimal places:

  (i) 100.00 °C

  (ii) 0 °C

  (iii) − 80.7 °C   [3]

 (b) Convert into °C, and to an appropriate number of decimal places:

  (i) 273 K

  (ii) 376.2 K

  (iii) 0.0 K   [3]

 (c) State and explain what is meant by *thermal equilibrium*.   [2]

2 Describe a method of calibrating a liquid-in-glass thermometer between 0° C and 100 °C.   [3]

3 Describe one advantage and one disadvantage of a thermistor used as a thermometer compared to a thermocouple.   [2]

4 **(a)** **(i)** State the two *fixed points* used for the absolute temperature scale.

  (ii) Describe the difference between the absolute temperature scale and other temperature scales.   [3]

 (b) State what is meant by the absolute zero of temperature.   [1]

 (c) A heater raises the temperature inside an incubator from 19.7 °C to 37.5 °C.

  Determine, in kelvin and to an appropriate number of decimal places:

  (i) the rise in temperature of the incubator

  (ii) the final temperature of the incubator.   [2]

5 A thermocouple produces an e.m.f. of 5.5 μV for each 1.0 °C temperature difference between the two junctions of the thermocouple. The cold junction is maintained at a constant temperature of 20.0 °C.

 (a) Calculate:

  (i) the temperature of the hot junction when the e.m.f. is 1.21 mV

  (ii) the e.m.f. when the temperature of the hot junction is 1200 °C.   [2]

 (b) State any assumptions you make.   [1]

6 **(a)** Define *specific heat capacity*.   [1]

 (b) Calculate:

  (i) the energy needed to raise the temperature of 8.0 kg of aluminium by 5 °C

   [Specific heat capacity of aluminium is 920 J kg⁻¹ K⁻¹.]

  (ii) the specific heat capacity of lead if 0.48 kJ are needed to raise the temperature of 200 g of lead by 18 °C.   [2]

7 A 3.0 kW electric heater is used to heat 140 kg of water in a tank from 20 °C to 65 °C. [Specific heat capacity of water is 4200 J kg⁻¹ K⁻¹.]

 (a) Determine the time taken to heat the water.   [2]

 (b) State, with a reason, whether your answer to (a) is an under-estimate or an over-estimate.   [2]

8 A continuous flow calorimeter is used to find the specific heat capacity of a liquid.

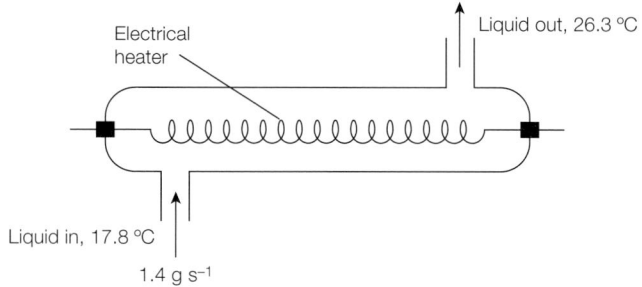

Liquid enters the tube at a constant rate of 1.4 g s⁻¹. The temperature of the liquid at the inlet is 17.8 °C. As it flows through the calorimeter, the liquid is heated by an electrical heater of output power 37.4 W. The liquid leaves the calorimeter at a temperature of 26.3 °C.

The flow rate is now doubled. The output power of the heater is increased until the temperature of the liquid leaving the calorimeter is again 26.3 °C. The output power of the heater is now 66.7 kW.

Calculate:

 (a) the specific heat capacity of the liquid   [2]

 (b) the heat energy lost each second to the surroundings.   [2]

Equation of state for an ideal gas

An **ideal gas** is one in which all the collisions between atoms or molecules are perfectly elastic and in which there are no intermolecular forces. Such a gas obeys the equation:

$$pV = nRT$$

where p is the pressure of the gas, V its volume, and T its absolute temperature measured in kelvin (K). n is the number of moles of the gas and R is the molar gas constant. The equation is called the **equation of state for an ideal gas**.

Real gases normally obey this equation except under extremely low values of temperature or very high pressures. For gases at moderate values of temperature and pressure it can be shown by experiment that:

- $p \propto \dfrac{1}{V}$ at constant temperature – Boyle's law (Figure 15.1a)

- $V \propto T$ at constant pressure – Charles' law (Figure 15.1b)

- $p \propto T$ at constant temperature – the pressure law (Figure 15.1c).

> **Remember**
>
> $$pV = nRT$$

> ★ **Exam tip**
>
> Don't forget that T in all the gas equations is the absolute temperature, in kelvin:
>
> $$T(\text{K}) = T(°\text{C}) + 273°$$
>
> The values of R and N_A are provided in Exam Papers 1, 2, and 4.

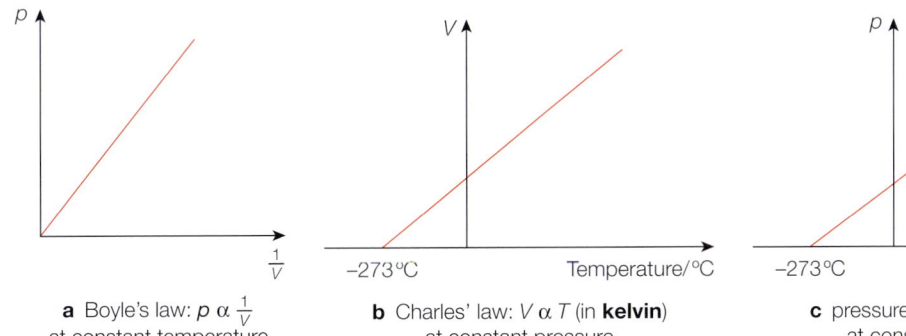

a Boyle's law: $p \propto \dfrac{1}{V}$ at constant temperature

b Charles' law: $V \propto T$ (in **kelvin**) at constant pressure

c pressure law: $p \propto T$ (in **kelvin**) at constant temperature

▲ **Figure 15.1** Gas laws

The three gas laws can be combined:

$$\frac{pV}{T} = \text{constant}$$

The size of the constant depends on the number of gas molecules (e.g., doubling the number of gas molecules in a sample doubles the number of collisions of the molecules with the walls of the container, and so doubles the pressure).

R is the value of the constant for one mole of gas molecules. A **mole** is just a number, **Avogadro's number** N_A (where $N_A = 6.02 \times 10^{23}\,\text{mol}^{-1}$), and so the constant for a different amount of gas is nR where n is the number of moles of gas present.

> 🔗 **Link**
>
> Avogadro's constant N_A is equal to the number of atoms in 12 g of ^{12}C.
>
> $$N_A = 6.02 \times 10^{23}\,\text{mol}^{-1}$$
>
> A *mole* (mol) of anything is N_A of it.

Internal energy of a gas

The molecules of a gas have both kinetic energy (because they are moving) and potential energy (because there are attractive forces between them). The sum total of all the molecules' potential and kinetic energies is the internal energy of the gas.

For an **ideal gas**, there are no intermolecular forces so the molecules have no potential energy – all the internal energy of an ideal gas is kinetic energy.

Kinetic theory of gases

The kinetic theory of gases is based on the model of a gas that has many identical particles moving about randomly, colliding with each other and the walls of their container, constantly changing speed and direction.

The theory makes a number of simplifying assumptions about gases:

- the gas molecules have negligible volume compared to the volume of the gas as a whole

- the forces between gas molecules are negligible, except when colliding with each other

- collisions between molecules, or between molecules and the walls of their container, are perfectly elastic

- for an individual gas molecule, the time for a collision is negligible compared to the time between collisions.

Imagine a single molecule of mass m, inside a box of side L, travelling at speed c towards the right-hand side of the box (Figure 15.2). When it collides with the wall it has a perfectly elastic collision and rebounds with the same speed in the opposite direction:

$$\text{change in momentum of the molecule} = mc - (-mc) = 2mc$$

The molecule moves towards the left-hand wall, rebounds, and returns to hit the right-hand wall again. The molecule will have travelled a total distance $2L$ at speed c between collisions with the right-hand wall:

$$\text{time taken between collisions with the right-hand wall} = \frac{2L}{c}$$

Using Newton's second law (force = rate of change of momentum):

$$\text{force on right-hand wall} = \text{rate of change in momentum}$$

$$= \frac{2mc}{\left(\dfrac{2L}{c}\right)} = \frac{mc^2}{L}$$

$$\text{pressure on right-hand wall} = \frac{\text{force}}{\text{area}} = \frac{\left(\dfrac{mc^2}{L}\right)}{L^2} = \frac{mc^2}{L^3}$$

For N molecules in the gas, all moving with different speeds, the pressure p is:

$$p = N\frac{m<c^2>}{L^3} = N\frac{m<c^2>}{V}$$

where $<c^2>$ is the mean of the squares of the velocities and V is the volume of the box. This assumes all the molecules are travelling in the same direction and colliding with the same two opposite faces of the box.

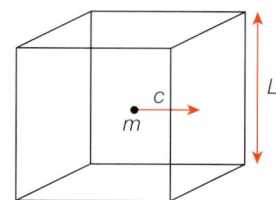

▲ **Figure 15.2** Molecule in a box

At any one time, as many molecules are moving up and down, or forward and backward, as left and right, and so we must divide by three to calculate the pressure in any direction:

$$p = \frac{1}{3} N \frac{m<c^2>}{V}$$

This equation is usually written:

$$pV = \frac{1}{3} Nm <c^2>$$

Re-arranging this equation gives:

$$p = \frac{1}{3}\left(\frac{Nm}{V}\right)<c^2> = \frac{1}{3}\rho <c^2>$$

where ρ is the density of the gas.

Worked Example

Estimate the r.m.s. speed of air molecules at room temperature.

Answer

Estimates: atmospheric pressure $\approx 1.0 \times 10^5\,$Pa

density of air at room temperature $\approx 1.29\,$kg m^{-3}

Rearranging $pV = \frac{1}{3} Nm<c^2>$ gives:

$$p = \frac{1}{3}\left(\frac{Nm}{V}\right)<c^2> = \frac{1}{3}\rho <c^2>$$

so $<c^2> = \dfrac{3p}{\rho} = \dfrac{3 \times 1.0 \times 10^5}{1.29} = 2.3 \times 10^5$

The root-mean-square speed $(c_{r.m.s.}) = \sqrt{<c^2>} = 480\,$m s^{-1}.

Kinetic energy of a gas molecule

Comparing $pV = \frac{1}{3} Nm <c^2>$ with the equation of state for an ideal gas:

$$\frac{1}{3} Nm <c^2> = nRT$$

$$\frac{1}{2} m <c^2> = \frac{3}{2}\left(\frac{n}{N}\right)RT$$

kinetic energy of a gas molecule $= \frac{1}{2} m <c^2> = \frac{3}{2}kT$

where k is the Boltzmann constant ($k = 1.38 \times 10^{23}\,$J K^{-1}).

So the average kinetic energy of a gas molecule is directly proportional to the absolute temperature of the gas.

↑ Raise your grade

(a) State two assumptions made in the kinetic theory of gases. [2]

The atoms have elastic collisions with each other and the walls of the container. ......

The volume of individual atoms is negligible compared to the volume of the container. ✔✔ ......

A good answer. Other answers include: no inter-molecular forces/molecules in random motion/ time of collision negligible compared to time between collisions/large number of molecules.

(b) Using the kinetic theory, it can be shown that the pressure p and the volume V of a gas are related

by the equation: $pV = \dfrac{1}{3}Nm<c^2>$

State the meaning of:

(i) Nm

The mass of the gas ✔✗ ......

The answer is correct, but note that there are 2 marks available. The second mark is for identifying m as the mass of an individual atom/molecule and/or N as the number of atoms/molecules. [2]

(ii) $<c^2>$ [1]

The speed of the gas molecules squared ✗ ......

The symbol $< >$ denotes a mean (average) value. $<c^2>$ is the mean of the squares of the velocities of all the gas atoms/molecules.

(c) Use this equation to show that the density ρ of a gas is given by $\rho = \dfrac{3p}{<c^2>}$. [1]

$$pV = \frac{1}{3}Nm<c^2> \Rightarrow p = \frac{1}{3}\frac{Nm}{V}<c^2> \text{ but } \frac{Nm}{V} = \rho \text{ (the density)}$$

so $\rho = \dfrac{3p}{<c^2>}$ ✔

$\dfrac{Nm}{V}$ correctly identified as the density.

(d) (i) An ideal gas obeys the equation $pV=NkT$ where k is the Boltzmann constant. Combine this equation with the equation given in part (b) to show that the kinetic energy of a gas molecule is proportional to T. [1]

$$pv = \frac{1}{3}Nm<c^2> = NkT ✔$$

N cancels and equation re-arranged to find $\frac{1}{2}m<c^2>$.

so $\dfrac{1}{2}m<c^2> = \dfrac{3}{2}kT \rightarrow$ the kinetic energy of a gas molecule is proportional to T.

(ii) A gas, initially at a temperature of $50\,°C$, is heated to a temperature of $150\,°C$. Compare the kinetic energies of a gas molecule at these two temperatures. [2]

Kinetic energy of a gas molecule is proportional to T so kinetic energy at $150\,°C$ is ......

three times kinetic energy at $50\,°C$. ✗✗ ......

A frequent error. The temperature T is the **absolute** temperature, measured in kelvin. The answer is:

$$\frac{\text{kinetic energy at }150°C}{\text{kinetic energy at }50°C} = \frac{(273+150)}{273+50} = 1.31$$

Exam-style questions

1 **(a)** State what is meant by an *ideal* gas. [1]

   (b) An oxygen cylinder contains oxygen at a pressure of 1.37×10^7 Pa and a temperature of 20 °C. Calculate the pressure if the temperature rises to 70 °C. [2]

2 **(a)** What is the equation of state for an ideal gas? [1]

   (b) A weather balloon is released from ground level, where the temperature is 20 °C. The pressure inside the balloon is 1.2×10^5 Pa.

   The temperature of the atmosphere decreases by 5 °C for each kilometre the balloon rises.

   Calculate:

   (i) the temperature of the atmosphere at a height 3 km above ground level

   (ii) the pressure of the gas inside the balloon at this height. [Assume the volume of the balloon remains constant.] [3]

3 **(a)** Define *Avogadro's constant*. [1]

   (b) A gas cylinder contains 9.0×10^{-3} m³ of argon (^{40}Ar) at a pressure of 1.37×10^7 Pa and a temperature of 20 °C.

   Calculate:

   (i) the number of moles of argon

   (ii) the mass of argon in the cylinder. [3]

   (c) For an ideal gas, $pV = \frac{1}{3} Nm <c^2>$.

   (i) Explain what is meant by $<c^2>$.

   (ii) Calculate the root-mean-square speed of the argon atoms. [3]

4 **(a)** State what is meant by a *mole* of gas. [1]

   (b) An ideal gas at a temperature of 25 °C exerts a pressure of 2.0×10^5 Pa. The volume of the gas is 4.0×10^{-3} m³. Calculate the number of moles of gas present. [2]

   (c) The gas is heated at constant volume so that the pressure increases to 2.5×10^5 Pa. What is the new temperature of the gas? [1]

5 **(a) (i)** Describe what is meant by the *internal energy* of a gas.

   (ii) Explain how this differs from the internal energy of an ideal gas. [2]

   (b) Calculate the internal energy of:

   (i) 30 g of helium (^{4}He) at 50 °C,

   (ii) 0.5 g of krypton (^{84}Kr) at 200 °C. [3]

   [Both gases can be treated as ideal gases.]

6 The equation of state for an ideal gas with pressure p and volume V is:

   $$pV = nRT$$

   (a) State the meaning of the symbol:

   (i) n **(ii)** R **(iii)** T. [3]

   (b) A gas cylinder contains 5.0×10^{-4} m³ of xenon at a pressure of 20.0×10^5 Pa and a temperature of 27 °C. Determine the number of atoms of xenon that are in the cylinder. [3]

7 A piston of mass 0.400 kg can slide freely up and down a cylinder of inner diameter 12.0 cm. The temperature of the air trapped inside the cylinder is 20 °C. The piston rests at a height of 8.0 cm above the base of the cylinder.

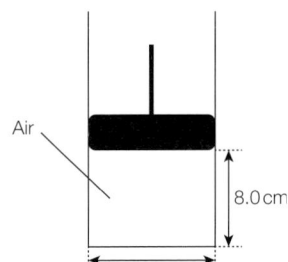

   (a) Calculate the pressure (above atmospheric pressure) of the air trapped in the cylinder. [1]

   (b) The temperature of the air is increased to 80 °C. Determine the new height of the piston in the cylinder. [2]

8 A canister of volume 2.50×10^{-4} m³ contains air at a pressure of 1.01×10^5 Pa. The temperature of the air is 20 °C.
   [Specific heat capacity of air at constant volume is 0.716 kJ kg⁻¹ K⁻¹. Molar mass of air is 29 g mol⁻¹.]

   (a) Determine how many moles of air are in the canister. [2]

   (b) The air in the canister is heated to 80 °C.

   Calculate:

   (i) the pressure of the air at the new temperature

   (ii) the change in internal energy of the air. [3]

16 Thermodynamics

Knowledge check

You should be able to:

- describe simple energy changes caused by heating and cooling

- understand how work is done by forces, electrical currents and heating

- describe energy transfer involving changes in kinetic and potential energy

- describe the behaviour of particles in a solid, liquid and gas.

Internal energy and the first law of thermodynamics

The **internal energy of a system** is defined as:

| internal energy of a system | = | sum of the random distribution of kinetic and potential energies of its molecules |

The internal energy of an object or system can be increased by:

- heating the object

- doing mechanical work on the object.

In practice, the total internal energy is very difficult to measure but changes in the internal energy can be found by analysing work done or heating effects.

If a gas is heated, its molecules move faster, and so have more kinetic energy. If the gas is compressed, the 'squashing' imparts kinetic energy to the molecules – when a bicycle pump is compressed quickly the end becomes hotter. The change in internal energy, the external work done, and the heat energy supplied to an object are linked together by the **first law of thermodynamics**.

The first law is really a statement of the **conservation of energy**. It states that the change in internal energy of a system (ΔU) is equal to the sum of the energy entering the system by heating (q) and the energy entering the system by work being done on it (W).

$$\Delta U = q + W$$

Examples of heating and cooling

- A positive value of q means that the system is being heated: for example, when ice is placed in contact with a warm surface, the ice will gain internal energy as it is heated by the surface.

- A negative value of q indicates that the system is losing heat: for example, a warm boiled egg gradually loses internal energy through conduction and radiation processes as it cools down.

Examples of doing mechanical work

- Internal energy is raised when work is done on the system by a force. When a spring in a toy is stretched, its internal energy increases; we describe this as an increase in elastic potential energy. This increase in internal energy is represented by a positive value of W.

- A negative value of W indicates a decrease in internal energy when the system does work on the surroundings. If the spring is released and the toy starts to move, then there is a transfer of the internal (potential) energy to kinetic energy of the moving toy through a force.

Internal energy of gases

Temperature changes

If a gas is heated, its molecules move faster, and so have more kinetic energy, which means that the internal energy has increased ($+q$). When a gas cools, its internal energy decreases ($-q$) as the average kinetic energy of the molecules decreases.

Work done

When a gas is compressed slowly by a force, work is being done on the gas. The internal energy of the gas increases ($+W$).

When a gas is allowed to expand, it does work on the surroundings, for example, by pushing the atmosphere out of the way. This gives a change in energy represented by a negative value of W.

The work done (W) when a gas changes volume (ΔV) at a constant pressure (p) is given by:

$$W = p\Delta V$$

Rapid compression and expansion of a gas lead to temperature changes in the gas: squashing a gas quickly will increase its temperature (as in a diesel engine) while allowing it to expand rapidly will decrease its temperature (as in a fire extinguisher).

Worked Examples

1 A litre of water at 100 °C is left to evaporate. When it has all evaporated, the water vapour occupies a volume of 1.7 m³. Calculate the change in internal energy of the water.

   [atmospheric pressure = 1.01×10^5 Pa; specific latent heat of vaporisation of water = 2.26×10^6 J kg⁻¹; density of water = 1.0×10^3 kg m⁻³]

   Answer

   Heat energy supplied to the water to evaporate it, q

   $$q = (1.0 \times 10^{-3}) \times (1.0 \times 10^3) \times 2.26 \times 10^6 = 2.26 \times 10^6 \, \text{J}$$

   Work done on atmosphere, W

   $$I = -1.01 \times 10^5 \times (1.7 - 1.0 \times 10^{-3}) = -1.72 \times 10^5 \, \text{J}$$

   W is negative because work is done on the atmosphere.

   Change in internal energy $\Delta U = q + W = 2.26 \times 10^6 - 1.72 \times 10^5$

   $$\Delta U = 2.09 \times 10^6 \, \text{J}$$

2 When a rubber band is quickly stretched and released several times, its temperature increases. Describe what happens to the rubber band in relation to the first law of thermodynamics. State whether each of the following is positive, negative, or zero: ΔU, q, and W.

   Answer

   To stretch the rubber band a force is moved in the direction of the force; that is external work is done to the rubber band (W is positive). No heat energy is supplied from an external source ($q=0$) so the change in internal energy $\Delta U = W$. As the rubber band becomes warmer than its surroundings it will start to emit heat energy (q is now negative).

   Eventually the external work done W will be equal to the heat energy lost q, and the internal energy of the rubber band will remain constant:

   $$\Delta U = W - q = 0$$

↑ Raise your grade

1   An electric motor is used to stir a container of liquid. Explain, in terms of the first law of thermodynamics, why the internal energy of the liquid increases. [3]

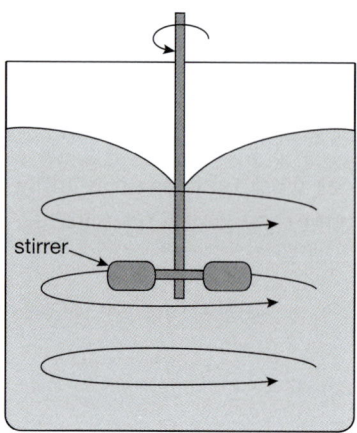

The motor gets hot due to the electric current in the wires and this heats the liquid. ✔ ✗ ✗

> This answer is acceptable for one of the three marks as electrical currents do cause heating effects and this would warm the liquid, but these processes need more description. The drag forces caused by the stirring action also do work on the liquid, so the student needs to mention 'work done on the liquid' by forces for the other two marks.

Exam-style questions

1   State what is meant by the *internal energy* of a system. [2]

2   (a) State the first law of thermodynamics.

Explain the meaning of any symbols you use. [3]

(b) A balloon bursts and all the air escapes rapidly. State and explain, using the first law of thermodynamics, what happens to the internal energy of the air that was inside the balloon. [2]

(c) A block of ice is removed from a freezer and placed in a warm room. The ice starts to melt.

(i) Explain why work is done on the ice as it melts.

(ii) Compare the work done on the ice with the change in internal energy of the ice as it melts. [3]

3   A cube of aluminium, with sides of length 2.0 cm, is heated so that its temperature increases from 20 °C to 450 °C. The cube expands so that its volume increases by 7×10^{-3} %.

[Density of aluminium = 2.7 g cm^{-3}; specific heat capacity of aluminium = 900 J kg^{-1} K^{-1}]

(a) The first law of thermodynamics states that

$$\Delta U = q + W$$

For the aluminium cube described, state whether:

(i) q is positive or negative

(ii) W is positive or negative. [2]

(b) Calculate the change in internal energy of the aluminium cube. [3]

Oscillations 17

Knowledge check

You should be able to:

- describe waves and oscillations in terms of frequency, period and amplitude

- describe energy changes including those involving kinetic, gravitational and elastic potential

- interpret graphs and their gradients

- use radians as a unit of angle.

Oscillations

Any to-and-fro motion about a fixed point, such as a pendulum clock (Figure 17.1) or a child bouncing up and down on a trampoline, is an example of an **oscillation**. Starting from the **equilibrium position**, a complete oscillation is the movement to the **maximum displacement** in one direction, back through the equilibrium position to the maximum displacement in the other direction and back again to the equilibrium position (Figure 17.2).

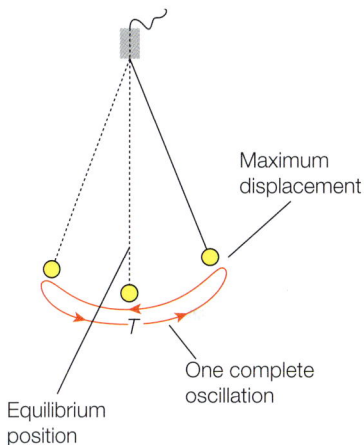

▲ **Figure 17.1** Simple pendulum

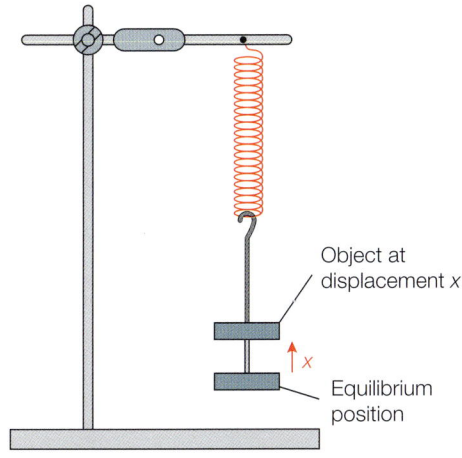

▲ **Figure 17.2** The oscillations of a loaded spring

Important terms

Amplitude is the maximum displacement from the equilibrium position. It can be measured in a variety of units (e.g., metres or degrees) depending on the type of oscillation. If the amplitude is constant, no energy is being lost and the oscillations are described as **free oscillations**.

Period T (s) is the time for one complete to-and-fro oscillation.

Frequency f Is the number of oscillations per second, measured in hertz (Hz). 1 Hz is one oscillation, or cycle, per second.

$$f = \frac{1}{T}$$

Angular frequency ω is defined as $\frac{2\pi}{T}$ ($= 2\pi f$), measured in rad s^{-1}.

$$T = \frac{1}{f} = \frac{2\pi}{\omega}$$

> ### Maths Skills
>
> For more on radians see the *Maths skills* section.

Worked Example

The graph in Figure 17.3 shows the displacement of a mass on a spring against time. Find the period, frequency, angular frequency, and amplitude of oscillation of the mass.

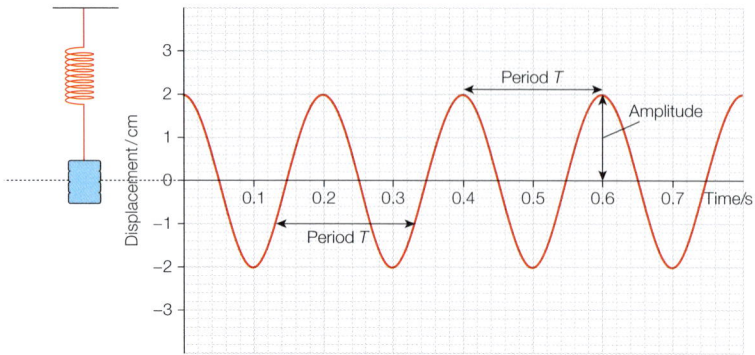

▲ **Figure 17.3** Displacement–time graph for an undamped oscillator

Answer

period T of the oscillation is 0.20 s

frequency $f = \dfrac{1}{T} = 5.0$ Hz

angular frequency $\omega = 2\pi f = 31.4$ rad s^{-1}

amplitude of the oscillation is 2.0 cm

Phase difference

The **phase difference** between two or more oscillations can be described in terms of fractions of a cycle, in radians or in degrees (see Table 17.1). Two oscillators that are out of step by half a cycle (in **antiphase**) are π radians out of phase; if one oscillator is at a maximum value when another is at zero displacement, the oscillators are one-quarter of a cycle out of phase or $\pi/2$ radians out of phase.

Figure 17.4 shows the displacement against time of two oscillators, A and B. Oscillator B is out of phase with oscillator A (oscillator A reaches its maximum displacement before oscillator B, so **leads** B (B **lags** behind A)).

The phase difference is $\dfrac{\Delta t}{T}$ cycles or $2\pi\dfrac{\Delta t}{T}$ radians or $360\dfrac{\Delta t}{T}$ degrees. In

Figure 17.4 the phase difference is one-quarter of a cycle (or $\pi/2$ rad or 90°).

▼ **Table 17.1** Phase difference

Cycles	Radians	Degrees
0	0	0
¼	$\dfrac{\pi}{2}$	90°
½	π	180°
¾	$\dfrac{3\pi}{2}$	270°
1	2π	360°
n	$2n\pi$	360n°

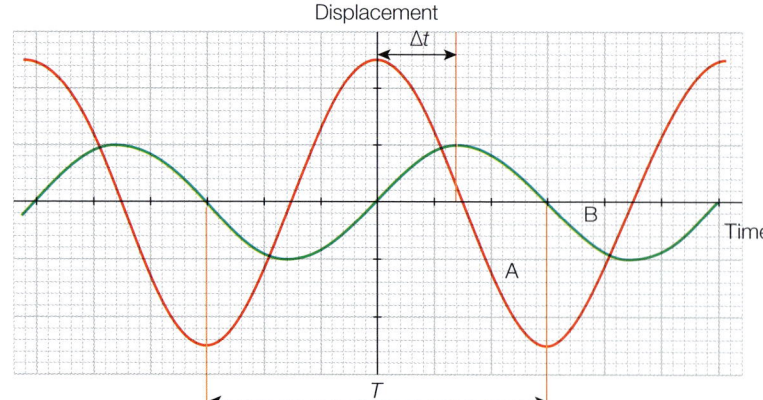

▲ **Figure 17.4** Phase difference

Simple harmonic motion (SHM)

Simple harmonic motion (SHM) is defined as motion in which the acceleration is:

* proportional to the displacement from a fixed point

* in the opposite direction to the displacement.

SHM can be expressed by the equation: $a = -\omega^2 x$

where a is the acceleration, x the displacement, and ω the angular frequency. The minus sign is important as ω^2 must be positive; it means that the acceleration will always have the opposite sign to the displacement. Many oscillations, including that of a mass on a spring and a simple pendulum, approximate to SHM.

A trolley of mass m is attached to a spring of stiffness k. It is pulled to one side and released (see Figure 17.5).

When the trolley is a distance x to the left of the equilibrium position, the spring exerts a force kx to the right. Using $F = ma$:

$$-kx = ma$$

Rearranging this equation:
$$a = -\frac{k}{m}x$$

As this is SHM, $a = -\omega^2 x$, so:
$$\omega^2 = \frac{k}{m}$$

Hence: $\omega = \sqrt{\frac{k}{m}}$ $f = \frac{1}{2\pi}\sqrt{\frac{k}{m}}$ $T = 2\pi\sqrt{\frac{m}{k}}$

The SHM equation $a = -\omega^2 x$ can be solved to find the velocity and displacement at time t. If the displacement is x and the velocity is v at time t:

$$x = x_0 \sin \omega t$$

and
$$v = v_0 \cos \omega t$$

where x_0 is the amplitude of the oscillation and v_0 is the maximum speed. It can also be shown that: $v = \pm \omega\sqrt{(x_0^2 - x^2)}$

> **Remember**
>
> For SHM the period T is independent of the amplitude of the oscillation.

> ★ **Exam tip**
>
> The equation $a = -\omega^2 x$ is provided in Exam Papers 1, 2, and 4.

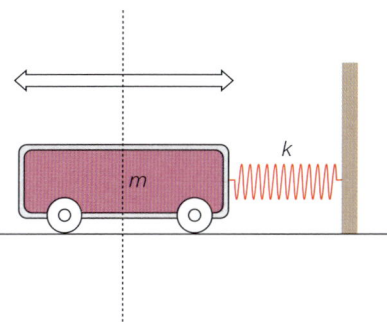

▲ **Figure 17.5** Simple harmonic motion

> ★ **Exam tip**
>
> You do not need to be able to derive the equations:
>
> $v = \pm \omega\sqrt{(x_0^2 - x^2)}$
>
> $v = v_0 \cos \omega t$
>
> $x = x_0 \sin \omega t$
>
> They are provided in Exam Papers 1, 2, and 4.

> **Remember**
>
> The maximum speed (when $x = 0$) is:
>
> $$v_0 = \omega x_0$$

Worked Example

A mass on a spring is oscillating with SHM. The amplitude is 0.30 m and the period of oscillation is 1.2 s. Calculate:

(a) the frequency

(b) the angular frequency

(c) the maximum speed

(d) the maximum acceleration.

Answer

(a) $f = \frac{1}{T} = \frac{1}{1.2} = 0.833\,\text{Hz}$

(b) $\omega = 2\pi f = 2\pi \times 0.833 = 5.23\,\text{rad s}^{-1}$

(c) $v_{max} = \omega x_0 = 5.23 \times 0.30 = 1.57\,\text{m s}^{-1}$

(d) $a = -\omega^2 x$, so $a_{max} = \omega^2 x_0 = 5.23^2 \times 0.30 = 8.2\,\text{m s}^{-2}$

Worked Example

A trolley, of mass $m = 0.90\,\text{kg}$, is suspended between two springs, each of spring constant $25\,\text{N}\,\text{m}^{-1}$, as shown in Figure 13.7. It is displaced to one side and released.

Calculate the angular frequency.

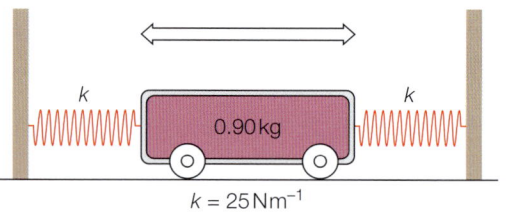

▲ **Figure 17.6** SHM oscillator

Answer

When the trolley is displaced a distance x, the restoring force trying to push/pull the trolley back towards its equilibrium position is $2kx$.

Using $F = ma$:

$$-2kx = ma$$

$$a = -\frac{2k}{m}x$$

$$\omega^2 = \frac{2k}{m} = \frac{2 \times 25}{0.90} \quad \rightarrow \quad \omega = 7.45\,\text{rad}\,\text{s}^{-1}$$

Using graphs to analyse SHM

Figure 17.7 shows the relationships between dispacement, velocity and acceleration during simple harmonic motion.

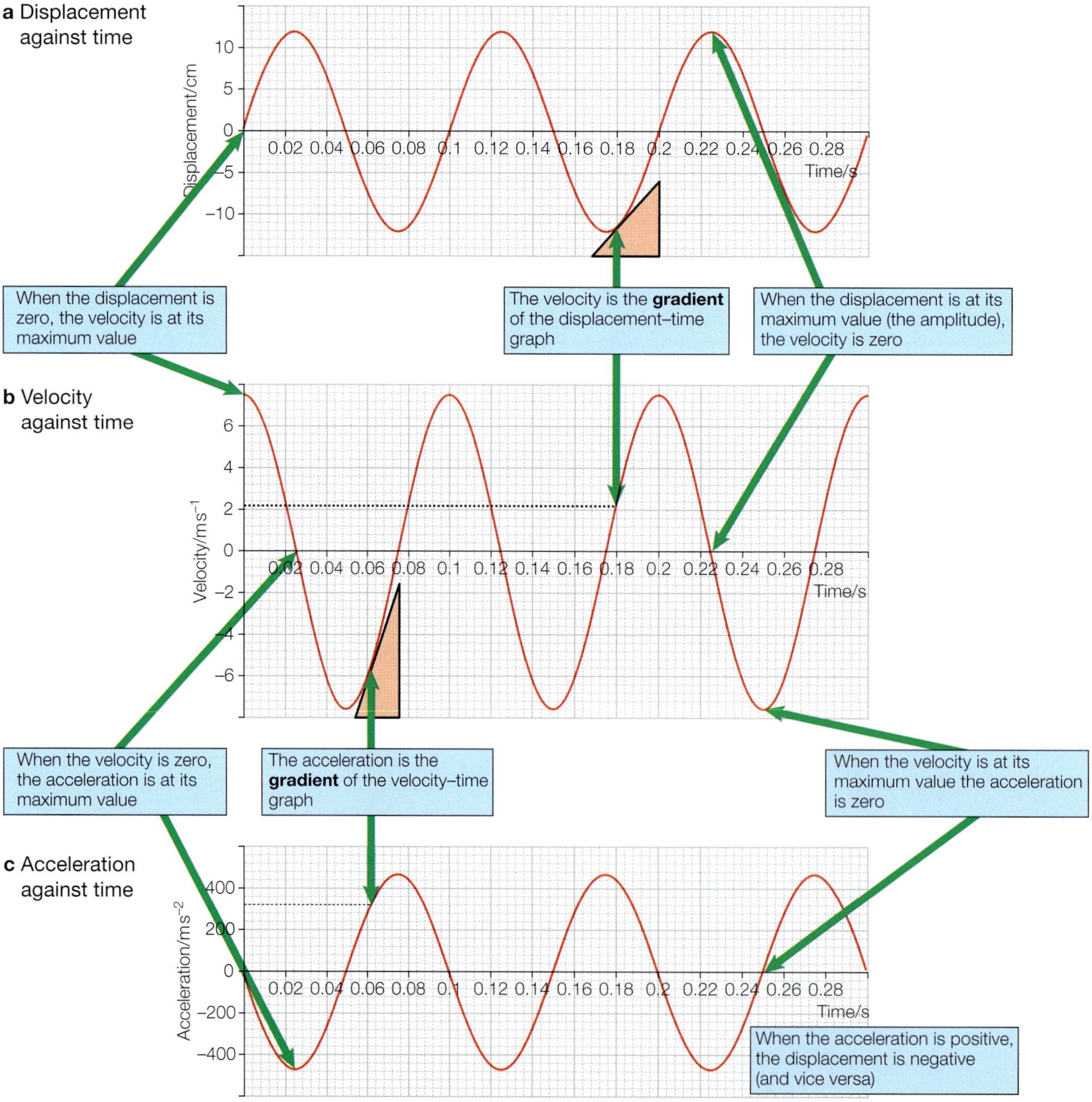

a Displacement against time

When the displacement is zero, the velocity is at its maximum value

The velocity is the **gradient** of the displacement–time graph

When the displacement is at its maximum value (the amplitude), the velocity is zero

b Velocity against time

When the velocity is zero, the acceleration is at its maximum value

The acceleration is the **gradient** of the velocity–time graph

When the velocity is at its maximum value the acceleration is zero

c Acceleration against time

When the acceleration is positive, the displacement is negative (and vice versa)

▲ **Figure 17.7** Displacement, velocity, and acceleration against time graphs for SHM.

SHM and energy

The energy of an object oscillating with simple harmonic motion changes from potential energy (P.E.) to kinetic energy (K.E.) and back to potential energy again (see Figure 17.8). For undamped SHM the total energy is constant.

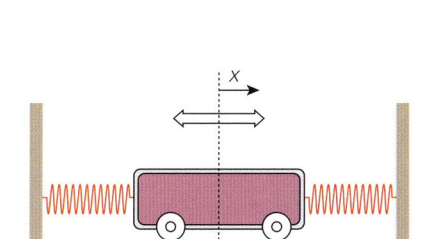

▲ **Figure 17.8** SHM and energy

Worked Example

A trolley of mass m is connected to a spring of spring constant k as shown in Figure 17.9. The spring is initially neither stretched nor compressed. The trolley is displaced to the left and released, and oscillates with SHM.

The displacement x of the mass from its equilibrium position at time t is:

$$x = x_0 \sin \omega t$$

where x_0 is the amplitude of the oscillation and ω the angular frequency.

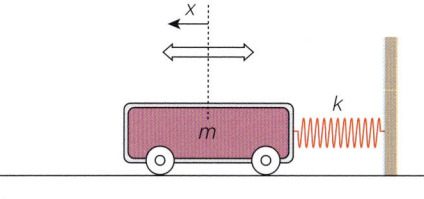

▲ **Figure 17.9**

Derive expressions for:

(a)   the potential energy E_p of the oscillator at time t

(b)   the kinetic energy E_k of the oscillator at time t

(c)   the **total** energy of the oscillator at time t.

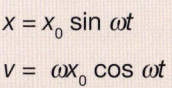

$x = x_0 \sin \omega t$

$v = \omega x_0 \cos \omega t$

Maths Skills

$\sin^2 \omega t + \cos^2 \omega t = 1$

Answer

(a)   potential energy $= \dfrac{1}{2} kx^2 = \dfrac{1}{2} kx_0^2 \sin^2 \omega t$

(b)   kinetic energy $= \dfrac{1}{2} mv^2 = \dfrac{1}{2} m(x_0 \omega \cos \omega t)^2 = \dfrac{1}{2} mx_0^2 \omega^2 \cos^2 \omega t$

(c)   total energy $= \dfrac{1}{2} kx_0^2 \sin^2 \omega t + \dfrac{1}{2} mx_0^2 \omega^2 \cos^2 \omega t$

but $\omega^2 = \dfrac{k}{m}$ so   total energy $= \dfrac{1}{2} kx_0^2 (\sin^2 \omega t + \cos^2 \omega t)$

$$= \dfrac{1}{2} kx_0^2$$

💡 **Remember**

- The total energy in SHM is constant.

- The total energy is proportional to the square of the amplitude.

Damped and forced oscillations, and resonance

Damped oscillations

Real oscillators (see Figure 17.10) lose energy over time, mainly as heat caused by friction and air resistance. If a mass on a spring is pulled down and released, it oscillates up and down, the amplitude gradually decreasing (exponentially) over time.

Critical damping

A **critically damped** oscillator returns to its equilibrium position in as short a time as possible without oscillating. The suspension of a car is designed to be critically damped – when the car goes over a bump in the road the suspension returns the car to its equilibrium position as quickly as possible without oscillating.

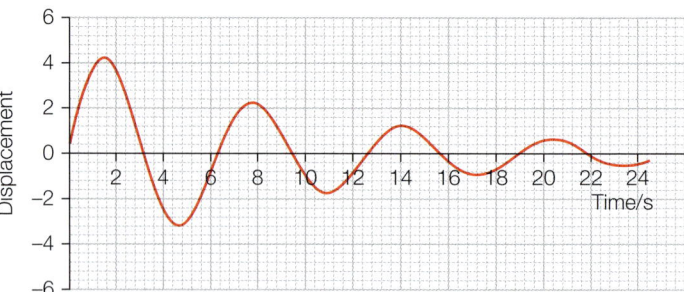

a An **underdamped** system oscillates several times as the amplitude gradually decreases to zero

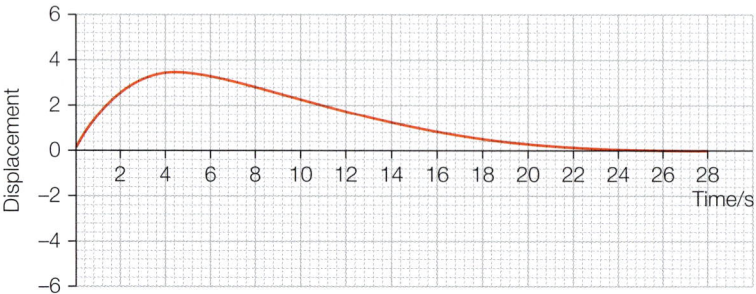

b An **overdamped** system very slowly returns to the equilibrium position without oscillating

▲ **Figure 17.10** Damped SHM

Forced oscillations and resonance

Oscillators can be driven by an external system, with energy being transferred from the external system to the oscillator. Figure 17.11 illustrates how a vibration generator can be used to drive the oscillations of a mass on a spring.

At very low driving frequencies, the mass and spring will oscillate with the same amplitude as the vibration generator (see Figure 17.12). As the driving frequency is increased, the amplitude of the oscillation gradually increases as more energy is transferred to the mass on the spring.

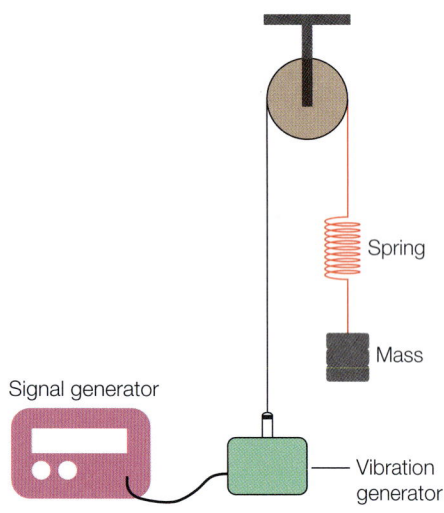

▲ **Figure 17.11** Forced oscillations and resonance

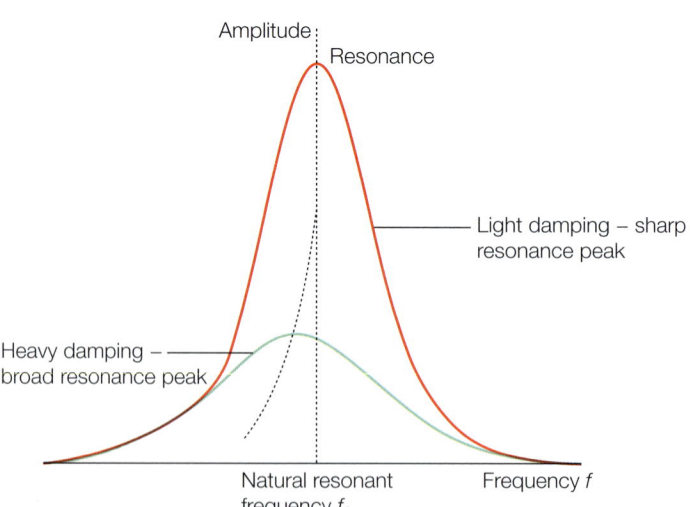

▲ **Figure 17.12** Forced oscillations with damping

Resonance

When the frequency of the forced oscillations is the same as the natural frequency of the mass and spring (the **natural frequency** f_0), the mass oscillates with its maximum amplitude. This is known as **resonance**. The size of the amplitude at resonance depends on the amount of damping.

Resonance effects can be both useful and damaging. Oscillations can build up until the amplitude is so large that it causes structural damage or failure when being shaken by earthquakes. To avoid this, structure are built to that different components have different natural frequencies. Resonance is a key feature of many musical instruments, amplifying particular frequencies of sound so that they can be heard more clearly. Sounds can also drive objects to resonate at their natural frequencies; certain frequencies can shatter glasses.

> 💡 **Remember**
>
> For heavily damped systems the resonant frequency is slightly lower than the natural frequency.

Worked Example

A mass is suspended vertically between two stretched springs. The lower spring is attached to a vibration generator connected to a signal generator. When the vibration generator is switched on the mass oscillates vertically, as shown in Figure 17.13.

(a) In the context of the apparatus shown, explain what is meant by:

    (i) natural frequency   (ii) forced oscillations   (iii) resonance.

(b) A student measures the amplitude of oscillation for different frequencies. She then plots a graph of amplitude against frequency. Sketch the graph she is likely to obtain. Label this graph A.

(c) She replaces the mass with an equal mass in the form of a thin disc. Sketch a second line on your graph showing how the amplitude will vary with frequency for the thin disc. Label this graph B.

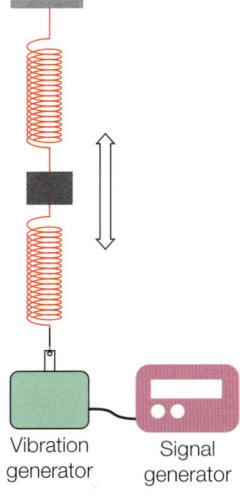

Vibration generator   Signal generator

▲ **Figure 17.13**

Answer

(a) (i) If the mass is pulled down and released, the frequency of vibration would be the natural frequency.

    (ii) The vibration generator forces the mass and springs to oscillate at the frequency of the generator (the driving frequency).

    (iii) As the driving frequency approaches the natural frequency, the amplitude of the oscillations becomes very large, reaching a maximum value when the driving frequency is equal to the natural frequency. The system is then resonating.

(b),(c)

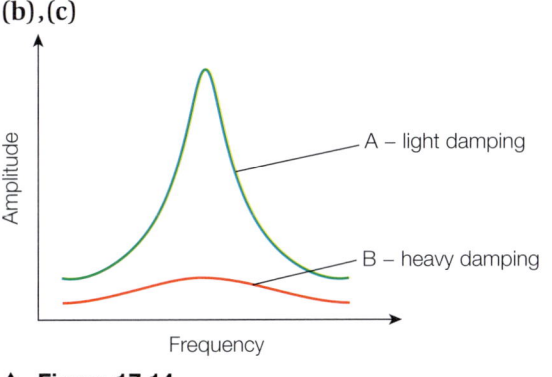

▲ **Figure 17.14**

↑ Raise your grade

1 **(a)** *Define simple harmonic motion.* [2]

> SHM is where the acceleration is proportional to the displacement. ✔ ✗

This is a correct answer but not a complete definition of SHM. It should also state that the acceleration is always in the opposite direction to the displacement.

(b) An aeroplane is moving at a constant speed *u*, flying at a level height. It experiences some turbulence and is displaced vertically, causing it to oscillate in the vertical direction.

Theory shows that the vertical acceleration *a* of the aeroplane is given by the equation:

$$a = -\frac{2g^2}{u^2}x$$

where *x* is the vertical displacement.

(i) Explain how it can be deduced from the equation that the aeroplane oscillates with simple harmonic motion. [2]

> *g* and *u* are both constants, so the acceleration *a* is proportional to the displacement. ✔ ✗

This is a correct answer but again not complete. The candidate has ignored the significance of the minus sign – this shows the acceleration is always in the opposite direction to the displacement.

(ii) Calculate the period of oscillation of an aircraft travelling at 300 km h⁻¹. [Use *g* = 9.81 m s⁻².] [2]

> $u = 300 \text{ kmh}^{-1} = \dfrac{300 \times 1000}{60 \times 60} = 83 \text{ m s}^{-1}$ ✔

The candidate has converted the speed of the aeroplane into SI units correctly, giving an answer to an appropriate number of significant figures.

> $\omega^2 = \dfrac{2g^2}{u^2} = \dfrac{2 \times 9.81^2}{83^2} = 0.0279 \rightarrow \omega = \sqrt{0.0279} = 0.167 \text{ rad s}^{-1}$ ✔

The candidate has equated the angular frequency *ω* to *g* and *u* correctly.

> $T = \dfrac{2\pi}{\omega} = \dfrac{2\pi}{0.167} = 37.6 \text{ s}$ ✔

The final calculation is correct.

period of oscillation = ..37.6.. s [3]

(iii) The oscillations of the aeroplane are *lightly damped*. Explain what *lightly damped* means. [1]

> The aeroplane does not complete a full oscillation before the vertical movement ends. ✗

This answer is incorrect – the candidate has described **heavily damped** oscillations. Lightly damped oscillations mean several oscillations will be completed before the amplitude has fallen to zero.

Exam-style questions

1 A ball of mass 200 g is suspended from a spring. When the ball is pulled down 6.0 cm and released, it oscillates with a period of 0.70 s.

Calculate:

(a) the frequency of oscillation [1]

(b) the maximum speed of the ball [1]

(c) the acceleration when the ball is 2.0 cm above the equilibrium position. [2]

2 The graph shows the variation of displacement with time for a mass–spring oscillator.

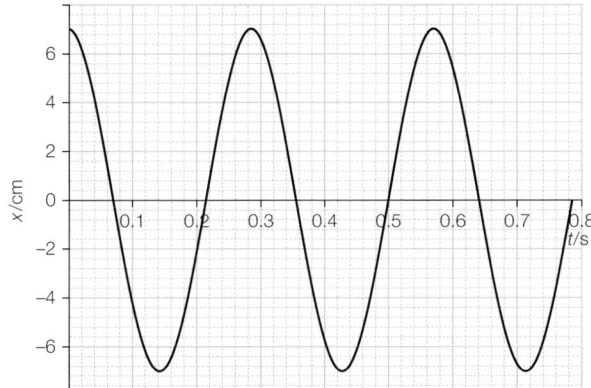

(a) Use the graph to determine:

(i) the period of oscillation

(ii) the frequency

(iii) the angular frequency. [4]

(b) Use the graph to calculate the speed of the mass at time $t = 0.20$ s. [2]

3 The displacement–time graphs of two oscillators, A and B, are shown in the graphs.

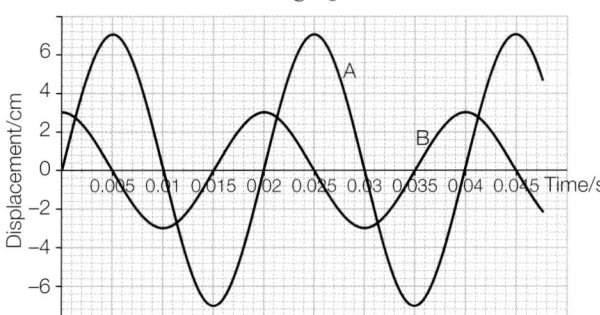

(a) State the phase difference between A and B:

(i) as a fraction of a cycle

(ii) in radians. [2]

(b) Determine the ratio $\dfrac{\text{amplitude of A}}{\text{amplitude of B}}$. [1]

(c) Use the graph to find:

(i) the frequency of A

(ii) the maximum speed of A

(ii) the maximum acceleration of A. [4]

4 **(a)** Define *simple harmonic motion*. [2]

(b) A small mass oscillates vertically according to the equation:

$$y = 15\cos 20\,t$$

where y is the displacement in centimetres of the mass at time t.

Calculate:

(i) the frequency of the oscillation

(ii) the angular frequency

(iii) the maximum speed of the mass. [4]

5 A U-tube contains liquid of density ρ. The tube is briefly tilted to one side and then returned to the vertical position. The liquid oscillates back and forth with simple harmonic motion.

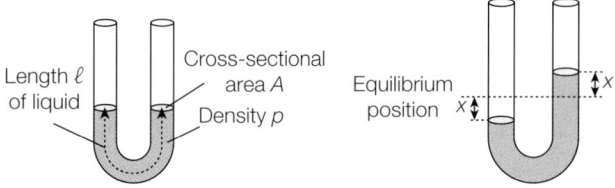

(a) Explain why the liquid oscillator satisfies the conditions for SHM. [2]

(b) Determine:

(i) the mass of liquid in the tube

(ii) the unbalanced force on the liquid at time t

(iii) the period of the oscillations. [4]

6 **(a)** In relation to oscillating systems, explain what is meant by:

(i) *forced vibrations*

(ii) *resonance*. [2]

(b) Describe one situation where resonance can be useful and one situation where resonance should be avoided. [2]

18 Electric fields

What is an electric field?

An **electric field** is a region where charged particles (both stationary and moving) experience forces. If a small positive charge q is placed at a point in the field and it experiences a force F, then the **electric field strength** E at that point is defined as:

$$E = \frac{F}{q}$$

This relationship is expressed in terms of the force acting on a charge:

$$F = qE$$

Electric field strength is a vector quantity – it has direction as well as magnitude. The direction of the electric field is the direction of the force on a unit positive charge and can be shown by electric field lines (see Figure 18.1).

> **Remember**
>
> $$E = \frac{F}{Q}$$
>
> SI units are $N\,C^{-1}$

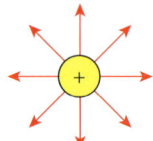

a Near an isolated positive charge

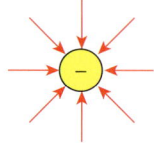

b Near an isolated negative charge

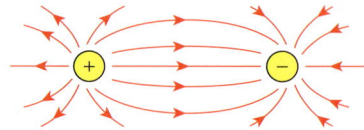

c Between two opposite charges

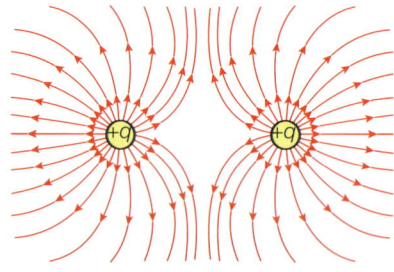

d Between two positive charges

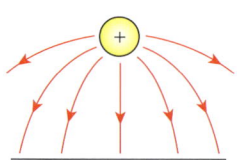

e Between a point positive charge, and an oppositely-charged plate

▲ **Figure 18.1** Examples of electric field lines

As positive charges repel other positive charges, the field lines point away from positive charges. The attraction between positive and negative charges is shown by the field lines pointing towards negative charges.

Electric fields surrounding isolated point charges (Figure 18.1a and b) are described as radial fields. These fields become weaker with increasing distance from the charge and this is shown by the field lines growing further apart.

The electric field at a point is the vector sum of all of the fields produced by charges near that point. This can give complex electric field shapes when point charges and charged surfaces interact (Figure 18.1d). However, electric field lines cannot cross or touch as they always indicate the direction of the resultant force produced by the fields at that point.

Uniform electric fields

Electric field between two parallel plates

If there is a potential difference across two large parallel plates, the electric field between the plates is **uniform** – it is constant both in magnitude and direction (see Figure 18.2a).

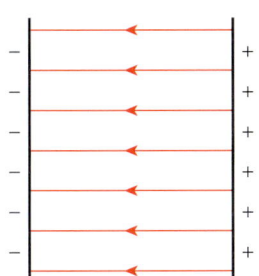

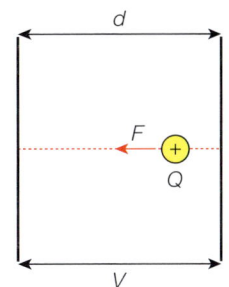

a Field between two parallel plates   **b** Force between two parallel plates

▲ **Figure 18.2** Electric field and force between two parallel plates

For two metal plates that are a distance d apart with a p.d. V between them, when a charge $+Q$ moves from the positively charged plate to the negatively charged plate (see Figure 18.2b), the work done on the charge is:

$$Fd = QV$$

$$E = \frac{F}{Q} = \frac{V}{d}$$

> **Remember**
>
> $E = \dfrac{V}{d}$ for parallel plates
>
> SI units are $V\,m^{-1}$
>
> (equivalent to $N\,C^{-1}$)
>
> This equation can also be written:
>
> $E = \dfrac{\Delta V}{\Delta d}$

Worked Example

1   An electron is between two parallel plates that are 2.0 cm apart. The p.d. between the plates is 5.0 kV. Calculate:

     (a) the electric field strength between the plates   **(b)** the force on the electron.

Answer

     (a) $E = \dfrac{V}{d} = \dfrac{5 \times 10^{3}}{2.0 \times 10^{-2}} = 2.5 \times 10^{5}\,V\,m^{-1}$ (or $N\,C^{-1}$)

     (b) $F = EQ$

          $= Ee = 2.5 \times 10^{5} \times 1.6 \times 10^{-19} = 4.0 \times 10^{-14}\,N$

2   A small charged ball, of mass 5.0 g, is suspended by a nylon thread between two parallel plates, and hangs at an angle of 40° to the vertical (see Figure 18.3). The p.d. between the plates is 2.0 kV, and the distance between the plates is 10.0 cm.

     How much charge is on the ball?

Answer

Resolving vertically: $T\cos 40° = mg = 5.0 \times 10^{-3} \times 9.81 = 4.91 \times 10^{-2}$

Resolving horizontally: $T\sin 40° = F_e = \dfrac{V}{d} \times Q = \dfrac{2.0 \times 10^{3}}{10.0 \times 10^{-2}} \times Q = 2 \times 10^{4} Q$

Dividing the second equation by the first:

$$\frac{2 \times 10^{4} Q}{4.91 \times 10^{-2}} = \frac{\sin 40}{\cos 40} = \tan 40$$

$$Q = \frac{4.91 \times 10^{-2} \times 0.839}{2 \times 10^{4}} = 2.1 \times 10^{-6}\,C \ (2.1\,\mu C)$$

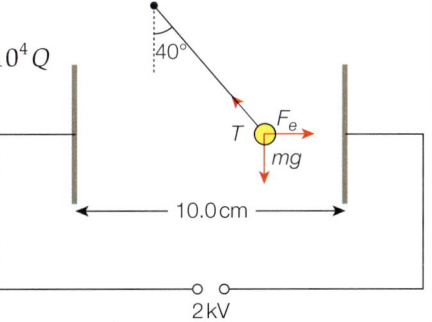

▲ **Figure 18.3**

Motion of charged particles in a uniform electric field

Uniform electric fields accelerate charged particles, changing their speed and direction of travel.

Worked Example

The p.d. between two metal plates, S and T, is 3.0 kV. The two plates are 6.0 cm apart. An electron is at rest next to the negatively charged plate S. It accelerates towards the positively charged plate T, a metal disc with a hole in the middle (see Figure 18.4).

Calculate:

(a) the electric field strength between the two plates

(b) the force on the electron

(c) the acceleration of the electron

(d) the speed of the electron when it reaches T.

Answer

(a) $E = \dfrac{V}{d} = \dfrac{3.0 \times 10^3}{6.0 \times 10^{-2}} = 5.0 \times 10^4 \, \text{N C}^{-1}$

(b) $F = Ee = 5.0 \times 10^4 \times 1.6 \times 10^{-19} = 8.0 \times 10^{-15} \, \text{N}$

(c) Using $F = ma$:

$a = \dfrac{F}{m} = \dfrac{8.0 \times 10^{-15}}{9.1 \times 10^{-31}} = 8.8 \times 10^{15} \, \text{m s}^{-2}$

(d) Using $v^2 = u^2 + 2as$:

$v^2 = 0^2 + 2 \times (8.8 \times 10^{15}) \times (6.0 \times 10^{-2}) = 1.1 \times 10^{15}$

$v = 3.2 \times 10^7 \, \text{m s}^{-1}$

This example shows how electrons are accelerated in, for example, cathode-ray oscilloscopes.

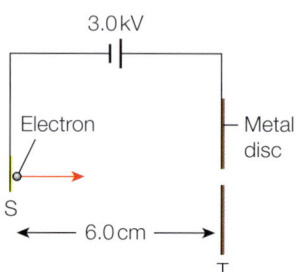

▲ **Figure 18.4**

> ★ **Exam tip**
>
> The mass of an electron m_e and the electronic charge e are provided in Exam Papers 1, 2, and 4.

Electric field strength between point charges

Coulomb's law

Coulomb's law says that if two **point charges**, Q_1 and Q_2, are a distance r apart (see Figure 18.5), the force F between them is:

$$F = \frac{Q_1 Q_2}{4\pi\varepsilon_0 r^2}$$

ε_0 is a constant, called the permittivity of free space. Its value is $8.85 \times 10^{-12} \, \text{F m}^{-1}$.

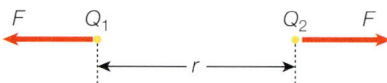

▲ **Figure 18.5** The force between two point charges

If the two charges are both positive, or both negative, the force is repulsive (by convention, a positive force). If one charge is positive, and the other negative, the force is attractive (by convention, a negative force).

Electric field strength of a point charge

The **electric field strength** at a point is the force on a unit positive charge at that point, and so for a point charge Q, by replacing Q_2 with '1' in Coulomb's law:

$$E = \frac{Q}{4\pi\varepsilon_0 r^2}$$

> ★ **Exam tip**
>
> If you're asked to define Coulomb's law, make sure that you refer to the force between two **point charges**.

> ★ **Exam tip**
>
> $\dfrac{1}{4\pi \varepsilon_0} = 8.99 \times 10^9 \, \text{N m}^2 \text{C}^{-2}$
>
> This value, and the value of ε_0, are provided in Exam Papers 1, 2, and 4.

> 💡 **Remember**
>
> Coulomb's law is defined for **point** charges, but a sphere of charge, with the charges equally spread over the sphere, acts as if all the charge were concentrated at the centre of the sphere.

Worked Example

1 Calculate the force between two protons 2.0×10^{-12} m apart. (The charge on a proton $= +e = 1.6 \times 10^{-19}$ C)

Answer

$$F = \frac{Q_1 Q_2}{4\pi\varepsilon_0 r^2} = 8.99 \times 10^9 \times \frac{(1.6 \times 10^{-19})^2}{(2 \times 10^{-12})^2} = 5.8 \times 10^{-5}\,\text{N}$$

2 A metal sphere of radius 50 mm carries a charge of $+200\,\mu$C.

   (a) What is the strength of the electric field 20 mm from the surface of the sphere?

   (b) In which direction is the electric field?

Answer

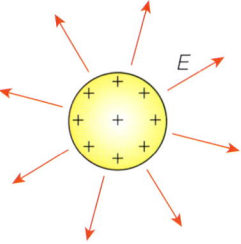

   (a) $E = \dfrac{Q}{4\pi\varepsilon_0 r^2} = 8.99 \times 10^9 \times \dfrac{200 \times 10^{-6}}{(70 \times 10^{-3})^2} = 3.7 \times 10^8\,\text{NC}^{-1}$

   (b) The direction is radially, away from the sphere (see Figure 18.6).

▲ **Figure 18.6**

Electric potential

The **electric potential** at a point is defined as the work done in bringing unit positive charge ($+1$ C) from infinity to that point. The work done is independent of the path taken (see Figure 18.7).

For a point charge $+Q$, the energy needed to bring $+1$ C from infinity to a point P that is a distance r from Q (i.e., the potential at point P) is inversely proportional to r (see Figure 18.8). The potential V is calculated using the equation:

$$V = \frac{Q}{4\pi\varepsilon_0 r}$$

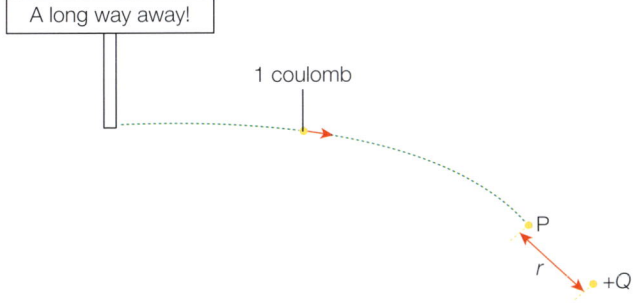

▲ **Figure 18.7** The meaning of electric potential

> **Remember**
>
> Electric force and electric field strength are both vectors because they have direction as well as magnitude.
>
> Electric potential is a scalar quantity – it only has magnitude.

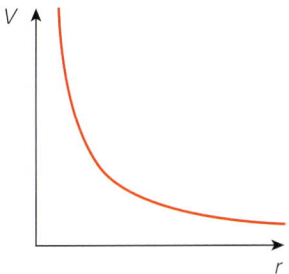

▲ **Figure 18.6** Electric potential V at a distance r from a point charge Q

Moving a unit positive charge towards $+Q$ requires work to be done as the positive charges are trying to push each other apart. The smaller the value of r (the closer the test charge gets) the greater the amount of work required. The unit positive charge gains electrical potential energy as it approaches Q, and so V is positive.

No work needs to be done to move the positive test charge towards a charge $-Q$. Instead, the unit positive charge is pulled towards $-Q$ and gains kinetic energy. The test charge loses potential energy as it approaches $-Q$, and so V is negative.

Electric potential is useful for calculating the energy needed or gained when charged particles move from one place to another. If the potential is known at two different points, then the energy needed for a unit positive charge to move from one point to the other is the potential difference.

> ★ **Exam tip**
>
> The formula
> $$V = \frac{Q}{4\pi\varepsilon_0 r}$$
> is provided in Exam Papers 1, 2, and 4, but remember that V has units of JC^{-1}.

Electric field and potential gradient

For two parallel plates, with a potential difference V between them, the potential V increases linearly from $0\,V$ to $+V$ (Figure 18.9). As shown earlier, the field strength is equal to $-\dfrac{\Delta V}{\Delta x}$ (the negative of the potential gradient), and is constant.

> **Remember**
>
> The negative sign is needed because the field strength E is in the opposite direction to the displacement Δx.

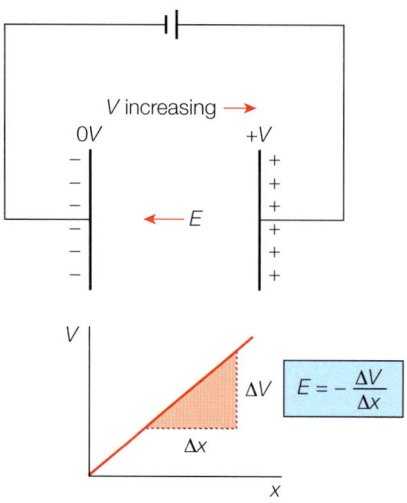

▲ **Figure 18.9** Electric field E between two parallel plates

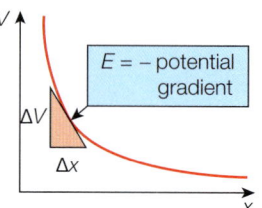

▲ **Figure 18.10** Field strength is the negative of the potential gradient

It can be shown that the equation:

electric field strength = − potential gradient

applies more generally. If a graph of electric potential against distance is plotted, the field strength at any point is the negative of the gradient of the graph at that point (Figure 18.10).

Comparing electrical and gravitational fields

The gravitational force between two masses and the electrical force between two charges both obey inverse-square laws (see Table 18.1).

▼ **Table 18.1** Comparison of electrical and gravitational fields

	Force	Field	Potential	Potential energy
Gravity	$F = \dfrac{Gm_1m_2}{r^2}$	$g = \dfrac{GM}{r^2}$	$\phi = -\dfrac{GM}{r}$	$\text{P.E.} = \dfrac{Gm_1m_2}{r}$
Electricity	$F = \dfrac{Q_1Q_2}{4\pi\varepsilon_0 r^2}$	$E = \dfrac{Q}{4\pi\varepsilon_0 r^2}$	$V = \dfrac{Q}{4\pi\varepsilon_0 r}$	$\text{P.E.} = \dfrac{Q_1Q_2}{4\pi\varepsilon_0 r}$

- The electrical force can be attractive or repulsive, depending on the sign of the charges, but the gravitational force is always attractive.

- Electrical potential can be positive or negative, depending on the sign of the charge. Gravitational potential must always be negative as work needs to be done to move a test mass to infinity where, by convention, it has zero potential energy.

- As two masses are moved apart, the potential energy of the two masses increases (work has to be done to separate the two masses).

- As two positive charges are moved apart, the potential energy of the two charges decreases (work has to be done to push the two positive charges together).

↑ Raise your grade

A beam of electrons is fired from an electron 'gun' inside a vacuum tube. The electrons pass between two metal plates, a distance d apart. A potential difference V is applied between the two plates, and the electron beam is deflected upwards, as shown.

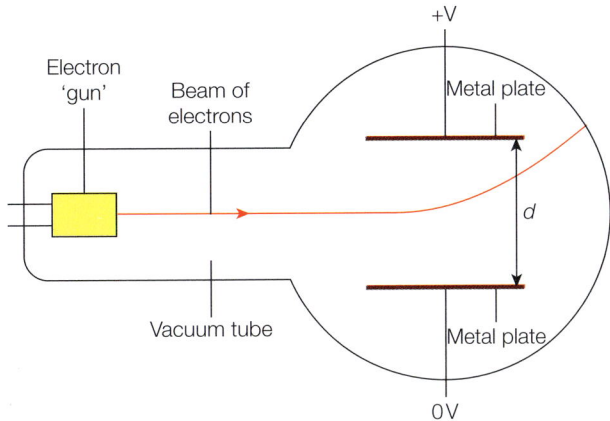

(a) Explain why the electrons are deflected upwards.   [1]

> Electrons are negatively charged. ✗

> Not sufficient explanation – should add 'and are attracted to the positive plate because opposite charges attract'.

(b) State, and explain, the effect on the deflection of the electrons for each of the following changes: **(i)** the distance d between the plates is decreased   [2]

> The electric field strength E (= V/d) will increase, increasing the upwards force on the
>
> electrons, so the electrons will be deflected more. ✔   A good answer.

(ii) the potential difference V between the plates is decreased   [1]

> The electric field strength E will decrease, decreasing the upwards force on the electrons,
>
> so the electrons will be deflected less. ✔   A good answer.

(iii) the speed of the electrons is decreased.   [1]

> The force on the electrons has not changed so the deflection will be the same. ✗

> An incorrect answer. The electrons spend more time between the two plates (because they are moving more slowly) so the upwards force acts on them for longer – the deflection would be larger.

(c) Describe, and explain, what would happen if the beam of electrons was replaced by a beam of α-particles travelling at the same speed as the electrons.   [2]

> Alpha particles are positively charged so they would be attracted to the negative plate.
>
> They would deflect downwards. ✔ ✗

> For the second mark the candidate should add that the deflection will be smaller because the mass of an alpha particle is much greater than the mass of an electron. The force acting on an alpha particle will be double that on an electron, but the mass of an alpha particle is several thousand times greater than the mass of an electron.

Exam-style questions

1   A charged oil drop, of mass 5.0×10^{-15} kg, is held stationary between two parallel, charged plates, 5.0 mm apart.

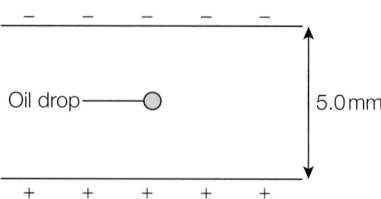

The p.d. between the two plates is 516 V.

What is the charge on the oil drop?

A 4.8×10^{-20} C   C 4.8×10^{-17} C

B 4.8×10^{-19} C   D 4.8×10^{-16} C   [1]

2   Two parallel metal plates, 20.0 cm apart, are connected to the terminals of a 5.0 kV supply.

What is the field strength between the plates?

A 0.25 N C^{-1}

B 25 N C^{-1}

C 2.5×10^4 N C^{-1}

D 2.5×10^6 N C^{-1} [1]

3   An electron is positioned midway between two oppositely charged, parallel plates connected to a d.c. supply. Which one of the graphs shows how the force F on the electron varies with the separation d of the plates?   [1]

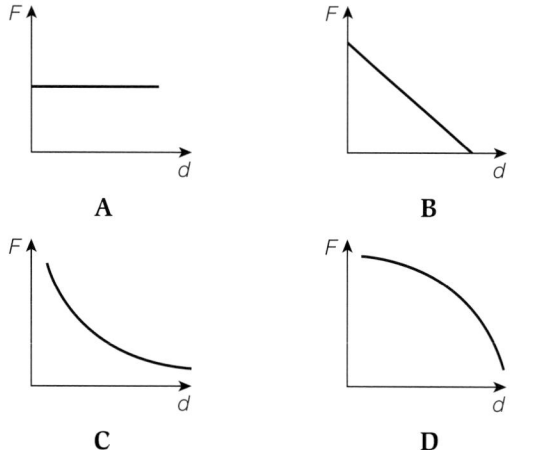

4   (a) Sketch the electric field pattern between two negatively charged particles. Each particle carries the same amount of charge.

[1]

(b) How would the diagram change if one of the particles had double the charge of the other?   [2]

5   A sphere of diameter 4.0 cm carries a negative charge of 6.0 nC.

(a) Calculate the electric field strength:
   (i) on the surface of the sphere
   (ii) at a point P, 10.0 cm from the centre of the sphere.   [3]

(b) Calculate the electric potential:
   (i) on the surface of the sphere
   (ii) at a point P, 10.0 cm from the centre of the sphere.   [3]

(c) Calculate the work done in moving a charge of $+4$ pC from the surface of the sphere to point P.   [2]

6   A metal sphere is suspended from an insulating wire. The sphere is connected to the positive output of a 5 kV d.c. supply.

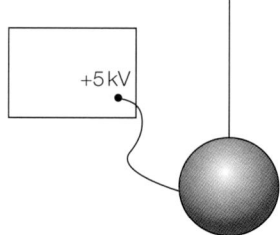

The potential 500 mm from the centre of the sphere is 3.0 kV.

Calculate:

(a) the charge on the sphere   [2]

(b) the radius of the sphere   [2]

(c) the potential 800 mm from the centre of the sphere.   [2]

7   The force between two charged particles is inversely proportional to the square of the distance between them – an example of an *inverse square law*. The variation of the gravitational force between two masses with their distance apart is also an inverse square law.

(a) Describe two ways in which gravitational and electrostatic forces are similar.   [2]

(b) Describe one way in which the two forces are not similar.   [1]

Knowledge check

You should be able to:

- describe the movement of charge in a circuit
- understand that the work done moving an electric charge through an electric field is given by $W = QV$
- describe the effect of resistance on the size of a current $\left(I = \frac{V}{R}\right)$.

Parallel-plate capacitors

Capacitors are devices for storing charge and energy in electric circuits. They usually consist of two metal plates a short distance apart, separated by an insulating material. Figure 19.1 shows the circuit symbol for a capacitor.

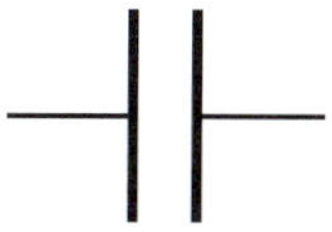

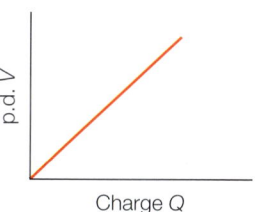

▲ **Figure 19.1** Symbol for a capacitor

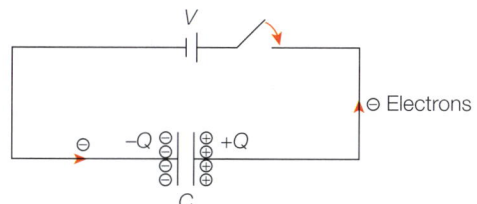

▲ **Figure 19.2** Charging a capacitor

In the circuit shown in Figure 19.2, when the switch is closed, electrons flow from the negative terminal of the cell onto the left-hand plate of the capacitor, and an equal number of electrons leave the right-hand plate and move to the positive terminal of the cell. The left-hand plate then has a negative charge $-Q$ and the right-hand plate has an equal positive charge $+Q$ (the **net** charge stored in a capacitor is zero).

▲ **Figure 19.3** $Q \propto V$

The greater the value of V, the potential difference (p.d.) across the capacitor, the greater the charge Q. Doubling V doubles the charge Q (see Figure 19.3):

$$Q \propto V$$

or

$$Q = CV \text{ therefore } C = \frac{Q}{V}$$

where C is the capacitance of the capacitor. The units of capacitance are **farads** (F). **Capacitance** is the charge stored per unit potential difference.

If a p.d. of 1V produces a charge of 1C (coulomb) on the plates of a capacitor (positive charge on one plate, negative on the other) the capacitor has a capacitance of 1F.

Most capacitors have capacitances very much smaller than 1F. Table 19.1 gives the prefixes often used for capacitances.

Remember

The capacitance C of a capacitor is the charge Q stored per unit potential difference:

$$C = \frac{Q}{V}$$

Worked Example

What are the SI base units of capacitance?

Answer

Units are: $\frac{C}{V} = \frac{\text{As}}{\text{JC}^{-1}} = \frac{\text{As}}{\text{kg m}^2\text{s}^{-2}\text{A}^{-1}\text{s}^{-1}} = \text{kg}^{-1}\text{m}^{-2}\text{s}^4\text{A}^2$

▼ **Table 19.1** Prefixes used with capacitance

Microfarads	µF	10^{-6} F
Nanofarads	nF	10^{-9} F
Picofarads	pF	10^{-12} F

Worked Example

1   A potential difference of 5.0 kV is connected across a 470 μF capacitor.

(a) What is the charge stored?

(b) How many extra electrons are there on the negative plate of the capacitor? $[e = 1.6 \times 10^{-19} \, \text{C}]$

Answer

(a) $Q = CV = 470 \times 10^{-6} \times 5 \times 10^{3} = 2.35 \, \text{C}$

(b) number of 'extra' electrons on negative plate $= \dfrac{2.35}{1.6 \times 10^{-19}}$

$= 1.47 \times 10^{19}$

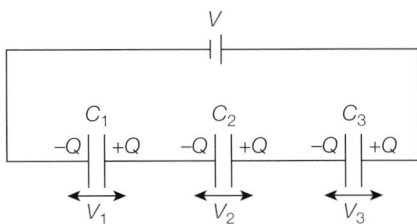

★ **Exam tip**

Take care not to confuse C (the symbol for the unit of charge, the coulomb) and C (the symbol for the capacitance of a capacitor).

This means there will be +2.35 C of charge on one plate, and –2.35 C of charge on the other plate.

Capacitors in series

Figure 19.4 shows three capacitors, with capacitances C_1, C_2, and C_3, connected in series to a d.c. supply with a terminal p.d. V. As the capacitors are in series, the charges on each plate of all the capacitors must be the same (charge is conserved).

From Kirchhoff's second law:

$$V = V_1 + V_2 + V_3$$

$$= \frac{Q}{C_1} + \frac{Q}{C_2} + \frac{Q}{C_3}$$

$$= Q \left(\frac{1}{C_1} + \frac{1}{C_2} + \frac{1}{C_3} \right)$$

▲ **Figure 19.4** Capacitors in series

The three capacitors are equal to one capacitor of capacitance C, where:

$$\frac{1}{C} = \left(\frac{1}{C_1} + \frac{1}{C_2} + \frac{1}{C_3} \right)$$

★ **Exam tip**

$$\frac{1}{C} = \left(\frac{1}{C_1} + \frac{1}{C_2} + \frac{1}{C_3} \right)$$

Capacitors in series add up like resistors in parallel.

Capacitors add up in series in the same way as resistors add up in parallel. Adding capacitors up in series makes a capacitor with **smaller capacitance**.

Worked Example

1   A 5 μF capacitor is connected in series to a 10 μF capacitor. The combination of capacitors is connected to a 12 V d.c. supply (Figure 19.5).

Calculate:

(a) the p.d. across the 10 μF capacitor

(b) the charge on the 5 μF capacitor.

Answer

(a) From Kirchhoff's second law: $V_1 + V_2 = 12$ (eqn 1)

Charge on both capacitors is the same:

$$Q = CV = 5 \times 10^{-6} \, V_1 = 10 \times 10^{-6} \, V_2$$

$$V_1 = 2V_2 \quad \text{(eqn 2)}$$

From eqn 1 and eqn 2: $V_1 = 8 \, \text{V}$, $V_2 = 4 \, \text{V}$

(b) $Q = CV = 5 \times 10^{-6} \times 8 = 40 \, \mu\text{C}$

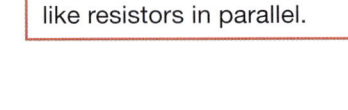

▲ **Figure 19.5**

Capacitors in parallel

In Figure 19.6 three capacitors, (C_1, C_2, and C_3) are connected in parallel to a d.c. supply with a terminal p.d. V.

The potential difference across each capacitor is the same as the potential difference across the d.c. supply (V), as they are all in parallel.

The total charge 'stored' is Q, where:

$$Q = Q_1 + Q_2 + Q_3$$
$$= C_1V + C_2V + C_3V$$
$$Q = V(C_1 + C_2 + C_3)$$
$$\frac{Q}{V} = C = C_1 + C_2 + C_3$$

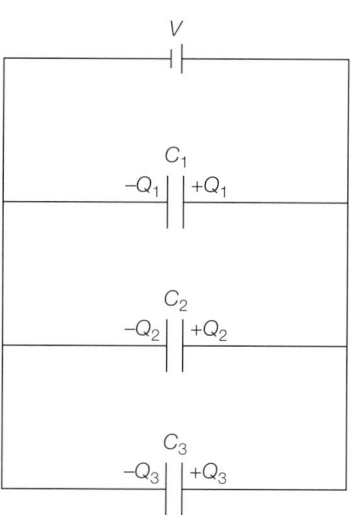

▲ **Figure 19.6** Capacitors in parallel

Capacitors add up in parallel in the same way as resistors add up in series. Adding capacitors up in parallel makes a capacitor with **larger capacitance**.

> ★ **Exam tip**
>
> $$C = C_1 + C_2 + C_3$$
>
> Capacitors in parallel add up like resistors in series.

1   Three capacitors, with capacitance $1000\,\mu F$, $2200\,\mu F$, and $4700\,\mu F$ are connected in parallel.

   (a) What is the total effective capacitance?

   (b) If the parallel combination of capacitors is connected to a $12\,V$ d.c. power supply, what is the total charge stored?

Answer

   (a) $C = C_1 + C_2 + C_3 = 1000 + 2200 + 4700 = 7900\,\mu F$

   (b) $Q = CV = 7900 \times 10^{-6} \times 12 = 95\,mC$

2   Three capacitors are connected as shown in Figure 19.7. The total effective capacitance is $20\,\mu F$. What is the value of C?

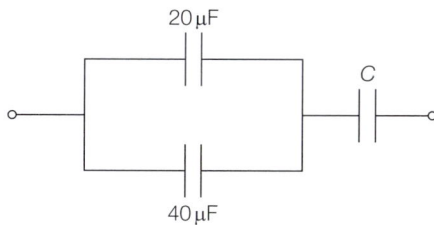

▲ **Figure 19.7**

Answer

$$\frac{1}{(20 + 40)} + \frac{1}{C} = \frac{1}{20}$$
$$\frac{1}{C} = \frac{1}{20} - \frac{1}{60} = \frac{2}{60}$$
$$C = 30\,\mu F$$

Energy stored in capacitors

When a capacitor is being charged, work is done in pushing charge onto one plate and 'pulling' it off the other. For a capacitor that is initially uncharged, the work done in pushing the first few charges onto one of the plates is very small, but as the charge on the plates builds up it gets harder and harder to push more charge onto the plate (because the charges already there are repelling them).

The energy stored W in a capacitor charged to a potential difference V is equal to the area under the graph of p.d. against charge (see Figure 19.8).

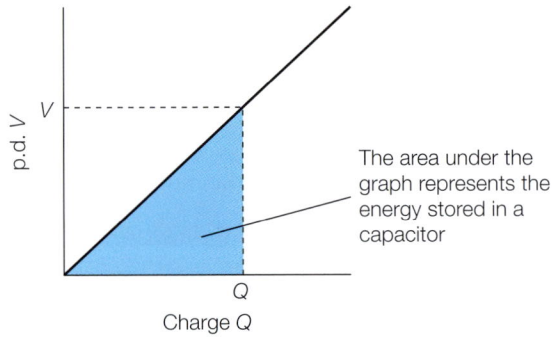

The area under the graph represents the energy stored in a capacitor

▲ **Figure 19.8** Energy stored in a capacitor

Calculating the energy stored

From Figure 19.8:

Energy stored: $\qquad W = \dfrac{1}{2}QV$

Using $Q = CV$: $\qquad = \dfrac{1}{2}CV^2$

> **Remember**
>
> Energy stored in a capacitor W
>
> $= \dfrac{1}{2}QV$
>
> $= \dfrac{1}{2}CV^2$

Worked Example

1 The graph (Figure 19.9) shows the charge Q on a capacitor as the potential difference V across it is gradually increased.

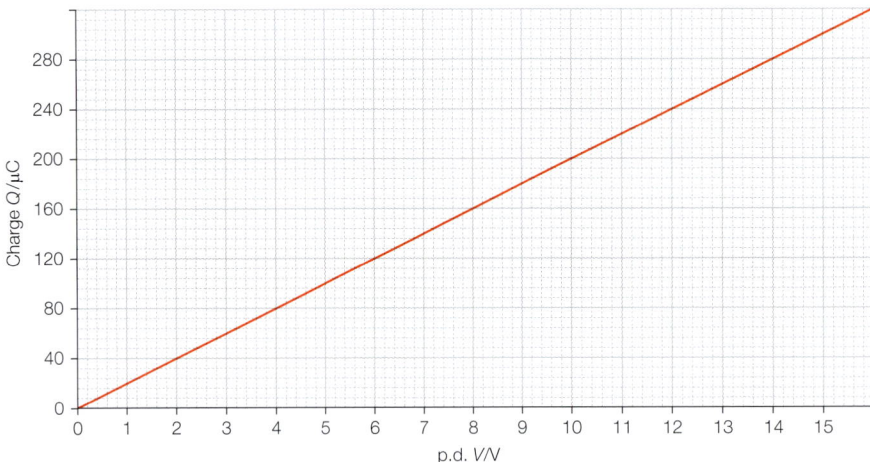

▲ **Figure 19.9**

(a) What is the capacitance of the capacitor?

(b) What is the energy stored in the capacitor when the p.d. across it is 14.0 V?

(c) The capacitor is partially discharged so that the p.d. across it falls from 14.0 V to 9.0 V. how much energy is lost by the capacitor?

Answer

(a) $C = \dfrac{Q}{V} = \dfrac{240 \times 10^{-6}}{12} = 20 \times 10^{-6}\,\text{F}\ (20\,\mu\text{F})$

(b) energy stored $= \dfrac{1}{2}CV^2 = 0.5 \times 20 \times 10^{-6} \times 14.0^2 = 1.96 \times 10^{-3}\,\text{J}$

(c) energy lost = area under graph between 14.0 V and 9.0 V

$= \dfrac{1}{2}(280 + 180) \times 10^{-6} \times (14.0 - 9.0) = 1.15 \times 10^{-3}\,\text{J}$

2 The top and bottom layers of a thundercloud can be treated as a capacitor of approximate capacitance $10^{-8}\,\text{F}$. Just before a lightning strike the p.d. between the top and bottom of the thundercloud is approximately $3 \times 10^9\,\text{V}$. Calculate the energy stored in a thundercloud.

Answer

$W = \dfrac{1}{2}CV^2 = \dfrac{1}{2} \times 10^{-8} \times (3 \times 10^9)^2 = 4.5 \times 10^{10}\,\text{J}$

Capacitor discharge

When the plates of a charged capacitor are connected through a resistor (Figure 19.10) the capacitor will discharge as electrons move around the circuit.

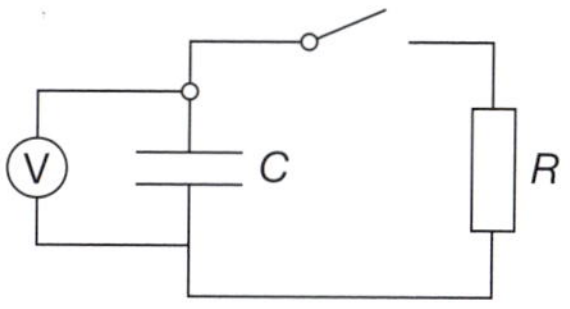

Exponential decay

The current in the discharging capacitor circuit is proportional to the resistance but also proportional to the potential difference across the plates, which, in turn, is proportional to the charge remaining on those plates. As the remaining charge falls the current in the circuit will fall in proportion, leading to an **exponential decay** in the charge remaining on the capacitor, as shown in Figure 19.11.

The charge remaining on the plates of a capacitor of capacitance C discharging through a resistance R after a time period t is given by:

$$Q = Q_0 e^{-(t/RC)}$$

As the current and p.d. are proportional to the charge remaining, similar exponential equations can be used to calculate them after time t, as shown in Table 19.2.

▼ **Table 19.2** Exponential decay equations for a discharging capacitor

Charge	Current	Potential difference
$Q = Q_0 e^{-(t/RC)}$	$I = I_0 e^{-(t/RC)}$	$V = V_0 e^{-(t/RC)}$

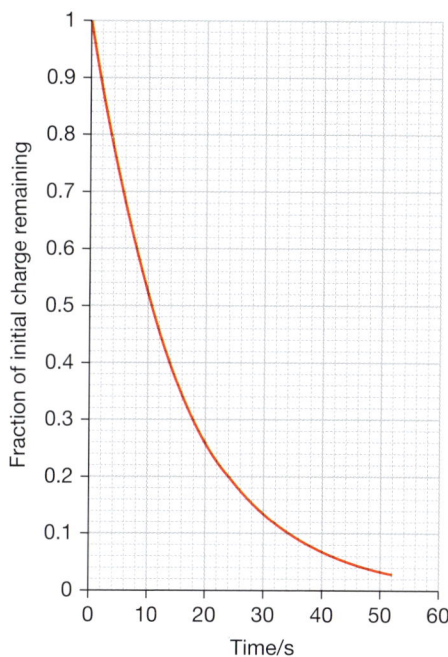

▲ **Figure 19.10**

▲ **Figure 19.11**

Time constant

The resistance and capacitance determine the rate of discharge of the capacitor through a resistor: high resistances decrease the discharge current and high capacitances store more charge per volt.

The product of the resistance and capacitance is known as the time constant τ.

$$\tau = RC$$

In a time of τ the charge of the capacitor plates will fall to $\dfrac{1}{e}$ of the original charge, where e is the base of natural logarithms, which is approximately 2.72. After 2τ the fraction of charge remaining will be $\dfrac{1}{e^2}$; after 3τ it will be $\dfrac{1}{e^3}$ and so on.

Worked Example

A capacitor of capacitance 600 μF is charged so that there is a p.d. of 4.0 V across its plates. The capacitor is allowed to discharge through a 150 kΩ resistor.

Calculate **(a)** the time constant for this circuit **(b)** the charge remaining on the capacitor after 30 s.

Answers

(a) $\tau = RC = 150 \times 10^3\ \Omega \times 600 \times 10^{-6}\ \text{F} = 90\ \text{s}$

(b) Initial charge on capacitor: $Q_0 = CV_0 = 600 \times 10^{-6}\ \text{F} \times 4.0\ \text{V} = 2.4 \times 10^{-3}\ \text{C}$

Charge remaining on capacitor:

$$Q = Q_0 e^{-(t/RC)}$$

$$Q = 2.4 \times 10^{-3}\ e^{-(30/90)}$$

$$Q = 1.72 \times 10^{-3}\ \text{C}$$

↑ Raise your grade

A capacitor of capacitance 200 μF is connected to a 12 V d.c. supply.

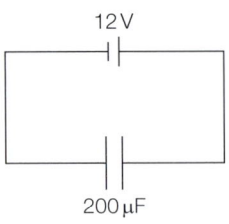

12 V

200 μF

(a) Define *capacitance*. [1]

Something has capacitance if it stores charge when a voltage is across it ✗

> This is a description of what capacitors do, rather than a definition. A good answer would be 'The capacitance of a capacitor is defined as the charge stored per unit potential difference'.

(b) (i) Show that the charge on the capacitor is 2.4 mC. [1]

$Q = CV = 12 \times 200 = 2400\,\mu C = 2.4\,mC$ ✔

A good answer.

(ii) Calculate the energy stored in the capacitor. [2]

Energy stored $= \dfrac{1}{2}CV^2 = 0.5 \times 200 \times 12^2 = 1.44 \times 10^4\,J.$ ✔ ✗

> The candidate has substituted the correct values into the correct equation (method mark), but has not taken account of the capacitance being in μF, not F. The answer should be $0.5 \times 200 \times 10^{-6} \times 12^2 = 1.44 \times 10^{-2}\,J$.

(c) The 200 μF capacitor, still charged, is disconnected from the power supply and connected to an uncharged 100 μF capacitor.

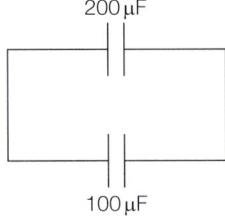

200 μF

100 μF

(i) Show that the potential difference across the 200 μF capacitor is now 8 V. [3]

The 2400 μC charge is 'shared' between the 2 capacitors so that the p.d. across each capacitor is the same (the capacitors are in parallel).

8 V across 100 μF means 800 μC; 8 V across 200 μF capacitor means 1600 μC.

Total charge = 800 + 1600 = 2400 μC (charge is conserved) ✔ ✔ ✔

(ii) Calculate the total energy stored in the two capacitors. [2]

A good answer.

Total energy stored $= \dfrac{1}{2}C_1V^2 + \dfrac{1}{2}C_2V^2 = \dfrac{1}{2} \times 100 \times 8^2 + \dfrac{1}{2} \times 200 \times 8^2 = 0.96 \times 10^4$

$= 0.96 \times 10^4\,J$ ✔ ✔

> Correct method and substitution. Calculation is correct, allowing for error carried forward – already penalised for using μF rather than converting to F. The answer should be $0.96 \times 10^{-2}\,J$.

(d) Explain why your answers to (b)(ii) and (c)(ii) are not the same. [1]

Total energy stored ✗

> Not enough for a mark. The candidate should have written 'Charge must flow from one capacitor to the other (an electric current) so energy is lost as heat because of electrical resistance'.

Exam-style questions

1. When a capacitor is charged to 400 V the charge stored on the capacitor is 800 mC.

   (a) What is the capacitance of the capacitor? [1]

   (b) Why is it ambiguous to refer to the charge *stored* on a capacitor? [1]

2. A 400 μF and a 600 μF capacitor are connected in parallel and the combination is connected in series to a 1000 μF capacitor. The three capacitors are then connected to a 10 V d.c. power supply.

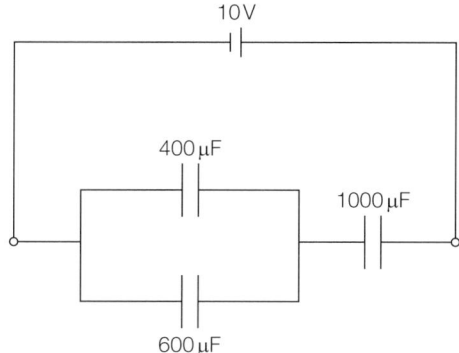

   (a) Determine the combined capacitance of:

   (i) the two capacitors in parallel

   (ii) the three capacitors in combination. [2]

   (b) Calculate the charge:

   (i) on the 400 μF capacitor

   (ii) on the 600 μF capacitor. [2]

   (c) Calculate the p.d. across the 1000 μF capacitor. [1]

3. A photoflash capacitor fitted to a camera has a capacitance of 500 μF. It is charged to a voltage of 300 V.

   (a) Determine the energy stored in the capacitor. [2]

   (b) The capacitor discharges completely in 3 ms. Calculate the average power output of the flash. [2]

4. A student has a 20 μF capacitor, a 47 μF capacitor, and an 82 μF capacitor.

   (a) (i) What is the largest capacitance that he can make from the three capacitors?

   (ii) Draw a diagram to show how the three capacitors should be connected. [2]

   (b) (i) How can the three capacitors be combined to have an overall capacitance of 50 μF?

   (ii) Draw a diagram to show how the three capacitors should be connected. [2]

5. A 1000 μF capacitor is connected to a 6 V d.c. supply, as shown.

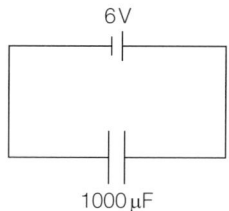

   (a) Calculate:

   (i) the charge on the capacitor

   (ii) the energy stored in the capacitor. [2]

   The 1000 μF capacitor, still charged, is disconnected from the battery. A 500 μF capacitor is now charged to 6 V by the d.c. supply and then disconnected. The two capacitors are then joined together, as shown below, the positively-charged plate of one capacitor being connected to the negatively-charged plate of the other.

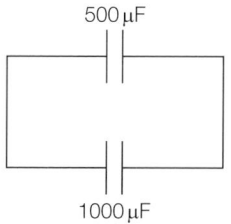

   (b) Calculate the charge on either plate of:

   (i) the 500 μF capacitor

   (ii) the 1000 μF capacitor. [2]

   (c) Determine the potential difference across:

   (i) the 500 μF capacitor

   (ii) the 1000 μF capacitor. [2]

   (d) (i) Calculate the total energy stored in the two capacitors.

   (ii) Explain why this is less than your value in (a)(ii). [2]

20 Magnetic fields

Magnetic fields

A **magnetic field** is the space around a magnet or a current-carrying wire in which magnetic forces (on a ferrous metal, another magnet or another wire carrying a current, for example) can act. The shape, size and direction of a magnetic field can be illustrated using magnetic field lines (see Figure 20.1).

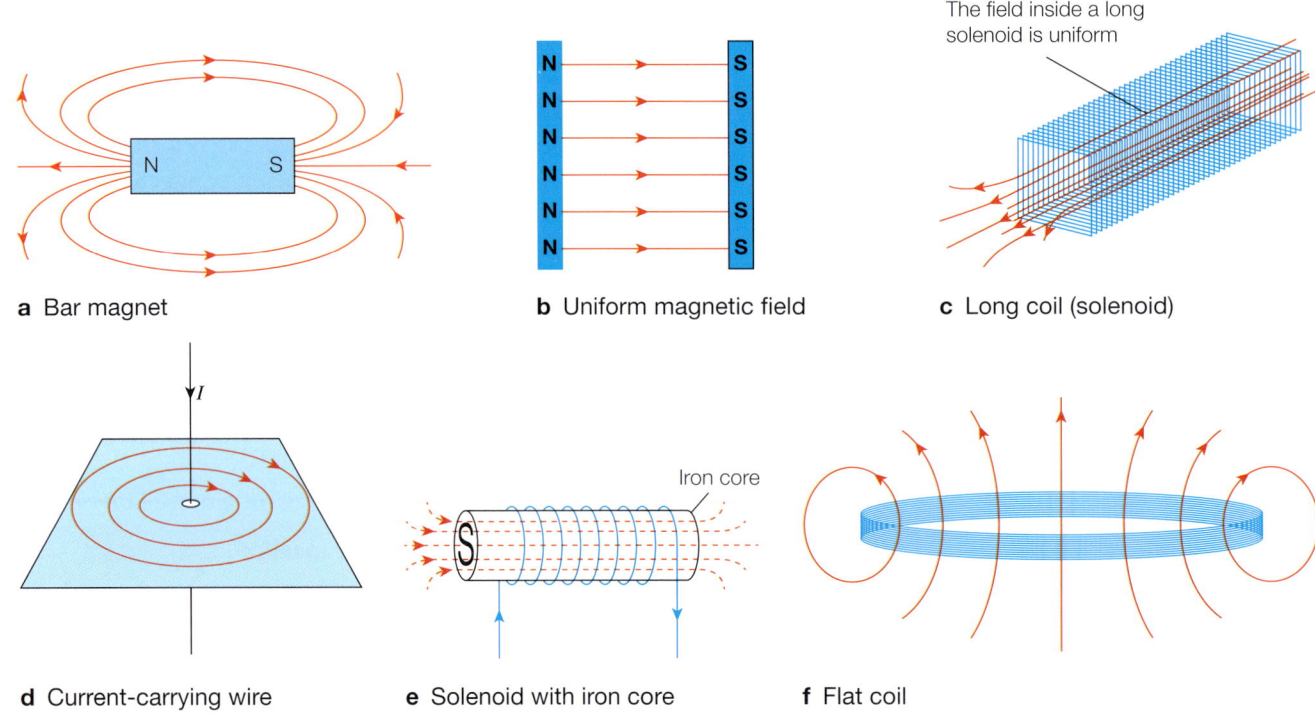

a Bar magnet   b Uniform magnetic field   c Long coil (solenoid)

The field inside a long solenoid is uniform

d Current-carrying wire   e Solenoid with iron core   f Flat coil

Iron core

▲ **Figure 20.1** Magnetic field patterns

> **Remember**
>
> The **corkscrew rule** can be used to find the direction of the magnetic field lines due to a current in a long straight wire. To make the corkscrew (current) move down, turn the corkscrew (field lines) clockwise (see Figure 20.2)

Key points to remember about magnetic fields:

- the closer together the field lines, the stronger the field

- the direction of a field line is the direction a small plotting compass would point (its north pole would point to the south pole of another magnet, so field lines always point from N to S)

- inserting a ferrous metal core into a coil or solenoid (e.g., iron) can increase the strength of the field many thousands of times.

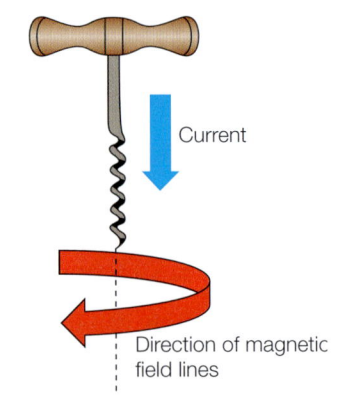

Current

Direction of magnetic field lines

▲ **Figure 20.2** Corkscrew rule

Force on a current-carrying conductor in a magnetic field

A wire of length l carrying a current I at right angles to a magnetic field B experiences a force F at right angles to both the direction of the current and the magnetic field, as shown in Figure 20.3.

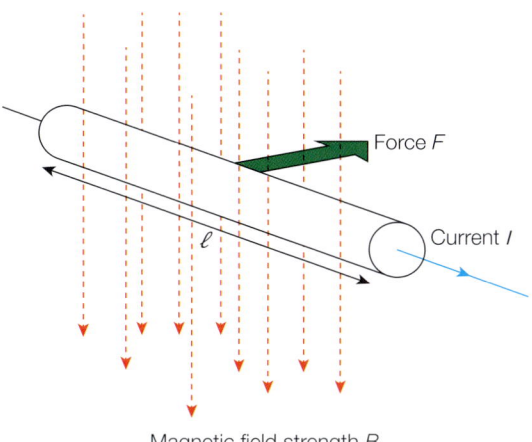

Force F

Current I

Magnetic field strength B

▲ **Figure 20.3** Force on a current-carrying wire

The force F is given by the equation: $F = BIl$

Rearranging this equation gives: $B = \dfrac{F}{Il}$

This equation defines the **magnetic field strength** B (also called the **magnetic flux density**) as **the force per unit current-length on a wire placed at right angles to the magnetic field**. The SI unit of magnetic field strength is the **tesla** (T). A magnetic field strength of 1 T is the strength of the magnetic field normal to a long straight wire carrying a current of 1 A that would produce a force of $1\,\text{N}\,\text{m}^{-1}$ on the wire.

The direction of the force can be found using **Fleming's left-hand rule** (see Figure 20.4). The thumb, first finger and second finger are first held at right angles to each other. The **F**irst finger is then aligned with the direction of the magnetic **F**ield and the se**C**ond finger made to point in the direction of the **C**urrent. The **Th**umb then automatically shows the force (**Th**rust) on the wire.

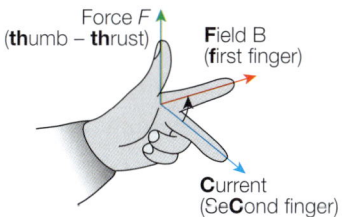

Force F
(**th**umb – **thr**ust)

Field B
(**f**irst finger)

Current
(Se**C**ond finger)

▲ **Figure 20.4** Fleming's left-hand rule

The equation $F = BIl$ only applies if the current and the magnetic field are at right angles to each other. If the magnetic field and the current are at an angle θ to each other, as shown in Figure 20.5, the magnetic field B can be resolved into two perpendicular components $B\cos\theta$ and $B\sin\theta$.

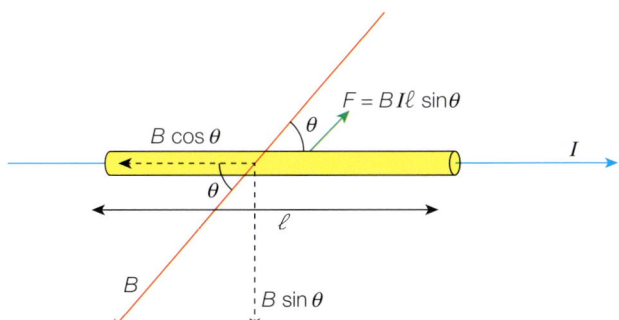

$F = BIl\sin\theta$

$B\cos\theta$

I

B

$B\sin\theta$

▲ **Figure 20.5** Force on a wire at an angle θ to the B-field

The component of the magnetic field parallel to the current has no effect. Only the component of the magnetic field perpendicular to the wire exerts a force. In this case:

$$F = BIl\sin\theta$$

Worked Example

1   In the UK, the magnetic field strength of the Earth is 1.7×10^{-4} T in a direction making 25° to the vertical (Figure 20.6).

Calculate the current needed in a copper cable of diameter 1 mm for the cable to be self-supporting in the Earth's magnetic field.

[The density of copper is 8900 kg m^{-3}.].

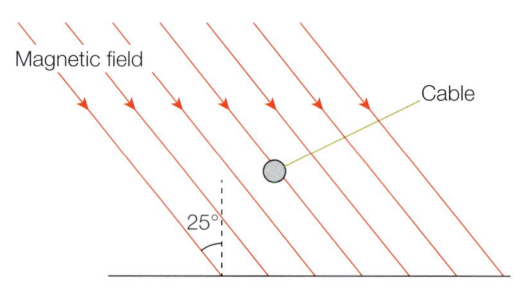

▲ **Figure 20.6**

Answer

Consider a 1.0 m length of the copper cable. Its weight would be:

$mg = \pi \times (0.5 \times 10^{-3})^2 \times 1.0 \times 8900 \times 9.81 = 0.069$ N

For the cable to be self-supporting:

$$BIl \sin \theta = 0.069$$

$$I = \frac{0.069}{1.7 \times 10^{-4} \times 1.0 \times \sin 25} = 960 \text{ A !}$$

> ★ **Exam tip**
>
> Notice it is the **horizontal** component of the magnetic field which produces a vertical force on the cable.

2   Two thin aluminium strips are suspended vertically and held close to each other. They are connected in series to a d.c. supply as shown in Figure 20.7.

Describe and explain what happens when the switch is closed.

Answer

When viewed from above, the current is descending ($\otimes$) in the left-hand strip of aluminium and ascending ($\odot$) in the right-hand strip. The magnetic field around the left-hand strip is circular, and (using the corkscrew rule) the field lines rotate clockwise (Figure 20.8).

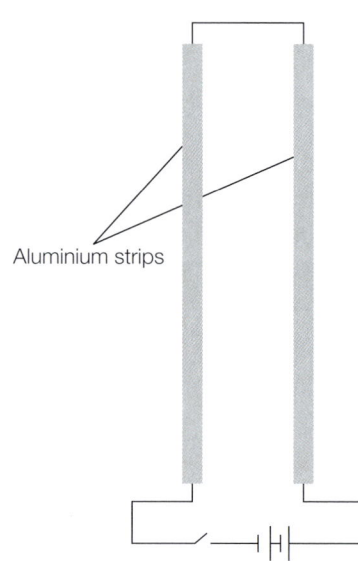

▲ **Figure 20.7**

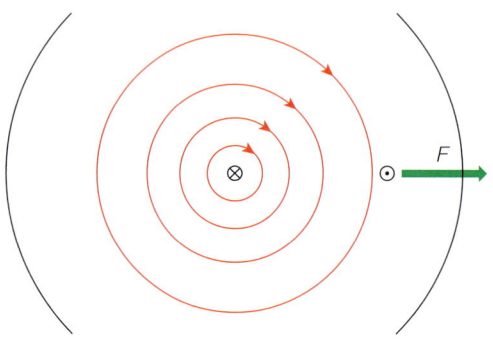

▲ **Figure 20.8**

Using the left-hand rule, the force on the right-hand strip, due to the magnetic field of the left-hand strip, is outwards, towards the right. Applying the same analysis to the left-hand strip shows it will be pushed out to the left – the two strips appear to repel each other. If the currents in the two strips are in the same direction, the strips will appear to attract each other.

Force on a charged particle moving in a magnetic field

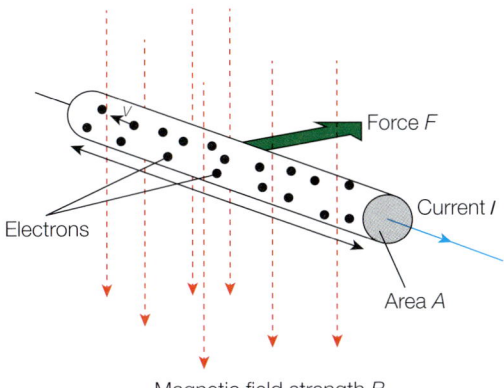

▲ **Figure 20.9** Force on a charged particle in a magnetic field ($F = Bqv$)

The magnetic force on a wire of length l carrying a current I perpendicular to a magnetic field B (Figure 20.9) is given by $F = BIl$. The current I is:

$$I = nAqv$$

where n is the density of charge carriers, A the cross-sectional area of the wire, q the charge on an individual charge carrier and v the drift velocity.

Combining these two equations:

$$F = B(nAqv)l = Bq(nAl)v$$

nAl is the total number of charge carriers, N, in the conductor of length l. The force on N charge carriers is:

$$F = BqNv$$

So the force on a single charge carrier is:

$$F = Bqv$$

If the angle between the field and the velocity is θ the force F is given by:

$$F = Bqv \sin\theta$$

Worked Example

A proton, travelling horizontally at a speed of $3.0 \times 10^7\,\text{m s}^{-1}$, enters a uniform magnetic field of strength 300 mT, at an angle of 40° to the field, as shown in Figure 20.10.

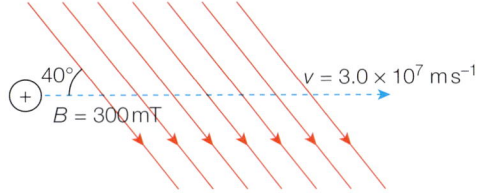

▲ **Figure 20.10** Force on a charged particle moving in a magnetic field

(a) Calculate the force on the proton.

(b) State the direction of the force.

Answer

(a) $F = Bqv = (300 \times 10^{-3} \times \sin 40°) \times 1.6 \times 10^{-19} \times 3.0 \times 10^7 = 9.3 \times 10^{-13}\,\text{N}$

(b) Using Fleming's left-hand rule, the first finger (the magnetic field) points downwards, the second finger (the current) goes from left to right, leaving the thumb (the thrust or motion) pointing towards the paper.

Notice that only the vertical component of the magnetic field ($B \sin\theta$) is used in the calculation.

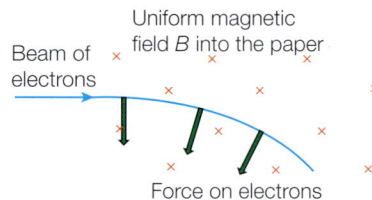

Deflecting charged particles in a magnetic field

A beam of electrons passing through a magnetic field, with a component of the field in a direction perpendicular to the path of the electrons, will experience a force at right angles to the direction in which they are travelling.

In Figure 20.11 the symbol × indicates a magnetic field into the paper (we are seeing the back of an arrow). The electron beam is moving left to right which means the conventional current flow is right to left. Using the left-hand rule, the force on the electron beam is downwards. As the beam deflects downwards the force alters direction so that it is always at right angles to the path of the electrons. The speed of the electrons does not change, only the direction.

For a beam of electrons, with each electron moving at speed v, and a uniform magnetic field B, the force on each electron will be Bev in a direction at right angles to the path of the beam. While the beam is in the magnetic field the electrons will follow a circular path.

▲ **Figure 20.11** Force on a beam of electrons in a magnetic field

Worked Example

Electrons are accelerated from rest across a p.d. of 5 kV in an electron 'gun' (Figure 20.12a). They then enter a region with a uniform magnetic field of magnetic flux density 4.0 mT and travel in a circular path. The direction of the magnetic field is perpendicular to the path of the electrons, as shown in Figure 20.12b.

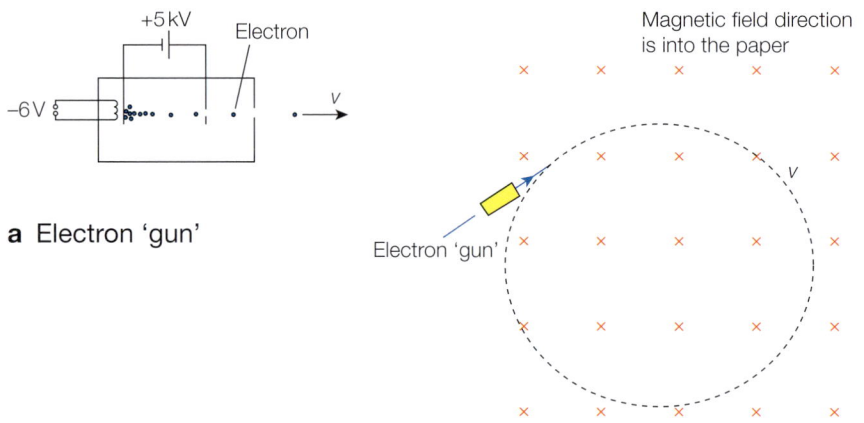

a Electron 'gun'

b Electrons in a uniform magnetic field

▲ **Figure 20.12** Electrons moving in circles

> **Remember**
>
> × This symbol indicates a magnetic field going into the paper (we see the back of an arrow).
>
> ⊙ This symbol indicates a magnetic field coming up out of the paper (the front of an arrow).

(a) Calculate the speed of the electrons leaving the electron 'gun'. [Charge on an electron is 1.6×10^{-19} C; mass of an electron is 9.11×10^{-31} kg.]

(b) Explain why the electrons travel in circles.   (c) Calculate the radius of the circle.

Answer

(a) $\dfrac{1}{2}mv^2 = eV$

$$v = \sqrt{\dfrac{2eV}{m}} = \sqrt{\dfrac{2 \times 1.6 \times 10^{-19} \times 5 \times 10^3}{9.11 \times 10^{-31}}} = 4.2 \times 10^7 \text{ ms}^{-1}$$

(b) The force on an electron is always at right angles to its direction of motion.

(c) Using $F = ma$, where a is the **centripetal acceleration** $\dfrac{v^2}{r}$:

$$Bev = \dfrac{mv^2}{r}$$

$$r = \dfrac{mv}{Be} = \dfrac{9.11 \times 10^{-31} \times 4.2 \times 10^7}{4.0 \times 10^{-3} \times 1.6 \times 10^{-19}} = 0.060 \text{ m} (6.0 \text{ cm})$$

> **Link**
>
> See Unit 12 *Motion in a circle* for more details about centripetal acceleration.

Separating particles with different velocities

A **uniform magnetic field** set at right angles to a **uniform electric field** can be used as a way of separating charged particles with different energies and velocities.

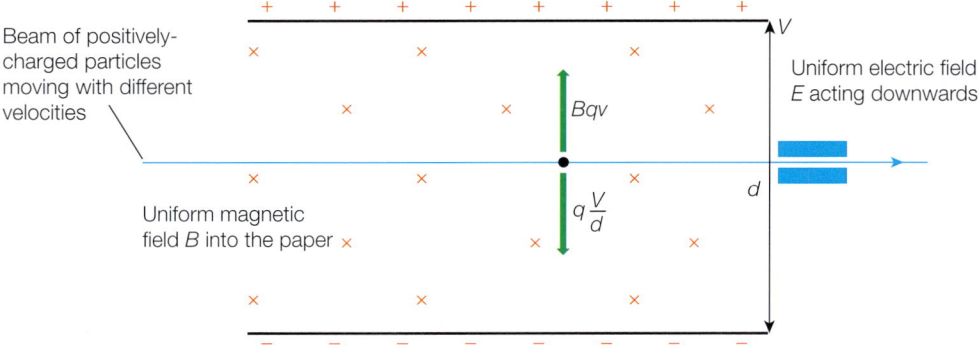

▲ **Figure 20.13** Separating charged particles moving with different velocities

Positively charged particles, moving with different velocities, enter a region with a uniform B-field and a uniform E-field, as shown in Figure 20.13. The magnetic field exerts a force Bqv upwards and the electric field exerts a force qV/d downwards. These forces will cancel out and the particle will travel in a straight line, undeflected by either field, if:

$$Bqv = q\frac{V}{d}$$

$$v = \frac{V}{Bd}$$

Particles travelling faster than this will be deflected upwards; particles slower than this will be deflected downwards. The experiment illustrates how particles moving with different speeds can be selected.

Hall effect

A thin slice of a semiconductor such as germanium is placed at right angles to a magnetic field of flux density B as shown in Figure 20.14.

The charge carriers in the thin slice are positively charged. If a current I passes through the slice the charge carriers will feel a force pushing them towards the back of the slice.

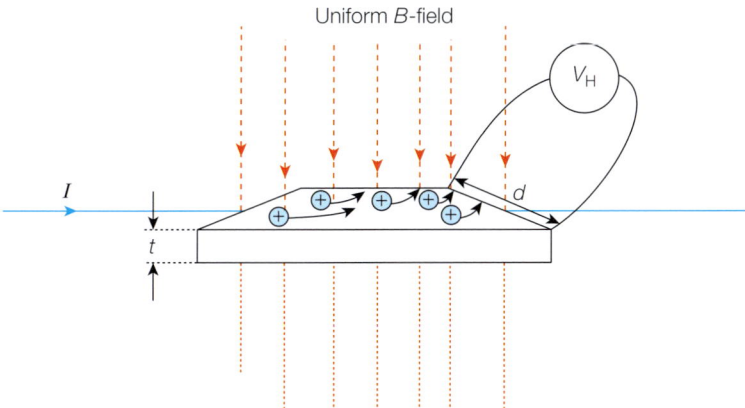

▲ **Figure 20.14** Hall effect

A potential difference V_H (the **Hall voltage**) builds up between the front and the back of the slice, creating an electric field similar to the electric field between two oppositely-charged parallel plates. This is known as the **Hall effect**.

The electric field tries to push the positively-charged particles back towards the front of the slice while the magnetic field continues to push the positively charged particles towards the back of the slice. A balance is achieved when the magnetic force on each particle (Bqv) is equal to the electric force (Eq), as shown in Figure 20.15.

$$Eq = Bvq$$

But $E = \dfrac{V_H}{d}$, so: $\qquad V_H = Bvd$

For a current-carrying conductor, $I = nAqv$. Combining these equations to eliminate v:

$$V_H = \frac{BId}{nAq} = \frac{BI}{ntq}$$

as the cross-sectional area of the slice, $A = dt$, where t is the thickness of the slice and d the depth.

This equation shows why the Hall effect is most easily observed using:

- a **thin** slice of material (so that the value of t is small)

- a semiconductor material such as germanium or silicon, as the value of n, the density of charge carriers, is much smaller for a semiconductor than a conductor like copper.

Using a Hall probe

A **Hall probe** uses the Hall effect to measure the strength of magnetic fields, the Hall voltage being directly proportional to the magnetic field strength B. Figure 20.16 shows a Hall probe being used to investigate how the magnetic field strength inside a solenoid varies with distance from the centre of the solenoid.

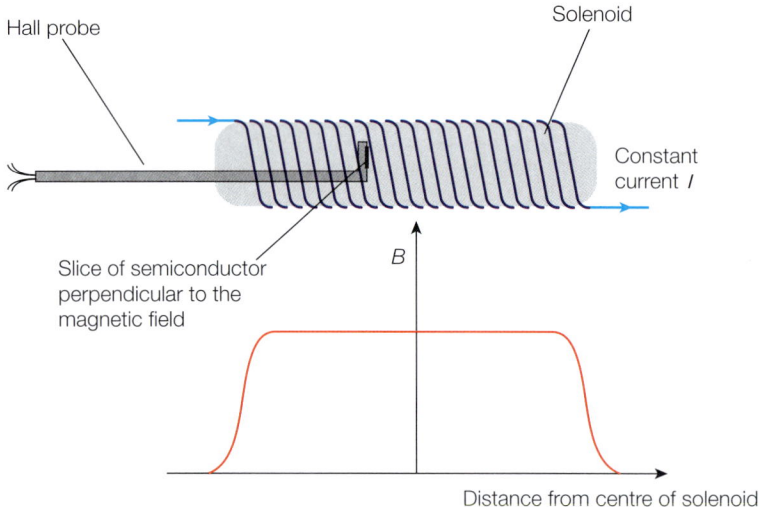

▲ **Figure 20.16** Using a Hall probe

In order to measure a magnetic flux density the Hall probe must first be calibrated using a known magnetic field. The probe is placed with the semiconductor slice perpendicular to a uniform magnetic field of known magnitude. A current I is passed through the slice and the Hall voltage V_H recorded. For a constant current I:

$$V_H = kB$$

The constant k can be found from the measurements taken.

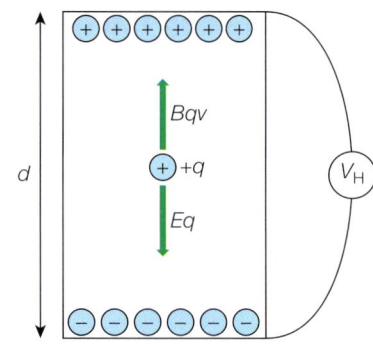

▲ **Figure 20.15** Hall voltage

> ★ **Exam tip**
>
> You may be asked to derive this equation.

The Hall effect can be used to measure the strength of magnetic fields and as magnetic field sensors; for example, some computer printers use Hall effect sensors to detect whether there is missing paper or the cover is open.

> 💡 **Remember**
>
> Key points to remember when using a Hall probe:
>
> - The semiconductor slice must be perpendicular to the magnetic field being investigated.
>
> - The probe can only measure magnetic fields which do not vary with time.

Comparing the forces in different types of field

Figure 20.17 compares the forces on mass, charge, and current in gravitational, electrical, and magnetic fields:

- All three forces are examples of 'action at a distance'.

- All three fields can be represented by field lines which show the direction of the force at points along the line.

- The density of the field lines indicates the strength of the field

- The field strength is defined as the *force per unit ... *** mass** (for gravitational fields), **charge** (electric fields), or **current-length** (for magnetic fields).

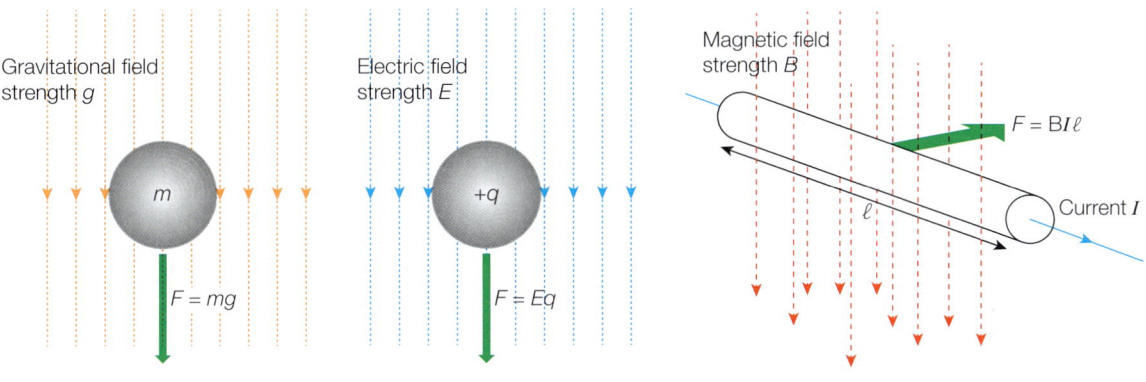

a   Gravitational field   b   Electric field   c   Magnetic field

▲ **Figure 20.17** Forces in gravitational fields, electric fields and magnetic fields

---

Worked Example

(a) Calculate:

    (i) the electric force

    (ii) the gravitational force

    between an electron and a proton in a hydrogen atom (Figure 20.18). Assume the radius of a hydrogen atom is 0.5×10^{-10} m.

(b) Calculate the ratio of these two forces.

(c) State what this ratio would be if the proton and electron were:

    (i) 1 m apart   **(ii)** 1 light-year apart.

[Charge on an electron = 1.6×10^{-19} C; mass of an electron = 9.1×10^{-31} kg; mass of a proton = 1.7×10^{-27} kg.]

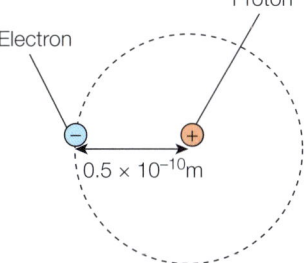

▲ **Figure 20.18** Hydrogen atom

Answer

(a) **(i)** Using Coulomb's law: $F_e = k\dfrac{q_1 q_2}{r^2} = 9 \times 10^9 \times \dfrac{\left(1.6 \times 10^{-19}\right)^2}{\left(0.5 \times 10^{-10}\right)^2} = 9.2 \times 10^{-8}$ N

    (ii) Using Newton's law: $F_g = G\dfrac{m_1 m_2}{r^2} = 6.67 \times 10^{-11} \times \dfrac{9.1 \times 10^{-31} \times 1.7 \times 10^{-27}}{(0.5 \times 10^{-10})^2} = 4.1 \times 10^{-47}$ N

(b) $\dfrac{F_e}{F_g} = \dfrac{9.2 \times 10^{-8}}{4.1 \times 10^{-47}} = 2.2 \times 10^{39}$

(c) **(i)** and **(ii)** In both cases the ratio would be the same. Both Coulomb's law and Newton's law are inverse square laws; the ratio of the two is independent of the distance between the two masses/charges.

↑ Raise your grade

(a) Define the *tesla*.

The tesla is the SI unit of magnetic flux density (magnetic field strength). ✔

A field of strength 1 tesla exerts a force of 1N on a wire carrying a current of 1A ✗ ✗

> A conductor carrying a current of 1 A **in a direction normal to the magnetic field** experiences a force of **1 N per metre length** of the wire.   [3]

An experiment is carried out to measure the strength of a uniform magnetic field between two magnets by measuring the force on a current-carrying wire. The magnets are repelling each other. The wire is supported by two clamps and connected to a d.c. power supply as shown.

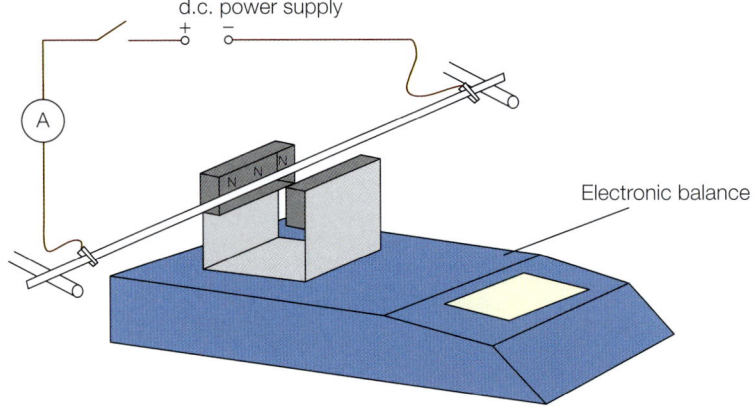

(b) (i) When the switch is closed a force acts on the wire. State the direction of this force.

downwards ✔   (Using Fleming's left-hand rule.)

(ii) The reading on the electronic balance decreases. Explain why.

From Newton's 3rd law, if the magnets exert a downward force on the wire, the wire ✔✔

exerts an equal and opposite (upwards) force on the magnets (which tries to 'lift' them). [3]

(c) The results of the experiment are shown in the table.

Length of wire in magnetic field	8.0 cm
Electric current	2.7 A
Initial reading on balance	95.4 g
Reading on balance when switch is closed	89.7 g

(i) Show that the force acting on the wire due to the magnetic field is approximately 5.6×10^{-2} N.

Change in force on balance = $(95.4 - 89.7) \times 10^{-3} \times 9.81 = 5.59 \times 10^{-2}$ N

✔   ✔

(ii) Calculate the strength of the magnetic field between the two magnets.

$$F = Bil \Rightarrow B = \frac{F}{il} = \frac{5.6 \times 10^{-2}}{2.7 \times 8.0} = 2.6 \times 10^{-3} \, T$$

✔ ✗

> The correct equation, re-arranged correctly to find B, but the value of ℓ should be converted to m. Correct answer is 0.26T.

strength of magnetic field = 2.6×10^{-3}  T [4]

Electromagnetic induction

When a wire 'cuts through' a magnetic field, an e.m.f. is induced across the ends of the wire. This is an example of **electromagnetic induction**.

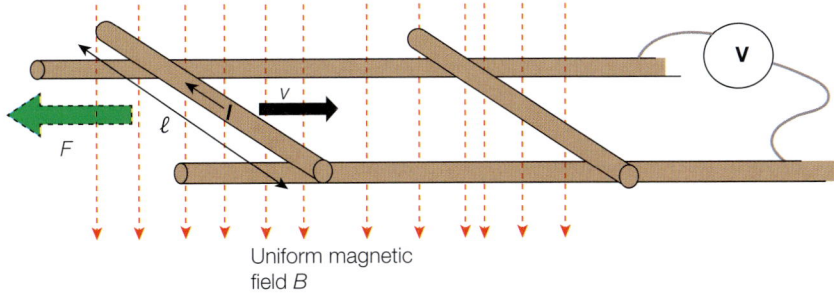

Uniform magnetic field B

▲ **Figure 20.19** Electromagnetic induction

A wire of length l moving perpendicular to a magnetic field B with velocity v (See Figure 20.19.) will have an e.m.f. E induced between the ends of the wire. Experiments show that the e.m.f. induced can be increased by:

- moving the wire faster

- increasing the magnetic field strength B

- increasing the length of wire in the magnetic field.

If a second wire is connected between the two rails to complete a circuit, a current I will flow and a force will act on the wire which is perpendicular to both the direction of the current and the magnetic field. The force will act in the opposite direction to the direction the wire is moving (i.e., it opposes the motion which produces it).

External work must be done against this force for the wire to move. From $F = BIl$, and using the principle of the conservation of energy,

electrical power generated = work done/second

$$IE = (BIl) \times v$$

$$E = Blv$$

> **Remember**
>
> The force F always acts in a direction which opposes the motion producing the force.

> **Remember**
>
> The **magnetic field strength B** (T) is also called the **magnetic flux density** ϕ (Wb m^{-2}).

> lv is the 'area swept out in one second'. Think of the wire as a brush – lv is the area of floor that would be swept in one second.

Worked Example

1　A rectangular loop ABCD of metal wire, of dimensions 20.0 cm × 40.0 cm and electrical resistance 50 Ω, is held perpendicular to a uniform magnetic field of strength 150 mT, as shown in Figure 20.20. The loop is pulled out of the field at a steady speed of 0.10 m s^{-1}.

Determine:

(a) the induced e.m.f.

(b) the current in the loop

(c) the force needed to pull the loop.

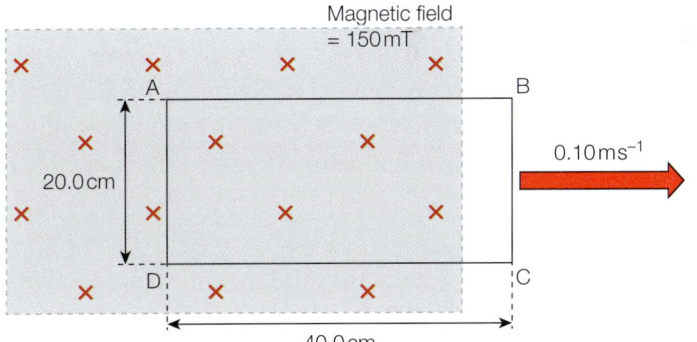

▲ **Figure 20.20** Induced e.m.f.s

Answer

(a) Using $E = Blv$: $E = 150 \times 10^{-3} \times 20.0 \times 10^{-2} \times 0.10 = 3.0 \, \text{mV}$

(b) $I = \dfrac{V}{R} = \dfrac{3.0 \times 10^{-3}}{50} = 6.0 \times 10^{-5} \, \text{A} \ (60 \, \mu\text{A})$

(c) Mechanical work done per second = electrical power generated

$$F \times v = V \times I$$

$$F \times 0.10 = 3.0 \times 10^{-3} \times 60 \times 10^{-6}$$

$$F = 1.8 \times 10^{-6} \, \text{N}$$

Faraday's law

The equation $E = Blv$ can be rewritten:

$$E = B\frac{dA}{dt}$$

where dA/dt is the 'area swept out' each second. The equation is a special case of Faraday's law of electromagnetic induction:

$$E = -\frac{d\phi}{dt} \qquad \text{where } \phi = BA$$

Faraday's law states that the induced e.m.f. in a circuit is proportional to the rate of change of flux linkage through the circuit.

Magnetic flux and flux linkage

Magnetic flux is defined by the equation:

$$\phi = BA$$

where A is the area perpendicular to the magnetic field B (see Figure 20.21). It is measured in webers (Wb)

Re-arranging this equation:

$$B = \frac{\phi}{A}$$

which explains why the magnetic field strength is also called the **magnetic flux density**, and can be measured in Wb m^{-2}. Note: $1\,\text{Wb m}^{-2} = 1\,\text{T}$

Magnetic flux can be thought of as the total number of magnetic field lines passing perpendicularly through an area. If the magnetic field lines are at an angle other than 90°, the component of the magnetic field perpendicular to the area is used.

The magnetic flux in this case is $\phi = BA\cos\theta$ (see Figure 20.22).

The greater the density of magnetic field lines, the greater the magnetic field strength. The greater the number of field lines enclosed by an area, the greater the magnetic flux.

If there were two wires moving in Figure 20.19, the induced e.m.f. would be twice as much; if there were N wires, the induced e.m.f. would be N times as much. The general form of Faraday's law is:

$$E = -\frac{\Delta(N\phi)}{\Delta t}$$

where $N\phi$ is the **magnetic flux linkage**, measured in webers (Wb). The symbol Δ means 'change in'.

The e.m.f. E induced is equal to the **rate** at which the flux linkage changes. The minus sign in the equation indicates that the induced voltage acts in such a way as to oppose the change producing the voltage (see Lenz's law, below).

For an e.m.f. to be induced, the magnetic flux must be **changing**. In the previous worked example, an e.m.f. only occurs when the total number of field lines enclosed by ABCD is increasing or decreasing – if all the loop is inside the magnetic field (or all outside) no e.m.f. is produced. The faster the magnetic flux is changing, the greater the e.m.f. induced. Faraday's law explains how generators, motors, and transformers work.

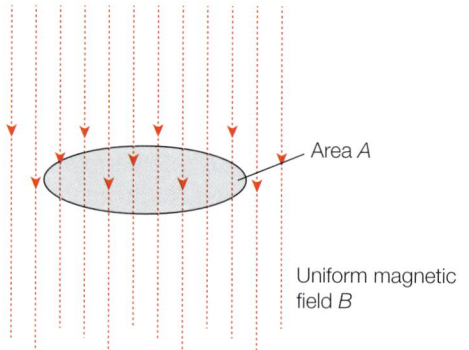

▲ **Figure 20.21** Magnetic flux $\phi = BA$

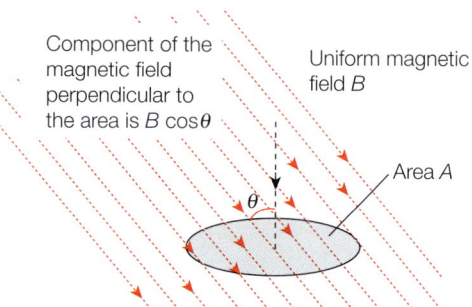

▲ **Figure 20.22** Magnetic flux $\phi = BA\cos\theta$

💡 **Remember**

magnetic flux $\phi = BA$

SI unit: Wb or T m^2

magnetic flux density

$$B = \frac{\phi}{A}$$

SI unit: Wb m^{-2} or T

magnetic flux linkage
$N\phi = NBA$

SI unit: Wb

Worked Example

A 20-turn circular coil of diameter 8.0 cm is held with its plane perpendicular to a uniform magnetic field of strength 1.4 T (see Figure 20.23). It is rotated 90° in 2.8 ms.

(a) Calculate the flux linkage:

    (i) before the coil is rotated

    (ii) after the coil is rotated.

(b) Determine the average e.m.f. between the ends of the coil while the coil is being rotated.

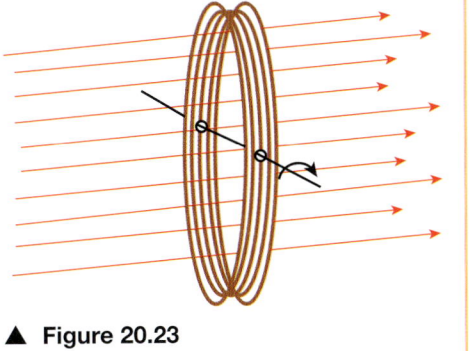

▲ **Figure 20.23**

Answer

(a) (i) flux linkage = $N\phi = 20 \times 1.4 \times \pi \times (4.0 \times 10^{-2})^2 = 0.14$ Wb

    (ii) flux linkage = 0

(b) Average e.m.f. = $\dfrac{\Delta(N\phi)}{\Delta t} = \dfrac{0.14}{2.8 \times 10^{-3}} = 50$ V

Lenz's law

Lenz's law states that the direction of the induced e.m.f. or current is such as to oppose the change that produces it.

It is a statement of the principle of conservation of energy applied to induced e.m.f.s.

In Figure 20.24a the north-seeking pole of a magnet is moving downwards and the flux inside the coil is increasing – more field lines are enclosed by ('link' with) the coil.

The induced voltage across the ends of the coil causes a current to flow anticlockwise when viewed from above. This makes the top of the coil a north pole, repelling the north pole of the magnet.

If the induced current flowed the other way, the top of the coil would become a south pole, attracting the magnet and causing it to move faster. The rate of change of flux linking the coil would increase, inducing a larger induced e.m.f. This in turn would create a larger current making the magnetic pole of the coil even stronger, causing the magnet to move even faster. The magnet would be gaining extra kinetic energy (more than it would gain by just falling without the coil of wire present); this would contradict the principle of conservation of energy.

In Figure 20.24b, as the north-seeking pole of the magnet moves up, the flux linking the coil decreases – fewer and fewer lines 'link' with the coil.

The e.m.f. induced causes a current to flow clockwise when viewed from above. This makes the top of the coil a south-seeking pole. This south pole tries to pull the magnet down; it tries to prevent the magnet moving up.

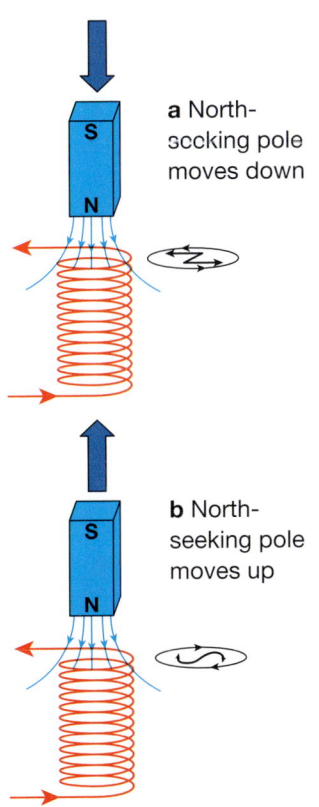

a North-seeking pole moves down

b North-seeking pole moves up

▲ **Figure 20.24** Lenz's law

> **Key terms**
>
> **Faraday's law**: the induced e.m.f. is proportional to the rate of change of flux linkage.
>
> **Lenz's law**: the direction of the induced e.m.f. or current is such as to oppose the change that produces it.

'Flux cutting', 'flux linking' and induced e.m.f.

As the magnet falls into the coil, the field lines are being 'cut' by the coil (more and more field lines 'link' with the coil). The total magnetic flux inside the coil is **increasing** in order to produce an e.m.f. across the ends of the coil.

As the magnet is pulled up from the coil, the number of magnetic field lines linking with the coil is **decreasing**, and so an e.m.f. is again induced, but in the opposite direction.

If the magnet is moved more quickly into the coil, the rate of change of flux linking with the coil would be greater, and so the e.m.f. induced across the ends of the coil would be larger.

If a stronger magnet is used (more field lines per unit area), and the magnet moved into the coil at the same speed, the number of field lines linking with the coil would increase more rapidly, inducing a greater voltage.

Worked Example

An aluminium disc of diameter 30.0 cm is connected to an electric motor and rotates at a constant speed of 120 r.p.m. in a vertical plane, as shown in Figure 20.25. A constant magnetic field of flux density 0.30 T acts horizontally and perpendicular to the plane of the disc. A voltmeter is connected to points O and P using sliding contacts.

(a) Calculate the potential difference between points O and P.

(b) Determine the p.d.:

    (i) between O and Q

    (ii) between P and Q.

    Justify your answers.

(c) Describe and explain what happens to 'free' conduction electrons in the aluminium disc.

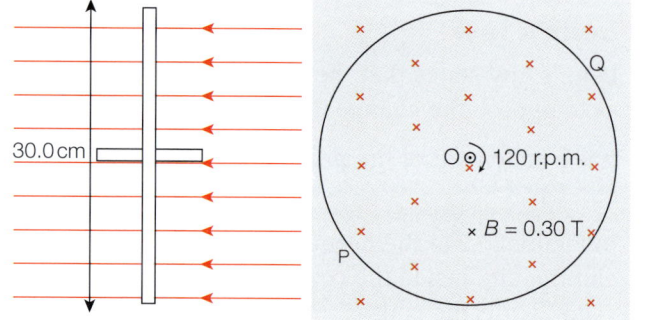

▲ **Figure 20.25** e.m.f. induced on a rotating aluminium disc

Answer

(a) Consider OP as a thin metal strip. It completes two rotations each second (120 r.p.m.) so 'sweeps out' an area of 2 × area of the disc in 1 s.

    e.m.f. between O and P = magnetic flux 'swept' per second

$$= 0.30 \times 2 \times (\pi \times 0.15^2) = 0.042 \text{ V}$$

(b) (i) OQ is 'cutting' or sweeping flux at the same rate as OP. The p.d. between O and Q will be the same as the p.d. between O and P.

    (ii) P and Q will be at the same potential, so the p.d. between P and Q will be zero.

(c) An electron moving downwards at one instant is equivalent to a conventional current moving upwards, as shown in Figure 20.26. Using Fleming's left-hand rule, the force on the electron is to the left (i.e., towards the centre). A similar argument applies to electrons in other positions on the disc – they all move towards the centre.

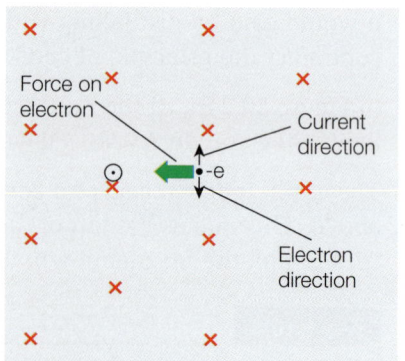

▲ **Figure 20.26**

Exam-style questions

[Charge on an electron is 1.6×10^{-19} C; mass of electron is 9.11×10^{-31} kg.]

1 A rectangular wire loop ABCD, with dimensions $10.0 \text{ cm} \times 8.0 \text{ cm}$, lies with its plane parallel to a uniform magnetic field of strength 0.50 T. The loop carries a current of 2.4 A, as shown.

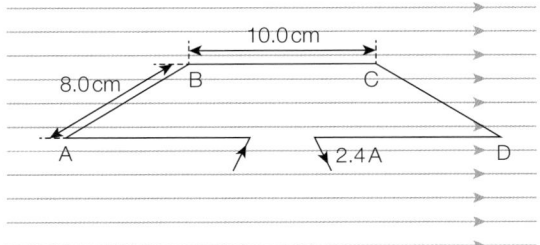

(a) State the direction of the force on:
   (i) AB **(ii)** CD. [2]

(b) Calculate the magnitude of the force on:
   (i) AB **(ii)** BC. [2]

(c) Calculate the total torque acting on the loop in this position. [2]

(d) Describe and explain what would happen if the loop was in a vertical plane. [2]

2 A β-particle, travelling at a speed v, enters a region with a uniform magnetic field of flux density 0.25 mT and a uniform electric field. The β-particle is undeflected, travelling in a straight, horizontal line, as shown.

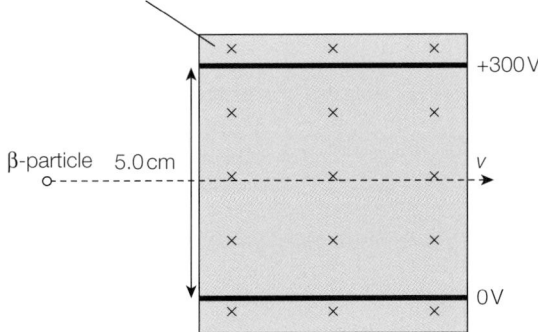

(a) Using information in the diagram, state the direction of:
   (i) the electric field
   (ii) the magnetic field. [2]

(b) Show that the electric field strength is 6 kV/m. [2]

(c) Calculate the velocity of the β-particle. [3]

(d) Describe and explain what would happen to the path of the β-particle if it was moving faster. [2]

3 A sliding rod AB, supported by two conducting rails, is pulled at a steady speed of 3.0 m s^{-1} through a magnetic field of flux density 40 mT, as shown.

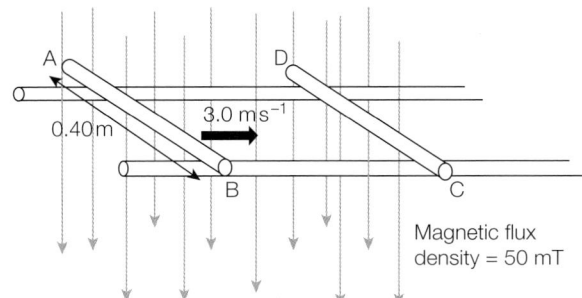

(a) Calculate:
   (i) the area 'swept' by the rod in one second
   (ii) the magnetic flux 'cut' by the rod each second
   (iii) the induced e.m.f. across the rod. [4]

(b) A second rod CD is placed across the rails and an electric current is induced in the loop ABCD.
   (i) State and explain the direction of the current in rod AB.
   (ii) Suggest a reason why the current in AB increases as it approaches rod CD. [3]

4 A horizontal flat circular coil of diameter 30.0 cm, consisting of 200 turns of wire, is perpendicular to a uniform magnetic field acting vertically downwards, as shown. The magnetic flux density increases linearly from 2.0 T to 5.0 T in 600 ms.

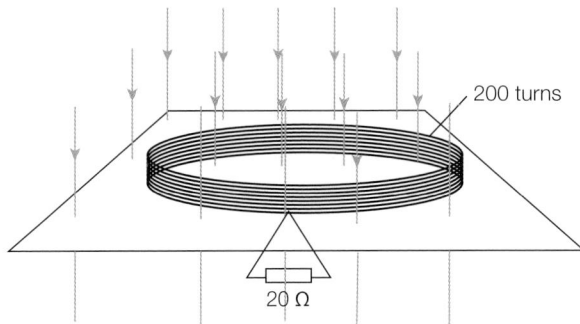

(a) Calculate the induced e.m.f. between the ends of the coil. [3]

(b) The ends of the coil are connected to a 20 Ω resistor.
   (i) Calculate the current in the resistor.
   (ii) Viewed from above, state the direction of the induced current in the coil. Explain your reasoning. [3]

21 Alternating currents (a.c.)

You should be able to:

- use the resistance ($V = IR$) and power ($P = IV = I^2R$) equations for d.c. circuits
- describe a sinusoidal oscillation with the relationship: $x = x_0 \sin \omega t$
- describe the functions of a diode and a capacitor.

Alternating currents and voltages (a.c.)

Alternating currents and voltages vary sinusoidally. They are used throughout the world when generating and transmitting electrical energy. Figure 21.1 shows an example of alternating current.

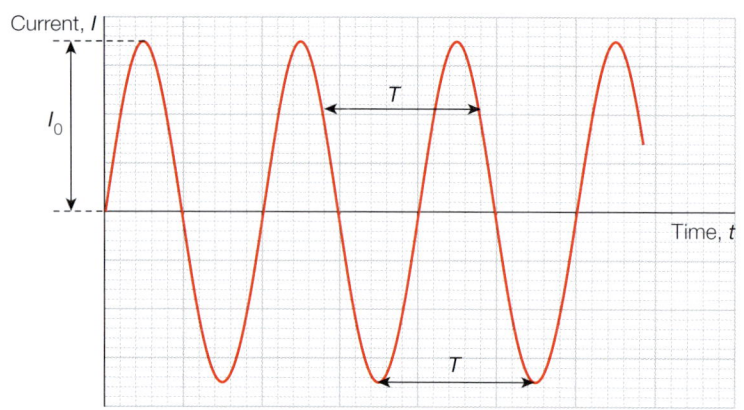

▲ **Figure 21.1** Alternating current (a.c.)

The **maximum (peak) value** of the current is I_0. The **frequency** f of the alternating current is $1/T$ where T is the **period**. The current I at time t is given by the equation:

$$I = I_0 \sin(2\pi f t) = I_0 \sin \omega t$$

where $\omega = 2\pi f$, and ω is the angular frequency in rad s^{-1}.

Similarly, for an alternating voltage:

$$V = V_0 \sin(2\pi f t)$$

> **Remember**
>
> Just as for SHM:
> $$\omega = 2\pi f = \frac{2\pi}{T}$$
> The general equation for a quantity x varying sinusoidally with time t is:
> $$x = x_0 \sin \omega t$$

> **Maths Skills**
>
> Alternative expressions are:
> $$I = I_0 \cos(2\pi f t)$$
> $$V = V_0 \cos(2\pi f t)$$
> for the case when I and V are at their peak values when $t = 0$.

Worked Example

Determine, from Figure 21.2:

(a) the period

(b) the frequency

(c) the peak voltage

(d) the peak-to-peak voltage.

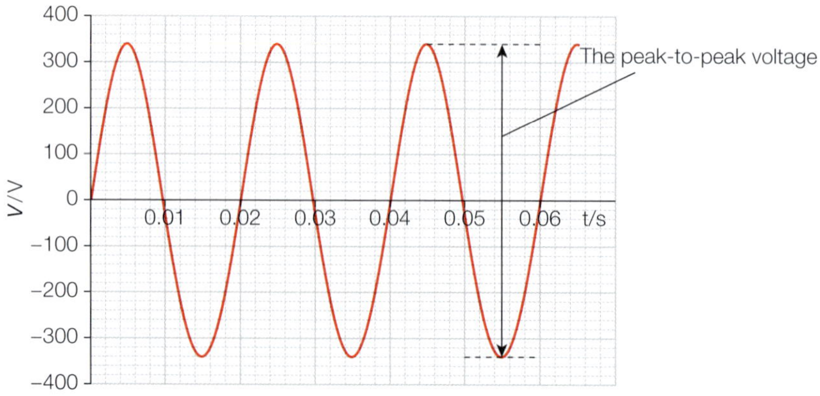

Answer

(a) period, $T = 0.02\,\text{s}$

(b) frequency $f = \dfrac{1}{T} = 50\,\text{Hz}$

(c) peak voltage $= 340\,\text{V}$

(d) peak-to-peak voltage $=$ range of voltages $= 680\,\text{V}$

▲ **Figure 21.2**

If an alternating voltage $V = V_0 \sin(2\pi ft)$ is connected across a resistor of resistance R (see Figure 21.3 a and b), an alternating current passes through the resistor. At time t the current is $I = I_0 \sin(2\pi ft)$, so P, the power dissipated in the resistor, is:

$$P = VI = V_0 I_0 \sin^2(2\pi ft) = \frac{V_0 I_0}{2}[1 - \cos(4\pi ft)]$$

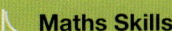

Maths Skills

$\cos 2\theta = 1 - 2\sin^2\theta$

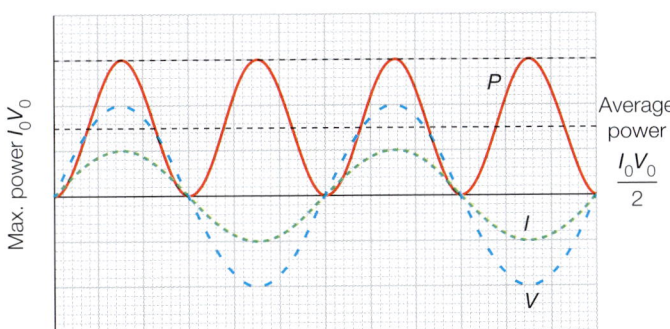

a Variation of current I, voltage V, and power P with time

▲ **Figure 21.3** Power dissipated in a resistor R

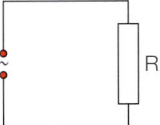

b a.c. current across a resistor

The maximum power is $I_0 V_0$.

The average power over a complete cycle is half this: $\dfrac{I_0 V_0}{2}$.

The direct current and voltage that would give the same power output are called the **root-mean-square** (r.m.s.) values.

$$I_{r.m.s.} = \frac{I_0}{\sqrt{2}} \qquad V_{r.m.s.} = \frac{V_0}{\sqrt{2}}$$

a.c. ammeters and voltmeters are calibrated to measure the r.m.s. values of current and voltage. The r.m.s. values of current and voltage can be used in the same equations already used for analysing d.c. circuits, such as the equation for power.

Remember

The **r.m.s. value** of an alternating current is the value of the direct current that would give the same heating effect when connected to the same resistor.

$$I_{r.m.s.} = \frac{I_0}{\sqrt{2}} \qquad V_{r.m.s.} = \frac{V_0}{\sqrt{2}}$$

For a.c. circuits:

$$P_{ave} = I_{r.m.s.} V_{r.m.s.}$$

$$= I_{r.m.s.}^2 R = \frac{V_{r.m.s.}^2}{R} = \frac{I_0 \times V_0}{2}$$

Worked Example

A heating element gives a power output of 24 W when connected to a 12 V d.c. power supply. What is the power output if the element is connected to

(a) a 12 $V_{r.m.s.}$ a.c. supply

(b) a 24 $V_{r.m.s.}$ supply?

Answer

Using the relationship $P = I_{r.m.s.}^2 R$

(a) The 12 $V_{r.m.s.}$ supply provides an r.m.s. current equivalent to that provided by the 12 V d.c. supply, so the power of the heating element would be the same: 24 W

(b) The 24 $V_{r.m.s.}$ supply provides an r.m.s. current equivalent to twice that provided 12 V d.c. supply, so the power of the heating would be *four* times as large: 96 W

Rectification

Half-wave rectification

Some electrical appliances can operate using either a.c. or d.c. – for example electric heaters, toasters or irons – but many other appliances require a direct current supply.

A single diode can be used to convert a.c. into d.c (see Figure 21.4). The diode is connected between the a.c. input and the load resistor R. In the first half of the a.c. cycle the diode is forward-biased, a current flows in the circuit, and there is a p.d. across the resistor R. In the second half of the a.c. cycle the diode is reverse-biased and no current flows. The p.d. across R is zero (see Figure 21.5).

This is called **half-wave rectification**. Although the output p.d. varies, it is never negative, though for half the time the output p.d. and the current are zero.

Full-wave rectification (the bridge rectifier)

Full-wave rectification can be achieved by combining four diodes in the form of a square, as shown in Figure 21.6.

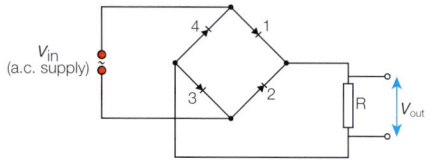

▲ **Figure 21.6** Full-wave rectification

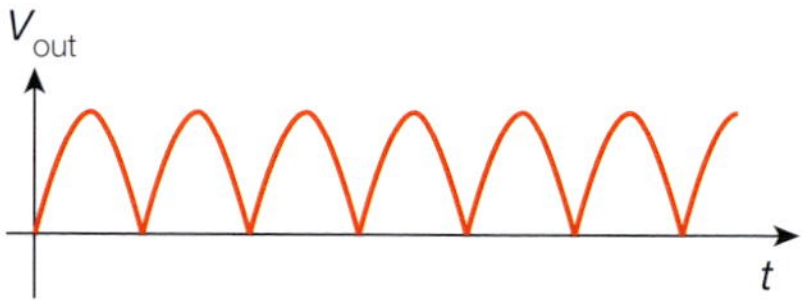

▲ **Figure 21.7** Full-wave rectification output

When V_{in} is positive, current flows through diode 1, downwards through resistor R and returns via diode 3; when V_{in} is negative, current flows through diode 2, downwards through resistor R and returns via diode 4. In both halves of the a.c. cycle the current is passing downwards through R. The output is shown in Figure 21.7.

Smoothing

Some electrical appliances, such as mobile phone chargers, require a steady d.c. supply to work properly. The 'bouncy' d.c. output from a half-wave or full-wave rectifier circuit can be 'smoothed' by the addition of a capacitor in parallel with the external load, as shown in Figure 21.8.

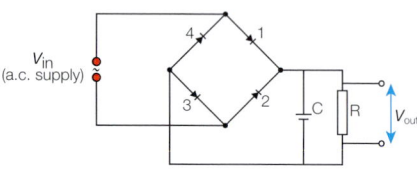

▲ **Figure 21.8** Smoothing

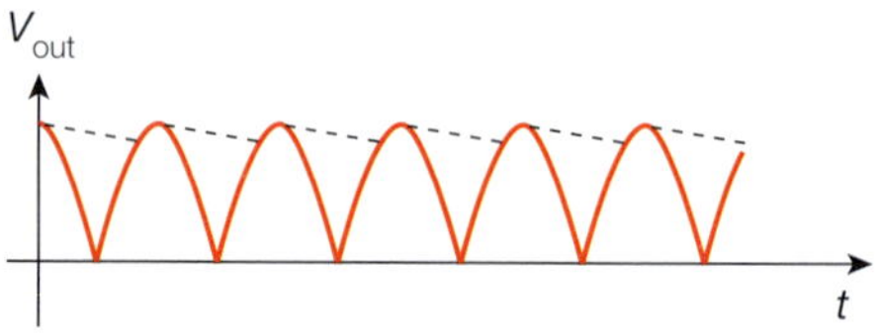

▲ **Figure 21.9** Ripple of smoothed output

When the p.d. across the load resistor is rising, the capacitor charges up. As the p.d. across the load resistor starts to fall, the capacitor maintains the output p.d. by only discharging slowly. When the p.d. from the rectifier rises again, the capacitor will charge up again and the process is repeated.

The output p.d. is not completely smooth – it has 'ripple', as shown in Figure 21.9. The amount of ripple depends on the value of CR (called the **time constant**) – the larger the value of CR, the smoother the output.

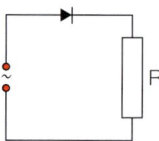

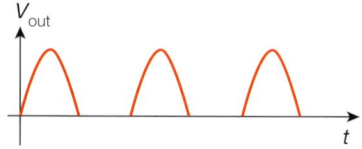

▲ **Figure 21.4** Half-wave rectification

▲ **Figure 21.5** Half-wave rectification output

Key terms

Frequency (f): The number of complete cycles of a.c. per second $\left(f = \dfrac{1}{T}\right)$

Full-wave rectification: Converting all of an a.c. input signal to d.c. (of varying amplitude)

Half-wave rectification: Converting half of an a.c. input signal to d.c. (of varying amplitude)

Period (T): The duration of one full cycle of an a.c. signal $\left(T = \dfrac{1}{f}\right)$

Root-mean-square: The effective 'average' current or voltage of an a.c. supply.

Remember

The value of the time constant CR should be much greater than the time period of the a.c. supply. If the load resistance is quite small the smoothing capacitor must be larger.

↑ Raise your grade

A student investigating electromagnetic induction sets up the experiment shown.

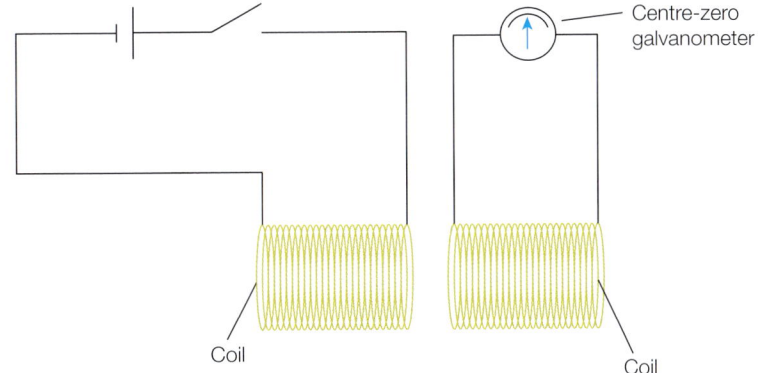

Coil   Coil

Centre-zero galvanometer

When the student closes the switch the galvanometer moves briefly to the right and then returns to zero.

(a) Explain, with reference to Faraday's law of electromagnetic induction, why the galvanometer briefly deflects one way and then returns to zero. [2]

The first coil becomes magnetised. This magnetises the second coil, which creates a current.

✗ ✗

When the switch is pressed the current in the first coil **increases** (from zero) causing an **increasing** magnetic field/flux in the first coil. This flux 'links' with the second coil. From Faraday's law, a **changing flux** in the second coil will induce a p.d. across the coil, so the galvanometer deflects. When the current in the first circuit is constant, there is no changing flux, and so no induced e.m.f. across the second coil.

(b) Describe and explain what will happen when the switch is opened. [2]

The first coil becomes demagnetised. This demagnetises the second coil – it is as if a magnet.

had been pulled out of the second coil. The changing magnetic field induces a voltage   ✔ ✗

The deflection of the galvanometer is in the opposite direction because the magnetic field inside the second coil is **decreasing** rather than increasing.

(c) The student now passes an iron bar between the two coils, as shown and repeats the experiment.

Explain why there is a much larger deflection on the galvanometer. [2]

The iron bar makes the magnetic field stronger. This makes the induced e.m.f. larger.   ✔ ✗

A better answer would be 'The iron increases the maximum magnetic field strength by several thousand times, so the **rate of change** of magnetic flux when the switch is opened or closed is much greater, inducing a much greater p.d. across the second coil'.

(d) The student now replaces the cell with a 'slow' a.c. supply from a signal generator, as shown. She decides to replace the centre-zero galvanometer with a double-beam oscilloscope.

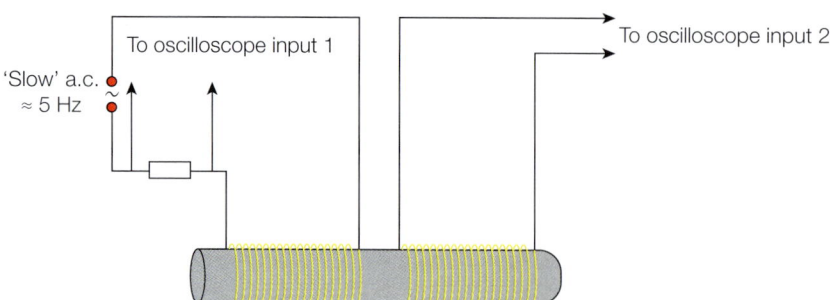

Suggest a reason why monitoring the p.d. across the second coil using an oscilloscope is preferable to using the centre-zero galvanometer. [1]

The voltage is changing direction too fast – the galvanometer cannot keep up. ✔

The needle of the galvanometer has **inertia** – it does not reach its maximum value in one direction by the time the induced p.d. changes direction.

(e) The output from the oscilloscope is shown.

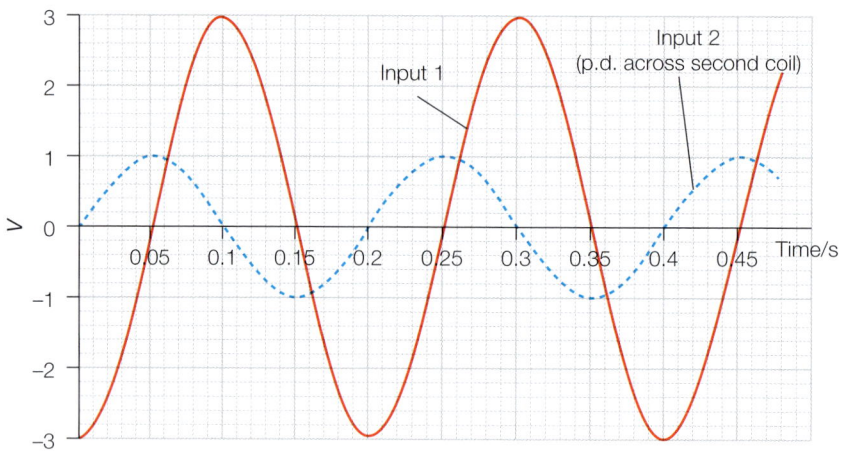

(i) Explain why input 1 is proportional to the current in the first coil. [1]

Input 1 is displaying the p.d. across a resistor, which is proportional to the current ✔

(ii) Explain why the potential difference across the second coil reaches a maximum when the current in the first coil is zero. [2]

When the current in the first coil is zero it is changing rapidly, so the magnetic field in the first coil is momentarily zero, but changing rapidly. This means that the magnetic field (flux) inside the second coil is changing rapidly so, from Faraday's law, there is a large e.m.f. induced across the second coil. ✔✔

A good answer.

Exam-style questions

1 An alternating voltage is displayed on an oscilloscope.

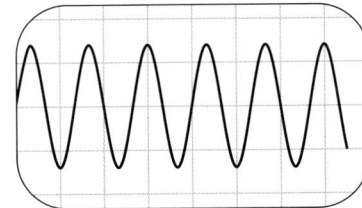

The *y*-gain of the oscilloscope is set to 0.1 V / div. The time-base setting is 100 μs / div.

(a) Determine:

    (i) the maximum voltage V_{max}

    (ii) the r.m.s. voltage $V_{r.m.s.}$

    (iii) the peak-to-peak voltage. [3]

(b) Calculate:

    (i) the period

    (ii) the frequency of the a.c. voltage. [2]

(c) Write an equation which describes the variation of voltage with time. [2]

2 An alternating potential difference has a peak value of 325 V and a frequency of 50 Hz.

(a) Determine the r.m.s. voltage $V_{r.m.s.}$. [1]

(b) The alternating p.d. is connected to a 100 Ω resistor. Calculate:

    (i) the r.m.s. current in the resistor

    (ii) the average current

    (iii) the average power dissipated in the resistor. [3]

(c) Sketch a graph of the power dissipated in the resistor against time for two complete cycles of the alternating voltage. [3]

3 An electric kettle operates from a 230 V mains supply. The kettle has a heating element of resistance 20 Ω.

Calculate:

(a) the r.m.s. current through the element [1]

(b) the power delivered by the heating element [1]

(c) the d.c. voltage needed to give an equivalent heating effect in the heating element. [1]

4 The diagram shows the electrical circuit of a voltage adaptor used for charging a laptop computer. The diagram has been divided into three sections.

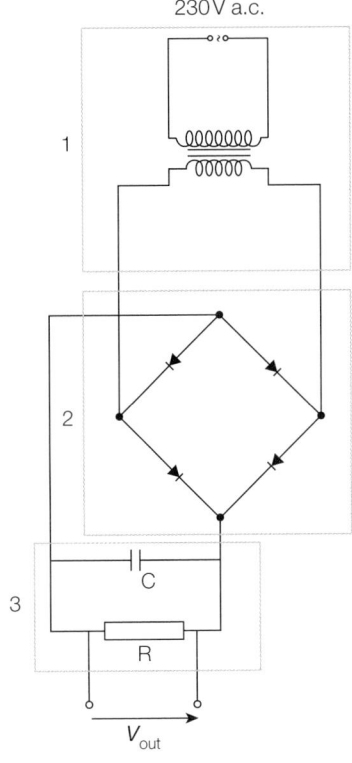

(a) Describe the purpose of section 1 of the circuit. [2]

(b) Section 2 of the circuit is a *bridge rectifier*.

    (i) State one advantage of using a bridge rectifier rather than a single diode

    (ii) Explain how the bridge rectifier works. [3]

(c) Section 3 of the circuit is responsible for smoothing the voltage output.

    (i) Describe what is meant by *smoothing*.

    (ii) Explain how the capacitor achieves smoothing. [3]

(d) The output p.d. will have some ripple.

    (i) Describe what is meant by *ripple*.

    (ii) Explain how the amount of ripple can be reduced. [3]

Photoelectric effect

Electromagnetic radiation incident on a metal can cause electrons (called **photoelectrons**) to be emitted from the surface. This is known as the **photoelectric effect**. For electrons to be emitted from the metal, the frequency of the electromagnetic radiation must be above a certain frequency, known as the **threshold frequency** f_0. The effect can be demonstrated using a gold-leaf electroscope as shown in Figure 22.1.

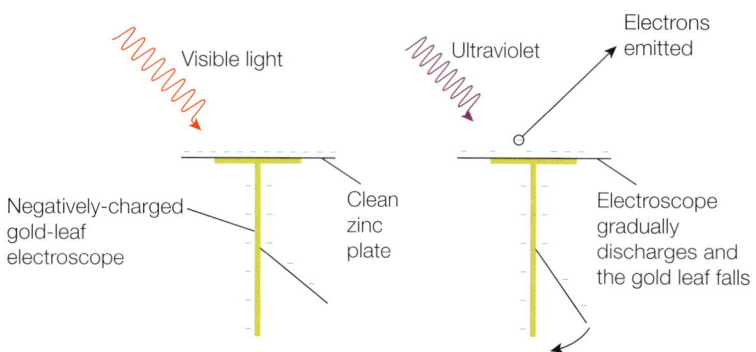

a Visible light incident on a zinc plate produces no photoelectrons, even at high intensity. The negatively-charged gold-leaf electroscope remains charged and the gold leaf does not fall.

b Ultraviolet light causes electrons to be emitted (called photoelectrons). The negatively-charged electroscope discharges and the gold leaf falls.

▲ **Figure 22.1** Photoelectric effect demonstrated using a gold-leaf electroscope

Key points about the photoelectric effect:

- Below the threshold frequency f_0, no electrons are emitted; this frequency depends on the metal being used.

- The greater the frequency of the radiation, the greater the *maximum* kinetic energy of the photoelectrons.

- A more intense (brighter) source of radiation of the same frequency produces *more* photoelectrons but does not change the maximum energy of the photoelectrons.

- Photoelectric emission occurs *instantaneously*, regardless of the intensity of the source of radiation provided that the frequency is above the threshold frequency for the metal.

These observations cannot be explained using the wave model of light. With the wave model, eventually enough energy would arrive to be able to free an electron from the surface, regardless of the frequency of the radiation or the type of metal being used.

To explain the photoelectric effect, Einstein proposed that light could be thought of as a stream of particles called **photons**, each with a packet or **quantum** of energy hf, where f is the frequency of the radiation and h is the Planck constant ($6.63 \times 10^{-34}\,\text{J s}$).

> 💡 **Remember**
>
> Energy of a photon $E = hf$
>
> The value of the Planck constant h is provided in Exam Papers 1, 2, and 4.

An incident photon delivers an amount of energy hf to an electron on the surface of the metal. The electron needs a minimum amount of energy ϕ (called the **work function**, which is different for different metals) to escape from the surface of the metal; the remaining energy appears as kinetic energy of the photoelectron.

The **maximum** kinetic energy of the photoelectrons (see Figure 22.2) is found from the equation:

$$hf = \phi + \frac{1}{2mv_{max}^2} \quad \text{or} \quad E_{k(max)} = hf - \phi$$

Some of the photoelectrons will have less than the maximum kinetic energy because they need more than the minimum energy ϕ to escape from the surface of the metal.

The threshold frequency f_0 which allows electrons to just escape from the surface without any additional kinetic energy is given by:

$$hf_0 - \phi = 0$$
$$f_0 = \frac{\phi}{h}$$

The wavelength corresponding to the threshold frequency f_0 is called the **threshold wavelength** $\lambda_0 = c/\lambda_0$ where c is the speed of light. *Above* the maximum wavelength λ_0 no photoelectrons will be emitted.

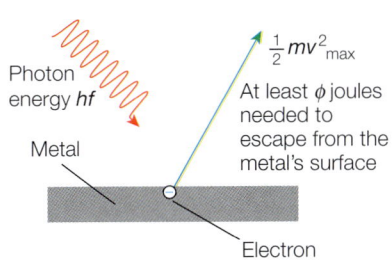

▲ Figure 22.2 $E_{k(max)} = hf - \phi$

Worked Example

The work function ϕ for aluminium is 6.85×10^{-19} J. Plank's constant is 6.63×10^{-34} J s.

Calculate the threshold frequency.

Answer

$$f_0 = \frac{\phi}{h} = \frac{6.85 \times 10^{-19}}{6.63 \times 10^{-34}} = 1.03 \times 10^{15} \text{Hz}$$

Light – particle or wave?

The photoelectric effect can only be satisfactorily explained by treating light as a stream of photons, delivering energy in 'lumps'; the diffraction and interference of light can only be explained by treating light as waves. Light exhibits either wave-like or particle-like behaviour, according to circumstances – it has a dual nature. This is described as **wave–particle duality**.

Matter waves

Light has particle-like properties, and matter has wave-like properties. Evidence for this includes a beam of electrons passing through a thin crystal of graphite – the electrons appear to 'diffract', producing a pattern of rings (see Figure 22.3).

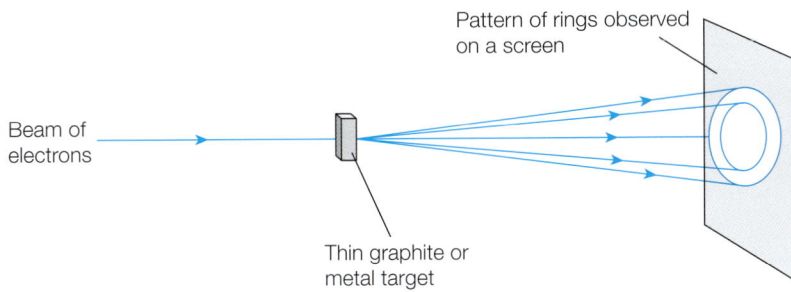

▲ Figure 22.3 Diffraction of electrons

Key terms

Photoelectric effect: The ejections of electrons from a charged metal surface due to absorption of light.

Photons: Particles of light which carry energy in proportion to their frequency.

Quantum: A fixed quantity, such as the energy of a photon.

Threshold frequency: The minimum frequency of light which causes electrons to be ejected from a charged metal surface.

Threshold wavelength: The minimum wavelength of light which causes electrons to be ejected from a charged metal surface.

Wave-particle duality: The wavelike and particle-like behaviour of particles such as photons and electrons.

Work function: The minimum about of energy required to eject a photon from a metal surface by the photoelectric effect.

The wavelength associated with the electrons is found from **de Broglie's equation**:

$$p = mv = \frac{h}{\lambda}$$

where p (or mv) is the momentum of the electron and h is Planck's constant.

When electrons are detected, they are detected as particles, with mass and velocity. The de Broglie wavelength associated with the electron enables the location of the electrons to be calculated. When the p.d. accelerating the electrons in the diffraction experiment is increased, the momentum of the electrons increases, and so the wavelength associated with the electrons decreases. The diameters of the rings observed when electrons are diffracted are seen to decrease.

Worked Example

Calculate the de Broglie wavelength for:

(a) an electron ($m = 9.11 \times 10^{-31}\,\text{kg}$)

(b) an alpha particle ($m = 6.64 \times 10^{-27}\,\text{kg}$)

moving at a speed of $1.5 \times 10^6\,\text{m s}^{-1}$.

Answer

From de Broglie's equation

$$\lambda = \frac{h}{mv} = \frac{6.6 \times 10^{-34}}{m \times 1.5 \times 10^6} = \frac{4.4 \times 10^{-40}}{m}$$

(a) $\lambda = \dfrac{4.4 \times 10^{-40}}{m} = \dfrac{4.4 \times 10^{-40}}{9.11 \times 10^{-31}} = 4.8 \times 10^{-10}\,\text{m}$

(b) $\lambda = \dfrac{4.4 \times 10^{-40}}{m} = \dfrac{4.4 \times 10^{-40}}{6.64 \times 10^{-27}} = 6.6 \times 10^{-14}\,\text{m}$

Electron energy levels

An electron in an isolated atom (e.g., an atom of hydrogen or helium gas) can only have certain specific amounts of energy; the energy of the electron is **quantised**. This can be represented on an energy level diagram – Figure 22.4 shows the different energy levels for the electron in a hydrogen atom. By convention, an electron which is completely free of the atom has zero energy. An electron inside the atom requires energy to 'escape' so it has less than zero energy (it has negative energy).

An electron with the lowest possible energy is in its **ground state** ($n = 1$). If an electron absorbs energy it can jump (*transition*) to a higher energy level – it is then in an **excited state**, as shown in Figure 22.5. After a short time, an electron in a higher energy state (E_1) will fall back down to a lower energy level (E_2). In doing so it emits an amount of energy $E_1 - E_2$ as a photon of frequency f, where:

$$hf = E_1 - E_2$$

Energy levels of electrons in atoms are usually measured in electronvolts (eV).

1 electronvolt is the energy gained by an electron in moving across a p.d. of 1 V. $1\,\text{eV} = 1.6 \times 10^{-19}\,\text{J}$

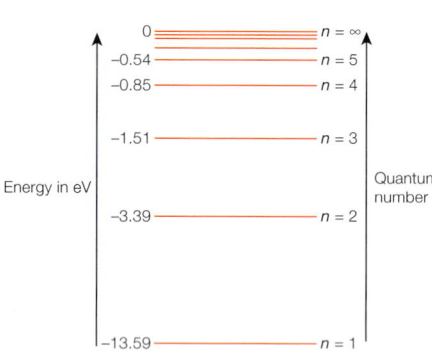

▲ **Figure 22.4** Electron energy levels in a hydrogen atom

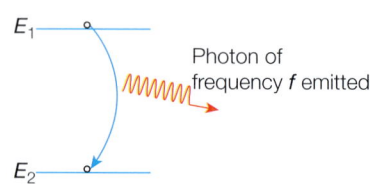

▲ **Figure 22.5** $hf = E_1 - E_2$

Worked Example

Determine the frequency of light emitted when the electron in a hydrogen atom falls from the $n = 4$ state to the $n = 2$ state.

Answer

Using Figure 22.4: $hf = E_1 - E_2 = [(-0.85) - (-3.39)] \times 1.6 \times 10^{-19}$

$$f = \frac{4.06 \times 10^{-19}}{6.63 \times 10^{-34}} = 6.12 \times 10^{14}\,\text{Hz}$$

★ **Exam tip**

Remember to convert energies in electronvolts to joules in any calculations.

If an electron absorbs enough energy to reach the zero energy level ($n = \infty$) it will have escaped the atom (the atom has been **ionised**).

Emission spectra and absorption spectra

The electron in an atom of hydrogen can make a number of different transitions from a higher energy level to a lower energy level. Each transition corresponds to a specific wavelength of electromagnetic radiation being emitted, as illustrated in Figure 22.6.

If the light emitted from excited hydrogen gas atoms passes through a diffraction grating or a prism, a series of bright lines is observed, called an **emission line spectrum** (Figure 22.7).

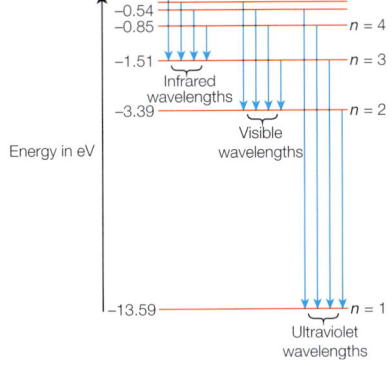

▲ **Figure 22.6** Electron transitions in hydrogen

Violet   Blue/Green

▲ **Figure 22.7** Emission line spectrum of hydrogen

If white light passes through hydrogen gas, for example, some frequencies of the white light will be absorbed by the gas. The frequencies are the same as those observed in an emission spectrum, but in this case electrons are absorbing energy in moving from a lower energy level to a higher energy level. The spectrum observed is called an **absorption spectrum** and the frequencies corresponding to the electron transitions appear as dark lines in an otherwise continuous spectrum (Figure 22.8).

410 434   486   656
Wavelength λ / nm

▲ **Figure 22.8** Absorption line spectrum of hydrogen

When photons interact with particles such as electrons, they transfer momentum. This means that the photons themselves must have momentum despite being massless. This means that the non-relativistic equation $p = mv$ cannot apply.

The momentum carried by a photon is related to its energy by the relationship: $p = \dfrac{hf}{c} = \dfrac{E}{c}$

Key terms

Absorption spectrum: The spectral pattern produced when white light passes through a cool gas.

Emission line spectrum: The spectral pattern produced by an excited gas.

Excited state: An electron is a higher energy level than its ground state.

Ground state: The lowest possible energy level for an electron.

Ionised: An atom which has lost and electron.

Quantised: Found in fixed amounts.

↑ Raise your grade

A student is investigating the photoelectric effect using a photocell.

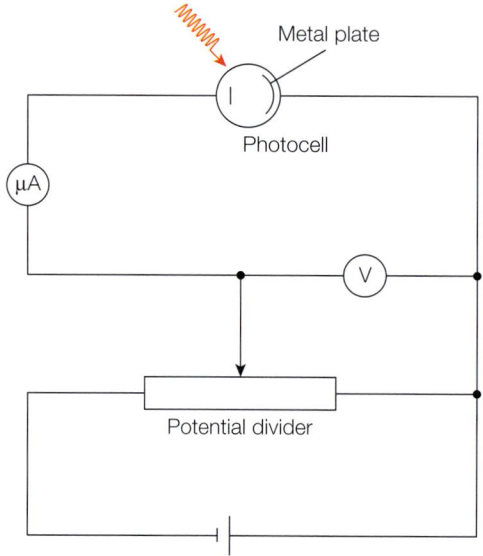

Electromagnetic radiation is incident on a metal plate inside an evacuated tube. Photoelectrons are emitted and a small current flows. A potential divider circuit provides a potential difference opposing this current. As the p.d. is increased, the current decreases and eventually falls to zero. The minimum potential difference needed to reduce the current to zero is called the stopping potential.

The student measures the stopping potential for different wavelengths of electromagnetic radiation:

Wavelength $\lambda/10^{-7}$ m	Stopping potential V_s/V	$1/\lambda$ / $\times 10^6$ m^{-1}
5.00	0.6 ± 0.05	2.00
4.29	1.0 ± 0.05	2.33
3.75	1.4 ± 0.05	2.67
3.33	1.8 ± 0.05	3.00
3.00	2.2 ± 0.05	3.33

✔ Column heading correct with quantity and unit separated by /.

✔ Calculations correct and recorded to same number of sig. figs. as the raw data.

(a) Complete the third column in the table by calculating values of $1/\lambda$. [2]

(b) (i) Plot a graph of V_s on the y-axis against $1/\lambda$ on the x-axis.

Include error bars for V_s. [2]

(ii) Draw the straight line of best fit and a worst acceptable straight line on your graph.

Both lines should be clearly labelled. [2]

(iii) Determine the gradient and the y-intercept of the line of best fit.

$$\text{gradient} = \frac{2.100 - 0.725}{(3.25 - 2.1) \times 10^6} = 1.20 \times 10^{-6}\ \text{Vm} \checkmark\checkmark$$

Read-offs correct and substituted into gradient calculation correctly. Hypotenuse of triangle for calculating gradient larger than half the length of the line drawn. Units correct.

Using point $(3.00 \times 10^6, 1.8)$ in $y = mx + c$

$$\text{y-intercept} = c = y - mx$$

$$= 1.8 - (1.20 \times 10^{-6}) \times (3.00 \times 10^6)$$

$$= -1.8\ \text{V} \checkmark\checkmark$$

gradient 1.20×10^{-6} Vm [2]

Read-off correct and substituted into $y = mx + c$ correctly.

y-intercept -1.8 V [2]

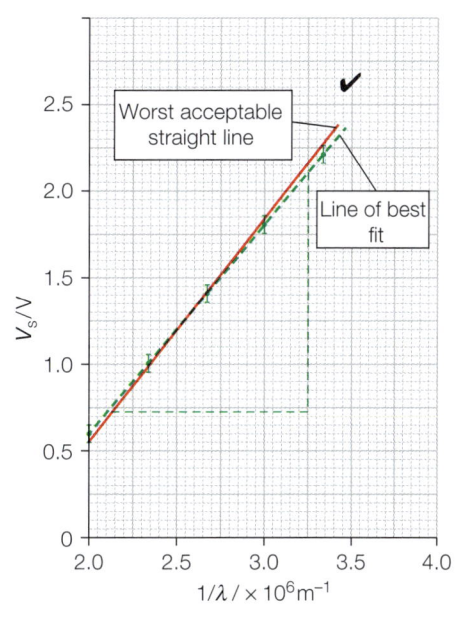

✔ Points plotted correctly.

✔

✔ Error bars drawn correctly.

The 'worst acceptable' straight line is the steepest (or the least steep) straight line which still passes through all the error bars.

(c) Theory predicts that:

$$V_s = \frac{hc}{e\lambda} - \frac{\phi}{c}$$

where e is the charge on the electron (1.6×10^{-19} C) and c is the speed of light ($3.0 \times 10^8\ \text{m s}^{-1}$).

Use your answers to (b)(iii) to determine h and ϕ.

$$\text{gradient} = \frac{hc}{e} = 1.20 \times 10^{-6} \checkmark$$

$$h = 1.20 \times 10^{-6} \times \frac{e}{c}$$

$$= 1.20 \times 10^{-6} \times \frac{1.6 \times 10^{-19}}{3 \times 10^8}$$

$$h = 6.4 \times 10^{-34} \checkmark$$

Gradient $= \dfrac{hc}{e}$.

Correct calculation of h.

$h = 6.4 \times 10^{-34}$ Js [2]

$$\text{y-intercept} = -\frac{\phi}{e} = -1.8\ \text{V} \checkmark$$

y-intercept $= -\dfrac{\phi}{e}$

$$\phi = 1.8 \times 1.6 \times 10^{-19} = 2.88 \times 10^{-19} ✗$$

$\phi = 2.88 \times 10^{-19}$ [2]

Units for ϕ omitted (J or V C).

Exam-style questions

$[c = 3.0 \times 10^8 \, \text{m s}^{-1}; \; e = 1.6 \times 10^{-19} \, \text{C};$
$h = 6.63 \times 10^{-34} \, \text{J s}]$

1 Electromagnetic radiation of wavelength
$4.5 \times 10^{-7} \, \text{m}$ is incident upon a metal surface
which then emits electrons with a maximum
kinetic energy of $4.2 \times 10^{-20} \, \text{J}$. Radiation of
wavelength $3.0 \times 10^{-7} \, \text{m}$ incident on the same
metal produces electrons with a maximum kinetic
energy of $2.63 \times 10^{-19} \, \text{J}$.

Calculate:

(a) the frequencies of the two radiations [3]

(b) the value of Planck's constant. [2]

2 Electromagnetic radiation of frequency $2.5 \times 10^{15} \, \text{Hz}$
is incident on a clean magnesium surface. The
work function for magnesium is $3.68 \, \text{eV}$.

(a) Calculate:

  (i) the maximum kinetic energy of the
  photoelectrons emitted

  (ii) the threshold frequency of magnesium. [3]

(b) Determine the *stopping potential* (the p.d.
needed to prevent the photoelectrons from
escaping). [2]

3 Electrons, initially at rest, are accelerated across a
potential difference of $40 \, \text{kV}$.

(a) Calculate:

  (i) the speed of the electrons

  (ii) their momentum. [3]

(b) State what is meant by the *de Broglie
wavelength*. [2]

(c) Calculate the de Broglie wavelength
associated with these electrons. [2]

4 The energy levels for electrons in a helium atom
are given by the equation:

$$E_n = \frac{-54.4}{n^2} \, \text{eV}$$

(a) Explain why electron energy levels are
negative. [1]

(b) Calculate the wavelength of the radiation
emitted by an electron falling from level $n = 5$
to $n = 2$. [3]

(c) State which region of the electromagnetic
spectrum this wavelength belongs to. [1]

5 The diagram shows the energy levels for the
electron in a hydrogen atom.

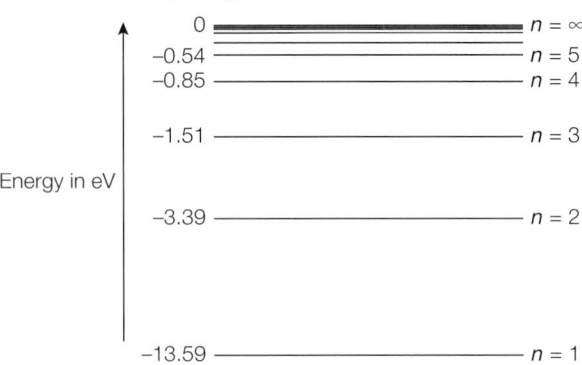

(a) Calculate the longest and shortest possible
wavelengths that could be produced by a
transition from an excited state to the ground
state ($n = 1$). [3]

(b) Determine the number of spectral lines
produced by transitions between the lowest
four states. [1]

6 An X-ray photon has a wavelength of
$2.00 \times 10^{-11} \, \text{m}$.

Calculate:

(a) the energy of the photon, in J [1]

(b) the energy of the photon, in eV [2]

(c) the momentum transferred by this photon. [2]

(d) Explain why X-rays photons can cause
ionisation in living tissue while radio wave
photons (wavelength ~10 cm) do not. [3]

7 Light can be described as showing both wave-like
and particle-like behaviours.

(a) Describe a situation where light acts in a
wave-like manner. [2]

(b) Describe a situation where light behaves
in a particle-like manner. [2]

Mass defect, mass excess, and binding energy

The mass of an atomic nucleus is slightly less than the total mass of the separate nucleons. For example, a helium-4 nucleus consists of two protons and two neutrons. Table 23.1 gives the masses of a proton, a neutron, and a helium-4 nucleus in u (atomic mass units).

> **Remember**
>
> The **unified atomic mass unit** (u) is defined as $\frac{1}{12}$ of the mass of a ^{12}C atom (including its electrons).
>
> $$1\,u = 1.661 \times 10^{-27}\,kg$$

▼ **Table 23.1** Masses of proton, neutron, and a helium-4 nucleus

	Mass / u
Proton	1.00728
Neutron	1.00867
Helium-4 nucleus	4.00150

The mass of a helium-4 nucleus (Figure 23.1) is less than the combined mass of 2 protons and 2 neutrons. The difference is called the **mass defect** Δm.

$$\Delta m = 2 \times 1.00728 + 2 \times 1.00867 - 4.00150 = 0.0304\,u$$
$$= 5.046 \times 10^{-29}\,kg$$

Einstein showed that mass m and energy E are related by the equation:

$$E = mc^2$$

where c is the speed of light ($3.0 \times 10^8\,m\,s^{-1}$). For a mass change Δm, the corresponding change in energy is:

$$\Delta E = c^2(\Delta m)$$

For a helium nucleus, $\Delta E = (3.0 \times 10^8)^2 \times 5.046 \times 10^{-29} = 4.54 \times 10^{-12}\,J$

This is the energy that is needed to completely separate a helium nucleus into individual protons and neutrons, and is called the **binding energy**. The value of the binding energy expressed in joule is very small, and so binding energies are usually expressed in electronvolts (eV).

Binding energy of a helium-4 nucleus in eV $= \dfrac{4.54 \times 10^{-12}}{1.6 \times 10^{-19}} = 2.84 \times 10^7\,eV$

The binding energy per nucleon $= \dfrac{2.84 \times 10^7}{4} = 7.1\,MeV$

Figure 23.2 shows how the binding energy/nucleon varies with nucleon number.

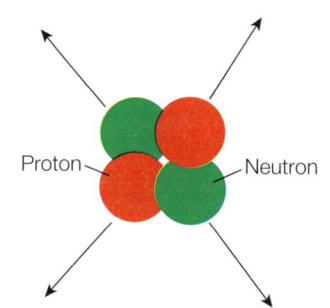

▲ **Figure 23.1** Binding energy of helium-4

> **Remember**
>
> $$\Delta E = c^2(\Delta m)$$

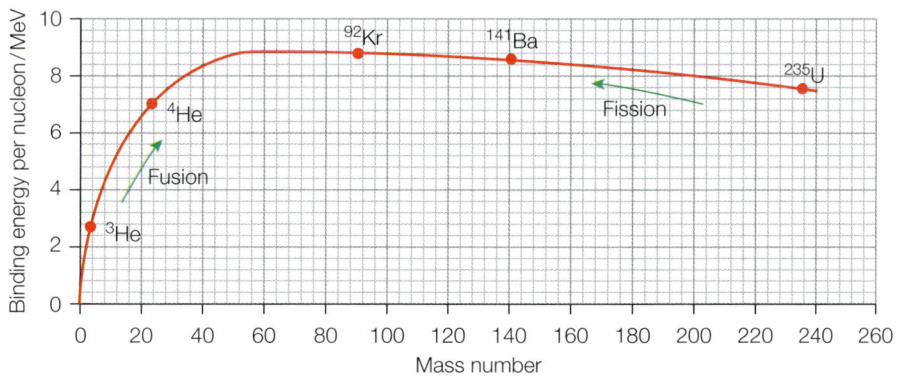

▲ **Figure 23.2** Binding energy per nucleon

> The greater the binding energy per nucleon, the more stable the nucleus because more energy is needed to remove a nucleon from the nucleus.

Nuclear fission

In **nuclear fission** a heavy nucleus splits into two lighter nuclei, together with a few individual neutrons. A large amount of energy is released, principally as kinetic energy of the fission fragments.

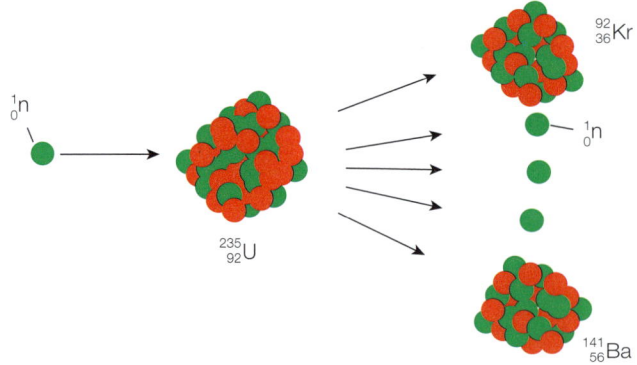

▲ **Figure 23.3** Nuclear fission

Figure 23.3 illustrates one example of nuclear fission. A ^{235}U nucleus absorbs a slow-moving ('thermal') neutron, momentarily becoming ^{236}U which is unstable, and splits into a nucleus of ^{92}Kr and ^{141}Ba, together with three neutrons. The binding energy/nucleon for ^{92}Kr and ^{141}Ba is greater than the binding energy per nucleon for ^{235}U. Each nucleon is now more tightly bound to the nucleus, requiring more energy to 'escape' from the nucleus than before the fission reaction, and so these nucleons must have lost energy (only the three 'free' neutrons have gained energy). This energy appears mainly as kinetic energy of the fission fragments (they are 'hot'!).

Energy released in nuclear fission

> ### Worked Example
>
> Calculate the energy released by one fission reaction in the example given in Figure 23.3, using the BE per nucleon values given in Table 23.2.
>
> ▼ **Table 23.2**
>
> | | BE per nucleon/ MeV |
> |---|---|
> | ^{235}U | 7.6 |
> | ^{141}Ba | 8.3 |
> | ^{92}K | 8.5 |
>
> **Answer**
>
> Energy released per fission reaction:
>
> $\Delta E = 141 \times (8.3 - 7.6) + 92 \times (8.5 - 7.6) - 2 \times 7.6 = 166\,\text{MeV}$

Mass excess

The mass of a nuclide in atomic mass units is very close to the nucleon number A. For example, the mass of U-238 is slightly more than 238 u (238.050788 u). The difference between the two is known as the **mass excess**.

$$\text{mass excess} = \text{mass (in u)} - \text{nucleon number}$$

Nuclear fusion

Nuclear fusion occurs when two light nuclei join together to form a larger, more stable nucleus. In one of the fusion reactions taking place in the Sun, two ^{3}He nuclei combine to form ^{4}He as shown in Figure 23.4.

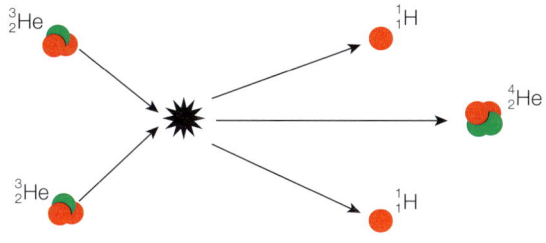

$$^3\text{He} + {}^3\text{He} \rightarrow {}^4\text{He} + 2{}^1\text{H} + \text{energy}$$

▲ **Figure 23.4** Nuclear fusion

Energy released in nuclear fusion

Four of the nucleons finish in a more stable nucleus so have lost energy; the other two have become single, 'free' protons so have gained energy.

Worked Example

Calculate the energy released per fission reaction in the example given in Figure 23.4, using the binding energy per nucleon values given in Table 23.3.

Answer

Energy released per fission reaction:

$$\Delta E = 4 \times (7.1 - 2.6) - 2 \times 2.6 = 12.8\,\text{MeV}$$

▼ **Table 23.3**

	BE per nucleon/ MeV
^{3}He	2.6
^{4}He	7.1

Radioactive decay

The exact time an unstable nucleus will decay cannot be predicted – radioactive decay is a **random** process – but the rate of decay of a very large number of atoms of any particular radioactive isotope can be calculated accurately. The activity A (the number of decays per second) is given by:

$$A = \lambda N$$

where N is the number of undecayed nuclei and λ is a constant called the **decay constant**. λ is the probability of a nucleus decaying in unit time (the fraction of atoms that decay in unit time) and has a different value for different radioactive isotopes.

The **activity** A is the number of nuclei decaying per second (the count rate) $\dfrac{dN}{dt}$, so:

$$\frac{dN}{dt} = -\lambda N$$

The minus sign occurs because the number of undecayed nuclei is decreasing with time. The solution to this equation is:

$$N = N_0 e^{-\lambda t}$$

where N_0 is the number of undecayed atoms at time $t = 0$ (see Figure 23.5).

> 💡 **Remember**
>
> The activity is measured in **bequerels** (Bq).
>
> 1 Bq = 1 disintegration per second

> **Key term**
>
> **Decay constant:** The probability of a nucleus decaying per unit time.

As the rate of decay is proportional to N:

$$A = A_0 e^{-\lambda t}$$

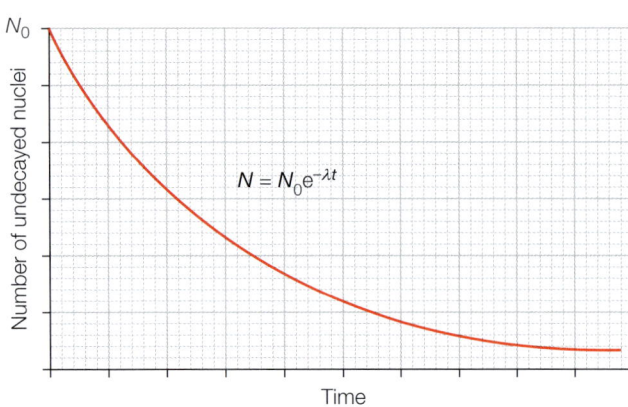

▲ **Figure 23.5** Exponential decay of the number of undecayed nuclei

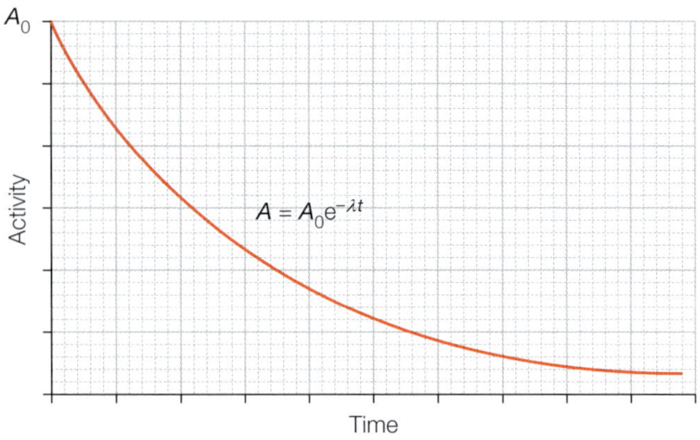

▲ **Figure 23.6** Exponential decay of activity

The activity A (the number of disintegrations per second) also decreases **exponentially** (Figure 23.6).

The count rate (C) on a detector such as a Geiger–Muller tube is proportional to the number of decays which gives a third exponential equation and a similar graph:

$$C = C_0 e^{-\lambda t}$$

A useful measure of the rate of decay is the **half-life** $t_{1/2}$ – the time it takes for the number of undecayed nuclei to fall by half (and for the activity to halve).

From $N = N_0 e^{-\lambda t}$:

$$\frac{N_0}{2} = N_0 e^{-\lambda t_{1/2}} \text{ so,} \qquad e^{\lambda t_{1/2}} = 2$$

$$\lambda t_{1/2} = \ln 2 = 0.693$$

$$t_{1/2} = \frac{0.693}{\lambda}$$

In Figure 23.7 the activity falls from 12 kBq to 6 kBq in 10 s. It then falls from 6 kBq to 3 kBq in the next 10 s, from 3 kBq to 1500 Bq in the next 10 s, and so on.

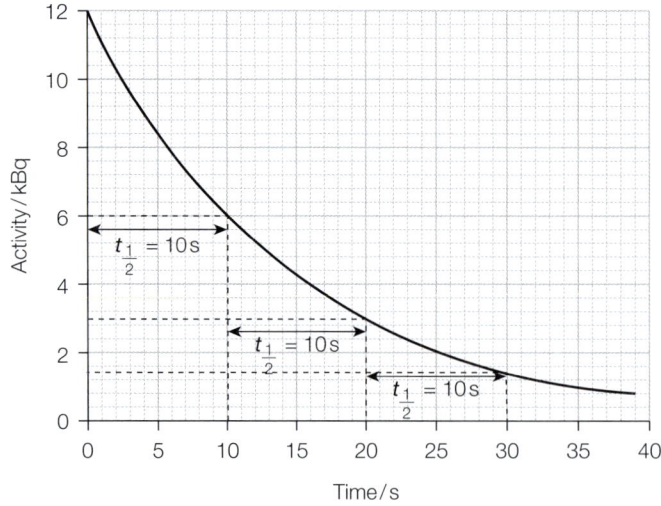

▲ **Figure 23.7** Half-life

Key terms

Half-life: The half-life $t_{1/2}$ is the average time it takes for half the radioactive nuclei to decay (and the time it takes for the activity to halve).

Mass excess: The difference between the mass of a nuclide in atomic mass unit and the nucleon number.

Nuclear fusion: This occurs when two light nuclei join together to form a larger, more stable nucleus.

Half-lives of radioactive isotopes vary from fractions of a second to many millions of years.

Worked Examples

1 An isotope of radon has a half-life of 3.83 days. If a sample of the gas has a mass of 12.0 g at time $t = 0$, determine how much of the sample remains after:

   (a) 1 day   **(b)** 1 week   **(c)** 1 year.

Answer

The decay constant $\lambda = \dfrac{0.693}{t_{1/2}} = \dfrac{0.693}{3.83} = 0.181$ days^{-1}

   (a) Fraction remaining after 1 day $= \dfrac{N}{N_0} = e^{-\lambda t} = e^{-0.181 \times 1} = 0.834$

   Mass of radon gas remaining $= 12.0 \times 0.834 = 10.0$ g

   (b) Fraction remaining after 1 week $= e^{-\lambda t} = e^{-0.181 \times 7} = 0.282$

   Mass of radon gas remaining $= 12.0 \times 0.282 = 3.4$ g

   (c) Fraction remaining after 1 year $= e^{-\lambda t} = e^{-0.181 \times 365} = 2.03 \times 10^{-29}$

   Mass of radon gas remaining $= 12.0 \times 2.03 \times 10^{-29} = 2.44 \times 10^{-28}$ g.

2 Uranium-235 has a half-life of 7.1×10^8 years. Calculate the activity of a sample of 1.0 g of U-235.

Answer

1.0 g of U-235 contains $1.0 \times \dfrac{6.02 \times 10^{23}}{235} = 2.56 \times 10^{21}$ atoms of uranium.

The decay constant $= \lambda = \dfrac{0.693}{7.1 \times 10^8} = 9.76 \times 10^{-10}$ years^{-1}

The activity $A = 2.56 \times 10^{21} \times 9.76 \times 10^{-10} = 2.50 \times 10^{12}$ disintegrations/year $= 8.1 \times 10^4$ Bq

↑ Raise your grade

The graph below shows how the activity of oxygen-14 decays over time.

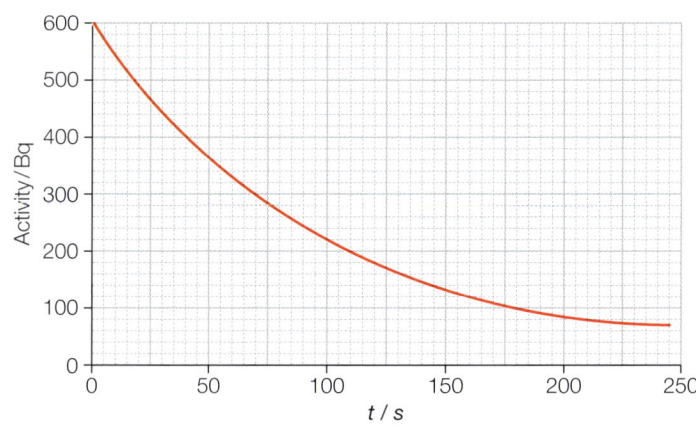

(a) Use the graph to determine the half-life of the sample.   [2]

70 ✔✔   Although the line is a bit rough the half-life is found.

(b) Determine the decay constant for oxygen-14.

$\lambda = \log 2 / t_{1/2} = 0.043$ ✗✗   The student should be using natural logarithms (ln 2), not base 10. The unit s^{-1} should also be given. The correct answer is 9.9×10^{-3} s^{-1}.

Exam-style questions

1 An isotope of uranium $\left(^{235}_{92}\right)$ undergoes an induced nuclear fusion to produce two smaller nuclei and some neutrons. Which of the following is possible? [1]

A $^{235}_{92}U + ^{1}_{0}n \rightarrow ^{141}_{56}Ba + ^{92}_{36}Kr + 1^{1}_{0}n$

B $^{235}_{92}U + ^{1}_{0}n \rightarrow ^{141}_{56}Ba + ^{92}_{36}Kr + 2^{1}_{0}n$

C $^{235}_{92}U + ^{1}_{0}n \rightarrow ^{141}_{56}Ba + ^{92}_{36}Kr + 3^{1}_{0}n$

D $^{235}_{92}U + ^{1}_{0}n \rightarrow ^{141}_{56}Ba + ^{92}_{36}Kr + 4^{1}_{0}n$

2 Fig. 1 shows the binding energy per nucleon for a range of isotopes.

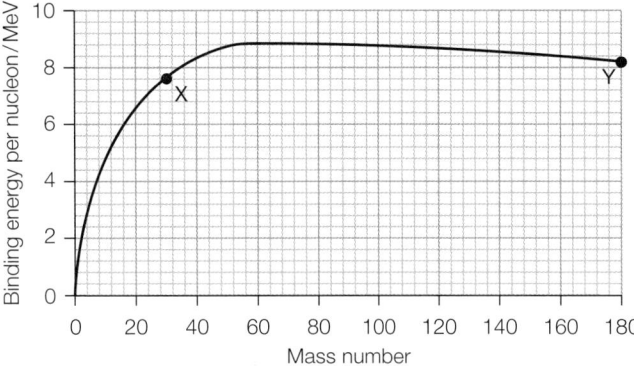

▲ Fig. 1

Which row in the table correctly describes the nuclear processes which would release energy for these two isotopes? [1]

Row	X	Y
A	Fission	Fusion
B	Fission	Fission
C	Fusion	Fusion
D	Fusion	Fission

3 Fig. 2 shows the number of atoms remaining over time for three different isotope samples.

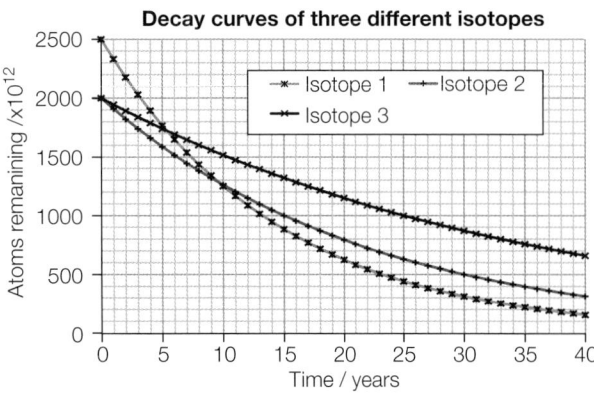

▲ Fig 2

(a) Which of the isotopes has the longest half-life? [1]

(b) Which of the samples has the greatest initial activity? [1]

(c) Determine the decay constant for Isotope 2. [2]

(d) Determine the activity of the sample of Isotope 2 after 30 years. [3]

4 The half-life of an isotope of radium is 1620 years.

(a) Explain what is meant by *isotope*. [2]

(b) Define radioactive *half-life*. [2]

(c) Show that the decay constant is 4.28×10^{-4} per year. [2]

(d) A sample of this isotope has an activity of 6.2×10^9 Bq. Calculate the number of radium atoms in the sample. [2]

5 A thermal neutron collides with a uranium nucleus $^{235}_{92}U$, which undergoes nuclear fission. The fission products are $^{93}_{38}Sr$ and $^{140}_{54}Xe$, together with two more neutrons:

$$^{235}_{92}U + ^{1}_{0}n \rightarrow ^{93}_{38}Sr + ^{140}_{54}Xe + 3^{1}_{0}n + \text{energy}$$

(a) Explain what is meant by a *thermal neutron*. [1]

(b) State what is meant by *nuclear fission*. [2]

(c)

	Mass/u
$^{235}_{92}U$	235.04393
$^{93}_{38}Sr$	92.91399
$^{140}_{54}Xe$	139.92162

Using the masses given in the table, calculate:

(i) the mass defect, in atomic mass units

(ii) the energy released in the reaction, in MeV. [5]

6 An isotope of rubidium, $^{87}_{37}Ru$, decays by β^--decay into $^{87}_{38}Sr$, which is a stable isotope. The half-life of $^{87}_{37}Ru$ is 4.9×10^{10} years.

The ratio of ^{87}Sr to ^{87}Ru in a sample of rock is found to be 0.0060.

Assuming that there was no ^{87}Sr when the rock was formed, calculate the age of the rock. [4]

Ultrasound

Ultrasonic waves are sound waves with frequencies above human hearing (typically 20 kHz). Unlike high-frequency electromagnetic waves such as X-rays, they are non-ionising, and so do not damage living tissue and are ideal for use in medical imaging. The frequencies used are usually between 1 MHz and 10 MHz (frequencies lower than 1 MHz are diffracted too much and frequencies higher than 10 MHz are absorbed too much by body tissue).

Ultrasonic waves are produced using a **piezo-electric transducer** in the shape of a disc. An alternating voltage applied between the faces of the disc causes it to vibrate. If the frequency chosen coincides with the natural frequency of the disc, resonance occurs, and the disc emits ultrasonic waves at the resonant frequency (see Figure 24.1).

In medical imaging, an ultrasound probe emits pulses of ultrasound into a body. The ultrasonic waves are partially reflected each time the ultrasonic waves pass from one material (medium) to another. The reflected waves cause the disc of the probe to vibrate, generating a small p.d. across the disc. The probe thus acts as both transmitter and receiver of the ultrasonic waves, the reflected pulses enabling a 'sound picture' to be constructed.

Transmission and reflection of ultrasound waves

When ultrasonic waves reach a boundary between two different materials, some of the wave energy is reflected and the rest is transmitted (and refracted). The proportion of energy that is reflected is determined by the **acoustic impedance** Z of each of the two materials, where:

$$Z = \rho c$$

ρ is the density of the material and c is the speed of sound in the material. Some typical values are given in Table 14.3.

The **intensity reflection coefficient** is the fraction of ultrasonic wave energy reflected, and is given by the equation:

$$\frac{I_R}{I_O} = \frac{(Z_1 - Z_2)^2}{(Z_1 + Z_2)^2}$$

where I_O is the intensity of the incident waves, I_R the intensity of the reflected waves, and Z_1 and Z_2 the acoustic impedances of the two materials (see Figure 24.2).

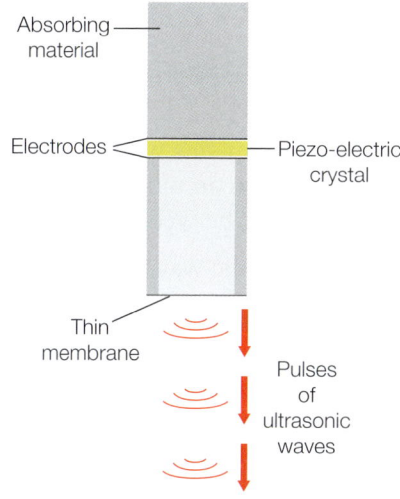

▲ **Figure 24.1** Piezo-electric probe

▼ **Table 24.1** Acoustic impedance

Material	Acoustic impedance $Z / \mathrm{kg\,m^{-2}\,s^{-1}}$
air	430
blood	1.59×10^6
bone	6.80×10^6
muscle	1.70×10^6
soft tissue	1.63×10^6
water	1.50×10^6

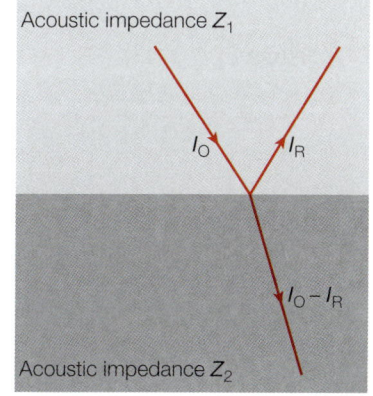

▲ **Figure 24.2** Transmission and reflection

Worked Example

Ultrasonic waves from a piezo-electric probe pass from bone into soft tissue. Calculate the fraction of ultrasonic wave energy that is reflected back towards the probe.

Answer

$$\frac{I_R}{I_O} = \frac{(Z_1 - Z_2)^2}{(Z_1 + Z_2)^2} = \frac{(6.80 - 1.63)^2}{(6.80 + 1.63)^2} = 0.376$$

Approximately a third of the incident wave energy is reflected.

Two key points to remember about the equation for the reflection coefficient:

- if $Z_1 \approx Z_2$ then $\frac{I_R}{I_O} \approx 0$ – almost all the wave energy is transmitted
- If Z_1 is very different from Z_2 (e.g., air and soft tissue) then $\frac{I_R}{I_O} \approx 1$ and almost all the wave energy is reflected.

This explains why a **coupling medium** such as a liquid gel is needed between an ultrasonic probe and soft tissue such as skin (see Figure 24.3). Any air between the probe and the soft tissue would mean virtually all the ultrasonic wave energy is reflected back off the soft tissue as $\frac{I_R}{I_O} \approx 1$. Placing the front of the probe in a gel (which has a similar acoustic impedance to soft tissue) means that almost all the wave energy will be transmitted into the body.

Absorption of ultrasonic waves

When a parallel beam of ultrasonic waves passes through a substance, the intensity of the waves decreases exponentially with distance.

The intensity I of a wave after passing through a distance x of a material is given by:

$$I = I_0 e^{-\mu x}$$

Where I_0 is the incident intensity, and μ the absorption coefficient of the substance (see Figure 24.4).

The absorption coefficient has the unit bel per metre $B\,m^{-1}$.

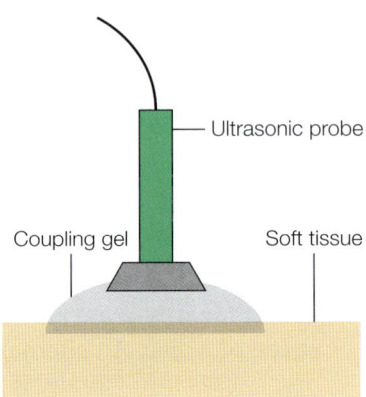

▲ **Figure 24.3** Use of coupling gel to limit reflection from soft tissue

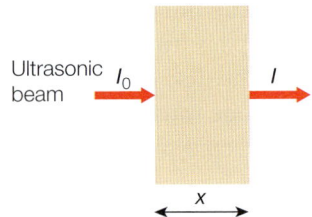

▲ **Figure 24.4** Exponential absorption

Worked Example

The absorption coefficient for $1\,MHz$ ultrasonic waves in fat is $6\,B\,m^{-1}$. What percentage of the energy of the wave will penetrate the fat to a depth of $10\,cm$?

Answer

$$I = I_0 e^{-\mu x}$$

$$\frac{1}{I_0} = e^{-\mu x} = e^{-6 \times 0.1} = 0.55 = 55\%$$

X-rays

X-rays are high-frequency electromagnetic waves, with wavelengths in the range 10^{-8} m to 10^{-11} m. They are produced by bombarding a metal target with high-energy electrons – the rapid deceleration of the electrons causes the emission of X-rays. Figure 24.5 illustrates how an X-ray tube works.

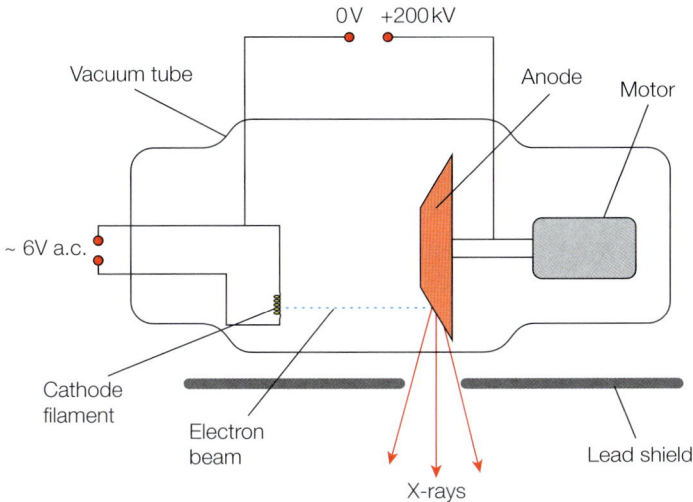

▲ **Figure 24.5** X-ray tube

Electrons are emitted from the hot filament (the **cathode**) by thermionic emission. They are then accelerated towards the **anode** by a p.d. of up to 200 kV. The anode is a small target made of tungsten (or other metal with a high melting point).

Approximately 1% of the kinetic energy of the electrons is converted into X-rays; the rest is converted to heat energy in the metal target. The tungsten target can be rotated rapidly by a motor so that a much greater area of tungsten is heated.

The X-rays are emitted through a thin 'window' surrounded by lead shielding. Metal tubes beyond the window **collimate** the beam (to make it parallel and not spread out like a fan).

The X-rays emitted have a continuous range of frequencies up to a maximum frequency f_{max}, determined by the magnitude of the accelerating potential. If an electron is accelerated across a p.d. V, the energy of the electron is eV, where e is the electronic charge. If all this energy is converted into a single X-ray photon:

$$hf_{max} = eV$$

$$f_{max} = \frac{eV}{h}$$

Worked Example

An X-ray machine has an accelerating potential of 50 kV. Calculate the shortest possible X-ray wavelength that can be emitted. [The speed of light is 3.00×10^8 m s^{-1}, e is 1.6×10^{-19} C, and Plank's constant is 6.63×10^{-34} J s.]

Answer

$$\lambda_{min} = \frac{c}{f_{max}} = \frac{hc}{eV} = \frac{6.63 \times 10^{-34} \times 3.0 \times 10^8}{1.6 \times 10^{-19} \times 50 \times 10^3} = 2.5 \times 10^{-11}\,\text{m}$$

Intensity and 'hardness' of X-rays

The **intensity** of an X-ray beam is a measure of the amount of energy emitted per second per unit area. The intensity can be increased by:

- increasing the accelerating potential

- increasing the number of electrons hitting the metal target. This can be achieved by increasing the current in the filament. The filament gets hotter, releasing more electrons each second.

The **hardness** of X-rays is a measure of their penetrating power. 'Hard' X-rays have higher energies (shorter wavelengths) than 'soft' X-rays, and so are more penetrating. When used to produce an X-ray image of an internal body structure, the soft X-rays are more easily absorbed, increasing the exposure of the patient to hazardous radiation – it is often better to use hard X-rays, using a metal filter to absorb the longer wavelength X-rays.

The hardness of X-rays can be increased by:

- using a filter to absorb the low energy, 'softer' X-rays

- increasing the accelerating potential. This increases the relative amounts of higher frequency (shorter wavelength) X-rays produced, as shown in Figure 24.6.

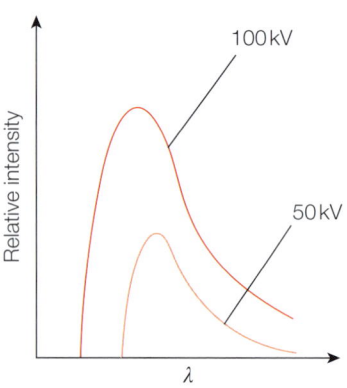

▲ **Figure 24.6** X-ray spectra

Absorption of X-rays

As an X-ray beam passes through matter it is **attenuated** – the intensity of the beam decreases. For a parallel beam with initial intensity I_o, the transmitted intensity I after passing through a thickness x of a material is given by:

$$I = I_o e^{-\mu x}$$

where μ is the attenuation coefficient of the material. (The SI unit of μ is m^{-1} though it is often given in cm^{-1}.)

The attenuation coefficient is proportional to the density of the material.

> 💡 **Remember**
>
> The intensity of an X-ray beam decreases exponentially with thickness of absorber.
>
> $$I = I_o e^{-\mu x}$$
>
> The **half-thickness** of a material is the thickness of the material needed to halve the intensity.

Worked Example

A metal plate of thickness 5.0 mm reduces the intensity of an X-ray beam by 40%. Determine:

(a) the attenuation coefficient   **(b)** the half-thickness of the metal.

> 📐 **Maths Skills**
>
> See Appendix: *Maths skills* for more about exponential functions.

Answer

(a) As the intensity has been reduced by 40%, the intensity must be $0.6I_o$, where I_o is the initial intensity:

$$0.6I_o = I_o e^{-0.005\mu}$$

Rearranging this equation and 'taking natural logs' of both sides:

$$-0.005\mu = \log_e 0.6$$

$$\mu = 100\,m^{-1}\ (1.0\,cm^{-1})$$

(b) When the thickness of the metal is the 'half thickness' $x_{1/2}$, $I = \dfrac{I_0}{2}$

so:

$$\frac{I_0}{2} = I_0 e^{-\mu x_{1/2}}$$

Rearranging this equation:

$$e^{\mu x_{1/2}} = 2, \text{so} \ldots \ldots x_{1/2} = \frac{\log_e 2}{\mu} = 0.69 \, \text{cm}$$

Computed tomography (CT) scanning

Ordinary X-ray images are two-dimensional 'shadow' pictures of a patient. The medical imaging technique called **CT scanning** (also known as CAT scanning – computed axial tomography) is a way of obtaining much more detailed images (see Figure 24.7), including 3D images.

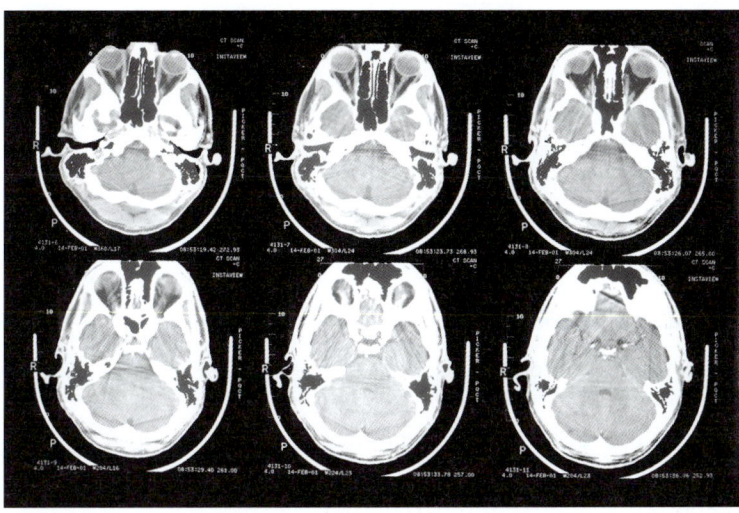

▲ **Figure 24.7** CT images of a brain

Figure 24.8 illustrates the main principles of CT scanning. An X-ray tube mounted on a gantry is able to rotate 360° around a patient. Several hundred X-ray sensors are mounted on a ring. The X-ray tube produces a narrow beam of X-rays which are detected by the sensors opposite the tube and the data sent to a computer. As the X-ray tube rotates, a detailed image is gradually built up and a cross-section of the patient can be displayed on a computer screen. If the patient is gradually moved along the axis of the ring, several 'slices' of the body can be obtained and combined to give a 3D image.

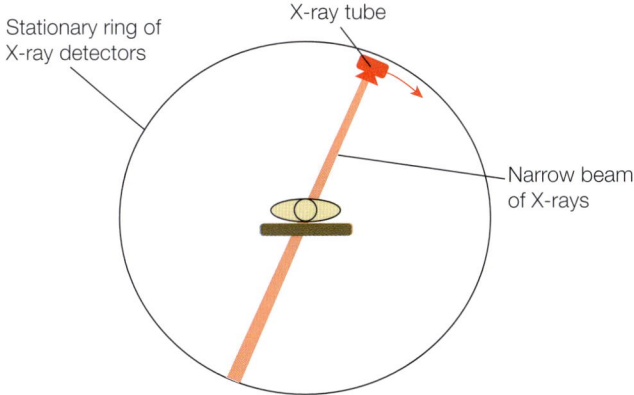

▲ **Figure 24.8** CT scanning

Positron emission tomography (PET)

Radiological tracers

In medical diagnosis, radioactive tracers can be injected to form images of tissues. Instead of X-rays being transmitted through the body and detected on the far side, the tracers produce gamma radiation which can be detected by external gamma cameras. The tracer contains a gamma emitting isotope which is chemically bonded to a molecule which will gather in particular tissue types in the body.

Positron emission tomography (PET) relies on tracers which have isotopes that decay by positron (β^+) emission. These positrons are the antiparticles to electrons and cannot travel very far through the body before encountering an electron from one of the surrounding atoms.

Annihilation

When a positron interacts with an electron, a process called **annihilation** occurs. The positron and electron particles cease to exist and are replaced by a pair of photons (see Figure 24.9).

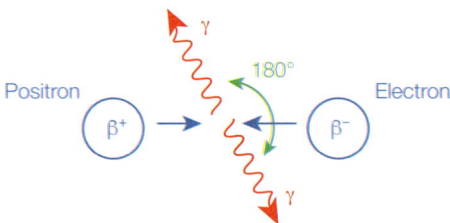

▲ **Figure 24.9** Electron–positron annihilation

During annihilation, conservation laws must be observed: these include conservation of charge, momentum and mass–energy. These rules mean that, when a positron and an electron annihilate, a pair of high-energy gamma photons with equal energy and opposite momentum is always produced. The mass and kinetic energy of the electron and positron are entirely replaced with the energy of the photons and conservation of momentum means that the photons travel in opposite directions.

Worked Example

What is the minimum energy released and the minimum frequency of the pair of photons produced in an electron–positron annihilation event?

Answer

The minimum energy will be when the electron and positron have no kinetic energy.

The mass of an electron (and positron) is 9.11×10^{-31} kg.

The energy can be found using $E = mc^2$.

$$E = 2 \times 9.11 \times 10^{-31}\,\text{kg} \times (3.00 \times 10^8\,\text{m s}^{-1})^2 = 1.64 \times 10^{-13}\,\text{J}$$

The frequency is found from $E = hf$; each photon has half of the total energy.

$$f = \frac{E}{h} = (\tfrac{1}{2} \times 1.64 \times 10^{-13}) / 6.63 \times 10^{-34} = 1.24 \times 10^{20}\,\text{Hz}$$

Detection

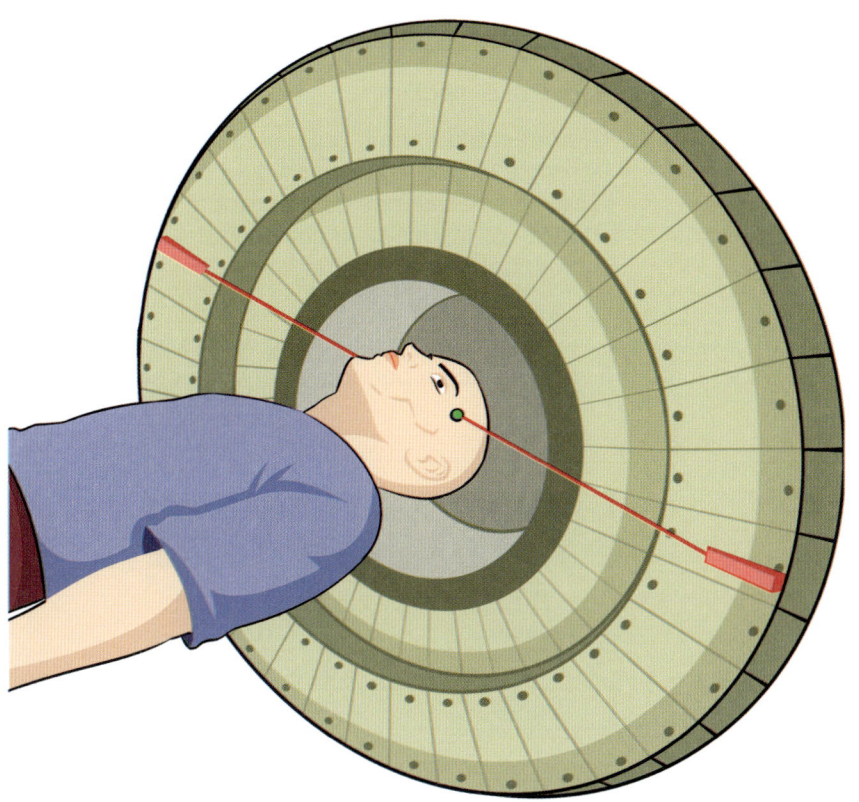

▲ **Figure 24.10** A PET detector ring

The two gamma photons produced by the annihilation event are detected by a ring of detectors surrounding the patient (see Figure 24.10). If the pair of photons are produced in the exact centre of the ring, then they will arrive at detectors which are positioned opposite each other at the same time. If they are produced off-centre, then one of the photons will be detected slightly earlier than the other, as it will travel a shorter distance to the detector. The detection can happen in different detectors (see Figure 24.11). This timing and detector position difference can be used to determine the position of the annihilation event very precisely.

The detectors feed information into a computer system which matches pairs of detection events (rejecting unmatched detections) and determines the positions where the annihilation happened. This data is used to build up a very detailed three-dimensional representation of the biological structures being studied.

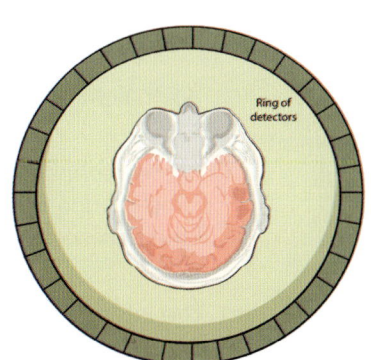

▲ **Figure 24.11** PET timing differences

Uses and hazards

PET scans are commonly used for cancer or brain imagery, where very high resolution and precise imaging is required. Synthetic glucose-like molecules are used as the tracer as this is readily absorbed into highly active tissues such as the brain or tumours. The tracer contains the positron emitting isotope ^{18}F which needs to be prepared in the hospital facility due to its short half-life of 100 minutes.

As positrons annihilate electrons in the molecules of the body, the effective dose of radiation given to the patient in a PET scan is equivalent to over 1000 chest X-rays or two CT scan series.

⬆ Raise your grade

(a) State what is meant by the *specific acoustic impedance* of a medium. [2]

> acoustic impedance = density × speed ✓✗

| A correct statement, but insufficient for 2 marks. A better answer would be 'specific acoustic impedance is the product of the density of the medium and the speed of ultrasound in the medium'.

(b) A vet is using a piezo-electric probe to examine an animal. Pulses of ultrasonic waves pass through the animal and are partially reflected each time the waves reach a boundary between one medium and the next.

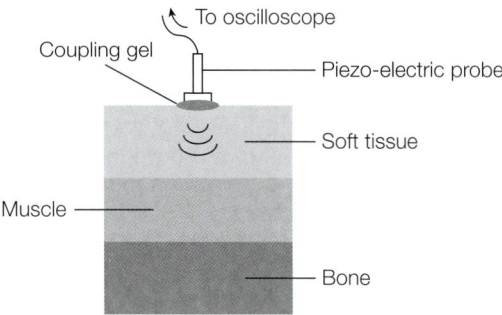

Material	Acoustic impedance Z / $kg\,m^{-2}\,s^{-1}$
air	430
muscle	1.70×10^6
soft tissue	1.63×10^6
bone	6.80×10^6

(i) Explain why the piezo-electric probe emits pulses of ultrasound. [4]

> So that the reflected pulses can be detected between emitting pulses ✓✗

A better answer would include '… which means that the time taken for each pulse to return can be calculated and hence the depth of each reflecting boundary'.

(ii) Explain the purpose of the coupling gel.

> So that most of the ultrasound pulse is transmitted through the soft tissue. ✓✗

A better answer would include '… because the coupling gel has a similar acoustic impedance to soft tissue/if there is air between the probe and the soft tissue most of the ultrasound would be reflected'.

(c) The intensity reflection coefficient at a boundary between two media is given by the equation:

$$\frac{I_R}{I_0} = \frac{(Z_1 - Z_2)^2}{(Z_1 + Z_2)^2}$$

where I_0 is the intensity of the incident ultrasonic waves, I_R the intensity of the reflected waves, and Z_1 and Z_2 the acoustic impedances of the two media.

Calculate the fraction of the ultrasonic wave energy transmitted when the ultrasonic waves reach the boundary between muscle and bone. [2]

> $$\frac{I_R}{I_0} = \frac{(Z_1 - Z_2)^2}{(Z_1 + Z_2)^2} = \frac{(1.63 - 6.80)^2}{(1.63 + 6.80)^2} = 0.376 \ (38\%) \quad ✓✗$$
>
> fraction transmitted = ...38%....

Correct substitution of values for Z_1 and Z_2 into the equation and correct calculation for the first mark, but the value obtained is the fraction reflected. The fraction transmitted is $1 - 0.376 = 0.624$ (62%)

Exam-style questions

1 A pregnant woman is having an ultrasound scan using a piezo-electric probe. A gel is spread over the patient's abdomen and the probe placed onto the gel.

   (a) Define *specific acoustic impedance.* [2]

   (b) Explain the purpose of the gel. [2]

   (c) Suggest a reason why only high frequency ultrasonic waves are used for this. [1]

2 Draw a diagram of a piezo-electric probe. Describe and explain how it produces and detects ultrasonic waves. [5]

3 (a) State the difference between a gamma-ray of wavelength 10^{-11} m and an X-ray of wavelength 10^{-11} m. [1]

   (b) Describe how X-rays can be produced. Include a labelled diagram in your answer. [4]

4 (a) Describe what happens to:

   (i) the hardness   (ii) the intensity

   of X-rays emitted by an X-ray tube if the accelerating potential is increased. [2]

   (b) Explain why longer wavelength X-rays are filtered out when using X-rays for medical imaging. [2]

5 A positron with kinetic energy 4.0×10^{-14} J annihilates a stationary electron.

   (a) What properties are conserved in this process? [3]

   (b) Calculate the frequency of one of the photons produced in this annihilation. [3]

6 (a) Show that the 'half-thickness' $x_{1/2}$ of an absorber of X-rays is related to the attenuation coefficient μ by the equation:

   $$x_{1/2} = \frac{\ln 2}{\mu}$$ [2]

   (b) The attenuation coefficient of bone for one frequency of X-rays is $0.60\,\text{cm}^{-1}$.

   (i) Determine the half thickness of bone for these X-rays.

   (ii) Calculate the thickness of bone needed to absorb 90% of these X-rays. [3]

7 A patient undergoes a medical diagnosis which involves positron emission tomography.

   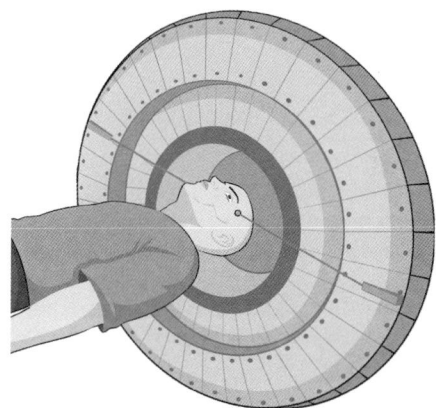

   (a) Describe how the process of positron emission can be used to locate structures in the body. [6]

   (b) Describe the risk and benefits associated with positron emission tomography. [4]

Knowledge check

You should be able to:

- describe inverse square relationships such as $I \propto \frac{I_0}{d_2}$.

The **luminosity** of a star (L) is the total radiant energy output across all wavelengths. This is a high number: for example, the luminosity of the Sun is 3.83×10^{26} W. The Sun is a fairly small, low luminosity star.

The radiation from a star spreads uniformly in all directions, its intensity decreasing with distance in accordance with an inverse square law: the area over which the radiation is spread is proportional to the distance squared (see Figure 25.1). At a distance d from the star, the radiant flux intensity (F) is given by:

$$F = \frac{L}{4\pi d^2}$$

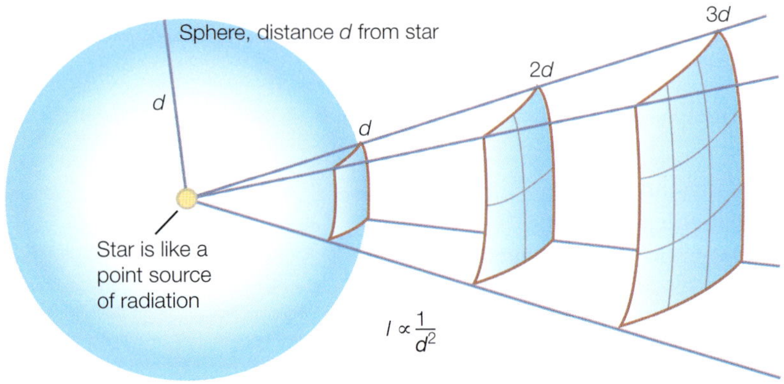

Key terms

Light year: The distance light travels in one year in a vacuum.

Luminosity: The total radiant energy emitted by a star.

Radiant flux intensity: The radiant energy passing through unit area. $F = \frac{L}{2\pi d^2}$

Parallax: The apparent angular movement of an object due to movement of the observer.

Standard candles: Distant objects (e.g. stars) with known intensity.

▲ **Figure 25.1** The inverse square law applied to radiant flux density and luminosity

Note that, although stars are very large, they can be considered point sources of radiation over very large distances.

The distances between stars and galaxies are vast and so are sometimes measured in light years. A **light year** (ly) is the distance light travels in one year (1 ly = 9.46×10^{15} m).

The distances to distant stars are measured by comparing their luminosity against the known luminosity of **standard candles** or events where the luminosity is known. One type of standard candle are cephid variable stars. These stars have varying luminosity; they grow brighter and dimmer over a cycle. Their luminosity is linked to the period of variation so, if the period is measured, the luminosity can be deduced.

If the period of a cephid variable in a local galaxy is measured, its luminosity can be calculated, and the **radiant flux intensity** is given by:

$$F = \frac{L}{4\pi d^2}$$

To the naked eye stars, with the exception of our Sun, appear as flickering dots. However, analysis of the light received from the stars, however faint, can give information about their luminosity, size and chemical composition.

Worked Example

The period of a cephid variable in a nearby galaxy shows that it has a luminosity 2200 times that of the Sun. The luminous flux measured at Earth is 1.20×10^{-12} W m^{-2}. What is the approximate distance between Earth and that galaxy?

$$d = \sqrt{\frac{L}{4\pi F}}$$

$$= \sqrt{\frac{2200 \times 3.83 \times 10^{26}}{4 \times \pi \times 1.20 \times 10^{-12}}}$$

$$= 2.36 \times 10^{20} \text{ m}$$

This is approximately 25 000 light years.

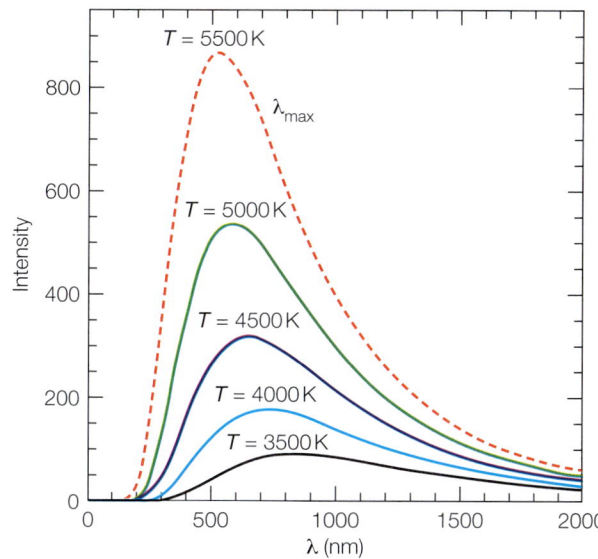

▲ **Figure 25.2** Radiation emitted by stars at different temperatures

A star can be considered to be a **black body** – a perfect absorber of radiation. None of the light entering a star is reflected back out. Black bodies such as stars emit radiation in a particular pattern (see Figure 25.2) and the surface temperature of the star is inversely proportional to the wavelength maximum on this graph (λ_{max}) in accordance with **Wien's displacement law**:

$$\lambda_{max} = \frac{1}{T}$$

$$\lambda_{max} = \frac{2.90 \times 10^{-3}\,\mathrm{m\,K}}{T}$$

For example, the peak wavelength for the Sun is 635 nm which gives a surface temperature of 4570 K.

As the temperature of a black body increases, its luminosity will increase as it radiates more power. The luminosity is linked to the absolute temperature of the body by the **Stefan–Boltzmann law** which, for a spherical object like a star, is:

$$L = 4\pi\sigma r^2 T^4$$

where $\sigma = 5.67 \times 10^{-8}\,\mathrm{W\,m^{-2}\,K^{-4}}$ is known as the Stefan–Boltzmann constant.

The Stefan–Boltzmann law and Wien's displacement law can be used together to estimate the radius of a distant star using the peak wavelength and the luminosity of the star.

When electrons transition between energy states in an atom they release photons of very specific frequencies and produce emission spectra. These spectra can be used to identify the elements and molecules in distant stars or galaxies.

However, when the spectra of most galaxies are analysed, they do not exactly match samples on Earth. The frequencies have been reduced and the wavelengths stretched so that the observed light is closer to the red part of the spectrum. This **redshift** is a result of the **Doppler effect**: the galaxies are

Worked Example

The luminosity of the star Alpha Centauri is 1.52 times that of the Sun. The peak wavelength of its spectrum is 500 nm. Use this information to estimate the radius of Alpha Centauri.

Answer

$$T = \frac{2.90 \times 10^{-3}\,\mathrm{m\,K}}{\lambda_{max}} = \frac{2.90 \times 10^{-3}\,\mathrm{m\,K}}{500 \times 10^{-9}} = 5.80 \times 10^3\,\mathrm{K}$$

$$r = \sqrt{\frac{L}{4\pi\sigma T^4}} = \sqrt{\frac{1.52 \times 3.83 \times 10^{26}}{4 \times 3.14 \times 5.67 \times 10^{-8} \times (5.80 \times 10^3)^4}} = 8.50 \times 10^8\,\mathrm{m}$$

moving away from our galaxy, the Milky Way.

The shift in the wavelength and frequency depend on the velocity with which the galaxy is moving relative to us (v) and is given by the relationship:

$$\frac{\Delta\lambda}{\lambda} \approx \frac{\Delta f}{f} \approx \frac{v}{c}$$

Edwin Hubble collected, collated and analysed the data from many galaxies and found that the distance a galaxy is from us (d) is proportional to its relative velocity (v):

$$v \approx H_0 d$$

The constant H_0 is known as the **Hubble constant** which has a value of approximately $2.2 \times 10^{-18}\,\text{s}^{-1}$.

Further measurements showed that the galaxies are all moving apart relative to each other; our galaxy is in no way the centre of the Universe. The volume of the Universe is increasing as the space itself expands in all directions, and the redshift is caused by the space between galaxies increasing. This can be hard to imagine but can be thought of as similar to bread dough (see Figure 25.3) with seeds spread throughout it. The dough will 'rise' over time and the seeds (the galaxies) will become more distant from each other because the medium they are in (the space) is expanding.

Link

See Chapter 22 *Quantum physics* for electron transitions.

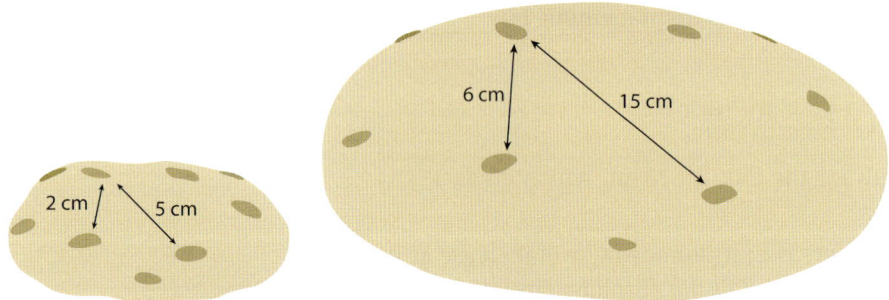

▲ **Figure 25.3** An analogy for the expanding Universe

The observation of cosmic redshift leads to the conclusion that, approximately 13.8 billion years ago, all of the observable galaxies occupied the same space. All of the mass–energy of the Universe occupied an infinitesimally small volume which was incredibly hot and dense. This was very unlike the current state; the temperature was too high for atoms or even protons to exist. The Universe expanded rapidly in a process called the **Big Bang** and, as it did so, it cooled leaving traces of the high temperature in the **cosmic microwave background radiation (CMBR)**.

Key terms

Big Bang: The early rapid expansion of the universe.

Cosmic microwave background radiation (CMBR): Radiation left over from the Big Bang which indicated that the initial state was at very high temperature.

Doppler effect: The change in wavelength of radiation due to relative movement of the source and observer or the expansion of space between them.

Hubble constant: A 'constant' of proportionality linking a galaxy's velocity relative to us and its distance away from us. The constant 'changes' over time.

Redshift: The increase in wavelength of light from an object moving away from us, which causes the light to shift towards the red part of the spectrum.

↑ Raise your grade

(a) The distance to a star in a nearby galaxy is determined using a standard candle technique. The star is found to be 2.9×10^{20} m away.

Describe what a standard candle is and give an example.   [3]

Standard candles are objects with a
known brightness ✗ used to measure
distances ✓ to stars. ✗

> A mark was awarded for the reference to measurement of distance, but brightness is not the same as luminosity so this could not be awarded a mark. No example was given, so another mark was missed.

(b) The peak wavelength of the light detected from the star is 400 nm and the radiant flux intensity is 1.3×10^{-12} W m^{-2}.

(i) Calculate the luminosity of the star.   [3]

$$F = \frac{L}{4\pi d^2}$$

$$L = 4\pi d^2 F \checkmark = 4 \times 3.14 \times (2.9 \times 10^{20})^2 \times 1.3 \times 10^{-12} = 1.37 \times 10^{30} \text{ W} \checkmark\checkmark$$

(ii) Estimate the radius of the star.   [7]

$$T = \frac{2.90 \times 10^{-3} \text{ m K}}{\lambda_{max}} \checkmark + \frac{2.90 \times 10^{-3} \text{ m K}}{400 \times 10^{-9}} = 7.25 \times 10^{-3} \checkmark \text{ }°C \text{ ✗}$$

> Although the calculation is correct, the student has given the units for temperature in degrees Celsius, but the answer should be in kelvin, so only two marks are given.

$$L = 4\pi\sigma r^2 T^4$$

$$r^2 = \frac{L}{4\pi\sigma r^2 T^4} \checkmark = \frac{1.37 \times 10^{30}}{4 \times 3.14 \times 5.67 \times 10^{-8} \times (7.25 \times 10^3)^4} = 6.963 \times 10^{20} \checkmark\checkmark$$

$$r = 2.64 \times 10^{10} \text{ m} \checkmark$$

> A very clear calculation given full marks. In calculations this long it is always worth checking three times.

Exam-style questions

1 A satellite in orbit around the Earth has a solar panel which powers the satellite by absorbing light in the visible part of the spectrum. This is only a small fraction of the radiant flux of the Sun so the panels are only 10% efficient. The satellite operates on a power of 1.0 kW.

The distance between the Sun and the Earth is 1.50×10^{11} m and the luminosity of the Sun is 3.83×10^{26} W.

(a) Calculate the power generated by each square metre of the satellite's solar panels.   [3]

(b) Calculate the area of solar panels required to power the satellite.   [1]

(c) The planet Neptune is 30 times further away from the Sun than Earth. Explain why probes sent to Neptune are nuclear powered instead of solar-powered. Justify your answer.   [3]

2 Observations of the galaxy Andromeda show a shift in the wavelength of a spectral line from a laboratory measurement of 656.3 nm to an astronomical measurement of 656.1 nm.

(a) Determine the approximate relative velocity of Andromeda to our galaxy, the Milky Way.   [3]

(b) Use the result of your calculation to explain what is unusual about the motion of Andromeda when compared with more distant galaxies.   [2]

3 The radius of the Sun is 6.96×10^8 m and the peak of the observable spectrum is 500 nm.

(a) Calculate the surface temperature of the Sun.   [2]

(b) Calculate the luminosity of the Sun.   [2]

26 Practical skills

Paper 3: Advanced practical skills

There are two questions on Paper 3 *Advanced practical skills*, each question lasting 1 hour and each worth 20 marks. The two questions are set in different areas of physics. They are designed to test your practical skills – no prior knowledge of the theory is needed.

Question 1

The first question usually requires you to set up some simple apparatus, such as an electrical circuit, or a beam supported by springs or an oscillating system. You will then make a number of simple measurements, such as measuring length using a rule with a millimetre scale, angle using a protractor, or time using a stopwatch.

Once you've taken your measurements (the **raw data**) you will be asked to calculate other quantities from them, and then to plot a graph. You will then find values from the graph, such as the gradient and the *y*-intercept, and use these values to find constants in an equation. See Table 26.1 for how marks are allocated for each skill.

▼ **Table 26.1** Mark allocation for Question 1

Question 1 (20 marks)		
Skill	**Mark***	**Skills needed**
Manipulation, measurement and observation	7 marks	Collecting data successfully, selecting a suitable range of values and Ensuring good quality data
Presentation of data and observations	6 marks	Compiling a table of results, recording data, observations, and calculations and drawing a graph
Analysis, conclusions and evaluation	4 marks	Interpreting the graph and drawing conclusions

*The remaining 3 marks are allocated across the skills in the table.

Question 2

The second question involves carrying out an experiment using apparatus that you may need to assemble, and taking a number of measurements. You will be asked to estimate the percentage uncertainty in one of these measurements and to consider the appropriate number of significant figures for any calculated values.

You will record a set of measurements for two values of the **independent variable**, and be asked whether the results you've obtained support a hypothesis. Some of the readings are designed to be difficult to measure accurately, and you will be asked to identify any limitations or sources of uncertainty in the experiment, and to suggest improvements. See Table 26.2 for how marks are allocated for each skill.

▼ **Table 26.2** Mark allocation for Question 2

Question 2 (20 marks)		
Skill	**Mark***	**Skills needed**
Manipulation, measurement and observation	5 marks	Collecting data successfully and ensuring good quality data
Presentation of data and observations	2 marks	Recording data, observations and calculations
Analysis, conclusions and evaluation	10 marks	Drawing conclusions, estimating uncertainties, Identifying limitations and suggesting improvements

*The remaining 3 marks are allocated across the skills in the table.

You could also be asked to measure force using a newton-meter, a volume of liquid using a measuring cylinder, or temperature with a thermometer.

💡 **Remember**

The **independent variable** is the one **you change** and control; for example, the length of a resistance wire or the number of masses hung on a spring.

The **dependent variable** is the one that is **altered** by the change you make; for example, the current in a circuit or the extension of a spring.

Key skills

Measuring length

Using vernier calipers

If the measurement to be made is less than 15–20 cm in length, then a pair of vernier calipers can be used to obtain a more accurate value. The value of a length measured using vernier calipers can usually be recorded to the nearest 0.1 mm.

To read a vernier scale (see Figure 26.2):

- first read the value on the main scale that is just before the zero line on the vernier scale

- then read the value on the vernier scale in line with a marking on the main scale and add this to the first reading.

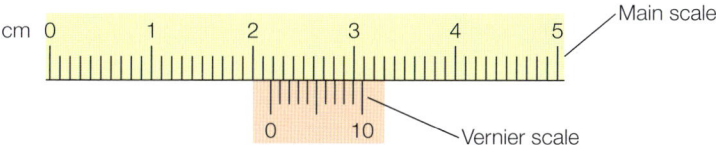

▲ **Figure 26.2** Using a vernier scale

Using a micrometer

A screw-gauge micrometer (see Figure 26.3) can be used to measure lengths up to 2–3 cm, to an accuracy of 0.01 mm.

- the reading on the barrel gives the length to the nearest 0.5 mm
- the reading where the centre line meets the thimble gives the value to be added on between 0.00 mm and 0.50 mm

Measuring angle

Readings of angle using a protractor should normally be recorded to the nearest degree (°).

Measuring time

Where possible time measurements should always be repeated and a mean value calculated. The precision of many stopwatches is ±0.01 s, but the accuracy depends on the reaction time of the experimenter (typically ±0.2 s).

The raw readings of time can be recorded to the precision of the stopwatch and then a mean value calculated.

Good practice

Repeating measurements

If you are measuring the period of oscillation of a pendulum, it is difficult to judge exactly when a complete oscillation begins and ends. It makes sense to measure the time for, say, 10 oscillations at least three times and then calculate a mean value. Likewise, any measurement which could vary significantly, such as the diameter of a metal wire, should be repeated two or three times and a mean value found.

Recording results

Results should always be collated and displayed in a clear table, with columns of readings (not rows). Key points to remember:

- **Range of values:** The range of values of the independent variable should be as large as possible, with the measurements at approximately equal intervals.

- **Column headings:** Each column heading should have the quantity that has been measured with the appropriate unit.

In Figure 26.2:

- main scale reading: 2.1 cm

- reading on vernier scale coinciding with reading on main scale: 0.8 mm (0.08 cm)

- final reading = 2.1 + 0.08 = 2.18 cm.

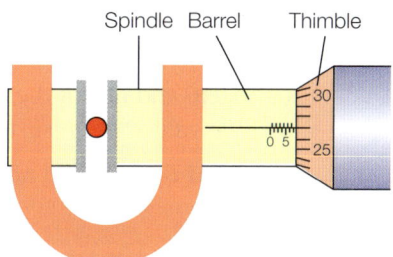

▲ **Figure 26.3** Reading a micrometer scale

In Figure 26.3:

- reading on the barrel is 7.00 mm

- reading on the thimble is 0.27 mm

- final reading = 7.00 + 0.27 = 7.27 mm.

The quantity and its unit should be separated by a '/' (called a 'solidus') or by placing the units in brackets (e.g., length / cm, time / s or θ / °, or length (cm), time (s), or angle θ (°)).

- **Consistency:** Raw readings (the measurements you actually make) should all be recorded to the same degree of precision. This sometimes means that a set of data will include values with different numbers of significant figures (e.g., 11.7 cm and 5.3 cm). The precision is determined by the instrument you are using.

Worked Example

This experiment investigates how the angle of tilt of a runway affects the time taken for a cylinder to roll down it (Figure 26.4).

Adjust the position of the clamp on the stand so that θ, the angle the runway makes with the horizontal, is 30°.

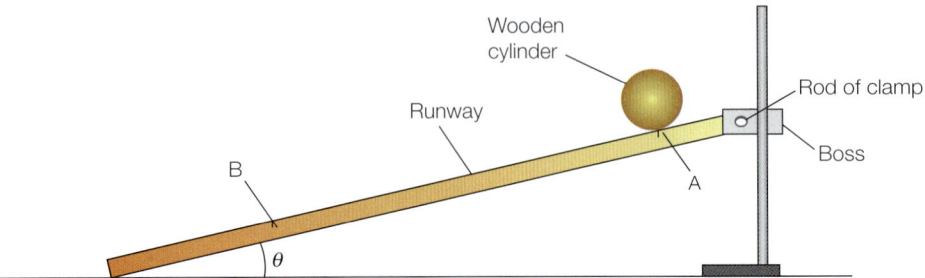

▲ **Figure 26.4**

Place the wooden cylinder at point A, release the cylinder, and measure the time T taken to reach point B.

Change θ and repeat the experiment until you have six values of θ and T. Record your results in a table and include values of $1/\sqrt{\sin\theta}$ in your table.

Values of θ should be no greater than 60°.

Answer

▼ **Table 26.4**

θ / °	T_1 / s	T_2 / s	T_3 / s	T / s	$\dfrac{1}{\sqrt{\sin\theta}}$
7	1.15	1.10	1.11	1.12	2.9
15	0.78	0.77	0.74	0.76	1.97
29	0.58	0.53	0.52	0.54	1.44
43	0.46	0.43	0.49	0.46	1.21
51	0.40	0.47	0.44	0.44	1.13
60	0.41	0.40	0.43	0.41	1.07

★ **Exam tip**

To achieve high marks:

- **Column headings:** All the quantities have the correct unit with the quantity and the unit separated by /. $1/\sqrt{\sin\theta}$ has no units; note also that the angle θ has a unit (°).
- **Collecting data:** Six sets of values of θ and T are recorded, showing the correct trend (T decreases as θ increases).
- **Range:** A good range of values of θ chosen, including a very small value and a very large value.
- **Consistency:** All the values of θ are recorded to the nearest degree; all the raw values of T are recorded to the same precision (the nearest 0.01 s).
- **Significant figures:** The mean values of T are recorded to the same number of significant figures as the raw values of T. The values of $1/\sqrt{\sin\theta}$ are recorded to one significant figure more than the significant figures of θ.

Drawing graphs

Choosing scales

When choosing scales for the *x*- and *y*-axes:

- choose scales so that the plotted points occupy more than half the grid of the graph

- scales do not have to start at zero – in many cases it is important that they do **not** start at zero, otherwise the plotted points will be compressed into a small area of the graph paper

- the scales must be linear

- the scales must be 'sensible' (e.g., increasing in twos, fives or tens). Scales that increase by a factor of three or seven, for example, make plotting the points accurately much more difficult and can lead to errors in taking read-offs for calculating a gradient or intercept value

- scale markings should occur no more than three large squares apart. A good rule is to put a scale value every two large squares

- the scales should be labelled with the quantities being plotted

- do not use 'springs' on your graphs (see Figure 26.5).

Plotting points

Points should be small, sharp pencil crosses (×), or dots with a circle around them (⊙). If one of the points appears to be **anomalous** – not following the trend of the other points – it is a good idea to repeat the measurements you made for this value to check, for example, that you didn't mis-read a meter.

Drawing lines of best fit

Drawing a good line of best fit (almost always a straight line) can be challenging, but can be achieved with practice. Look at the examples in Figure 26.6.

Always use a 30 cm rule in good condition and a sharpened pencil. Try to ensure that there is a good balance of points above and below the line. The first and last points are of no greater importance than the points in between – sometimes the best line does not go through any of the points that have been plotted.

If one point appears to be anomalous, draw a ring around it and then draw the line of best fit based on the remaining points.

Calculating a gradient

To calculate the gradient, mark two new points **on the line** you have drawn. The distance between the two points should be greater than half the length of the line drawn – the bigger, the better.

Calculating an intercept

The *y*-intercept can be found in one of two ways:

- **Directly from the graph:** but **only** if the scale on the *x*-axis does not have a false origin (i.e., the scale on the *x*-axis starts at zero).

- **Using $y = mx + c$:** Select a point on the line you have drawn. Read off the *x* and *y* values and substitute them into the equation for a straight line, together with your value for the gradient (*m*). Then rearrange the equation to find the intercept *c*.

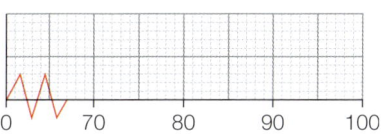

a Avoid a 'break' in the scale to include zero. This is an example of a **false origin**

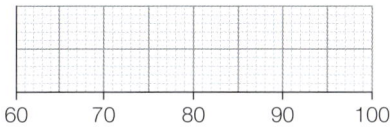

b The scale is linear and continuous. A 'zero' value at the origin is not necessary

▲ **Figure 26.5** Avoid 'springs' on your axes

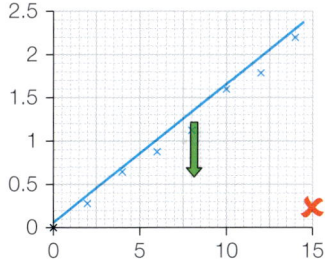

a A better line can be drawn by moving the candidate's line downwards

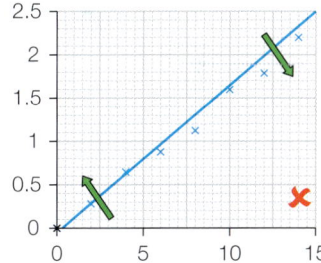

b A better line can be drawn by rotating the line clockwise. The candidate has joined the first and last points without taking into account the other points

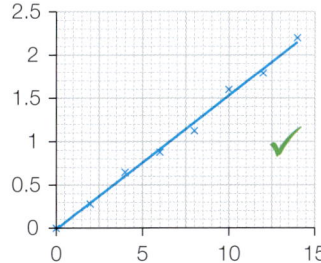

c A good attempt at drawing the straight line of best fit, with points above and below the line

▲ **Figure 26.6** Plotting the line of best fit

Answering Question 2

Identifying uncertainties

A physical quantity can never be measured exactly. **Uncertainties** arise due to the limitations of the instrument you are using (how accurate it is), your skill at using it, and changes in the environment (e.g., the effects of draughts or a change in temperature). Random fluctuations in what you are trying to measure (e.g. the activity of a radioactive source) will also give rise to uncertainty and the act of measuring can affect the measurement itself – when a thermometer at room temperature is placed in a warm liquid it will cool the liquid slightly. Physicists try to estimate these uncertainties to calculate the degree of confidence they can have in a calculated value.

Worked Example

(a) (i) Measure the diameter d of a small ball made of modelling clay, using a 30 cm rule.

(ii) Estimate the percentage uncertainty in your value.

(b) Measure the mass m of the ball (an electronic balance is available).

(c) (i) Calculate the density ρ of the modelling clay, using the equation:

$$\rho = \frac{6m}{\pi d^3}$$

(ii) Justify the number of significant figures you have given for your value of ρ.

Answer

(a) (i) d / mm: 26   27   30

Mean value for $d = 28$ mm

(ii) $\Delta d = 4$ mm, percentage uncertainty $= \frac{4}{28} \times 100 = 14\%$

> **Exam tip**
>
> The diameter of the ball is likely to vary in different directions, and so it is sensible to measure the diameter in three different directions and find a mean.

> **Remember**
>
> The precision of the rule is ±1 mm, but the accuracy of the measurement is much less than this. The accuracy is affected by:
>
> - parallax error due to the distance between the rule and the edges of the ball
>
> - variation in the diameter of the ball as it is a rolled piece of modelling clay which is not perfectly spherical.
>
> For these reasons, a better estimate of the absolute uncertainty is 2–5 mm. The percentage uncertainty is found from the equation:
>
> percentage uncertainty $= \dfrac{\text{absolute uncertainty}}{\text{mean diameter}} \times 100$

(b) $m = 21$ g

(c) (i) $\rho = \dfrac{6m}{\pi d^3} = \dfrac{6 \times 21}{\pi \times 2.8^3} = 1.8 \text{ g cm}^{-3}$

(ii) The value of ρ can be quoted to the same number of significant figures as the least number of significant figures in the raw data (or one more). In this example, ρ is calculated from the raw data of m (2 sig. figs.) and d (2 sig. figs.), and so ρ can be quoted to two or three significant figures.

> **Remember**
>
> Use consistent units in calculations.

> **Exam tip**
>
> It's not enough to say that the answer is consistent with the 'raw data'.

Testing hypotheses, sources of uncertainty and limitations of the procedure

The experiment in Question 2 will normally require you to obtain two sets of experimental results which are used to test the validity of a theoretical equation. The last section of the next worked example looks at sources of uncertainty and possible improvements.

Worked Example

A student investigates the terminal speed of a ball-bearing of diameter d falling through a liquid (see Figure 26.7). He measures the time t taken for each ball to fall from A to B, repeating each measurement and finding a mean value for t (see Table 26.5).

▼ Table 26.5

Diameter d of ball / mm	Time to fall from A to B / s
30	2.2
38	1.4

It is suggested that the relationship between t and d is:

$$t = \frac{k}{d^2} \quad \text{where } k \text{ is a constant}$$

(a) Using your data, calculate two values of k.

(b) Explain whether your results support the suggested relationship.

(c) Describe four sources of uncertainty or limitations of the procedure.

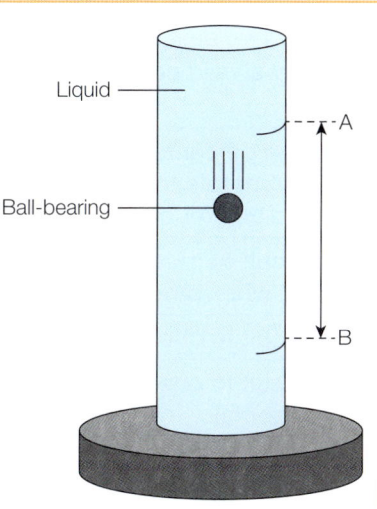

▲ Figure 26.7 Ball-bearing falling in liquid

Answer

(a) $k = t d^2$

so $\quad k_1 = 2.2 \times (30 \times 10^{-3})^2 = 1.98 \times 10^{-3} \, \mathrm{m^2 \, s}$

$\quad k_2 = 1.4 \times (38 \times 10^{-3})^2 = 2.02 \times 10^{-3} \, \mathrm{m^2 \, s}$

(b) percentage difference between k values $= \dfrac{2.02 - 1.98}{2.00} \times 100 = 2\%$

This is less than my estimate of the overall uncertainty in the experiment of 10%, and so the two results are consistent with the suggested relationship.

> ★ **Exam tip**
>
> Answers such as 'k values are quite close' or 'values of k are far apart' are not good enough.

(c) 1. While two readings may be consistent with a particular relationship, they are not conclusive proof. More readings are needed using ball-bearings with other diameters.

2. The diameters were measured with a 30 cm rule, leading to parallax error.

3. It is not certain that the ball-bearings had reached their terminal velocity by the time they reached point A.

4. The time taken to fall from A to B is very short, and so there is a large percentage uncertainty in the value of t.

Identifying improvements

The last section of Question 2 looks at improvements. These can address the problems identified earlier, but this is not essential.

Some of the limitations, sources of error and possible improvements are specific to a particular experiment (see Table 26.6).

▼ **Table 26.6** Sources of error and suggestions for improvements

Limitations and sources of error	Improvements
Only two sets of data recorded – not enough to draw a valid conclusion.	Take more readings (with different values of the independent variable), and either calculate further values of k and compare them, or plot a graph.
Large percentage uncertainty in using a 30 cm or 0.5 m rule to measure small distances such as the diameter of a wire or the thickness of a coin.	Use vernier calipers or a micrometer.
Difficult to measure a small change in length (e.g. when stretching a metal wire).	Use a travelling microscope.
Difficult to check whether a rule is vertical.	Hold a set-square on the bench against the rule.
Difficult to check whether a wooden strip is horizontal.	Use a spirit level, or measure the height of the strip above the bench at both ends and check they are the same.
Difficult to hold a rule or a protractor steady when making a measurement.	Clamp the rule or protractor using a clamp and stand.
Difficult to judge the start or end of an oscillation.	Place marker (e.g. a pencil mounted on a stand) at the centre of the oscillation (the equilibrium position). Start and stop a stopwatch when the oscillator passes the equilibrium position.
Difficult to time an event accurately because the time is short or there is a large percentage uncertainty in the value of time.	Video the experiment, then play back frame-by-frame and use the camera's clock (or a stopwatch filmed next to the experiment) to measure the time. It should be clear how the measurement is to be made (e.g. include a metre rule in the picture when filming so that changes in height can be measured).
Difficult to release an object (e.g. a ball) without exerting some external force.	Hold the object against a stop (e.g. a piece of card) and release.
Difficult to calculate the volume of a ball as its diameter varies.	Partially fill a measuring cylinder with water, place the sphere in the water and measure the change in volume.
Difficult to measure volume of liquid in beaker/ measuring cylinder as liquid is clear/ transparent/ difficult to see meniscus.	Add coloured dye to the liquid.

> ★ **Exam tip**
>
> - Avoid describing improvements that you should do as part of a good scientific technique, such as repeating measurements or reading instruments by viewing in a direction perpendicular to the scale (to reduce parallax errors).
>
> - Avoid 'inventing' problems such as a damaged rule, or 'it was difficult to drop the ball-bearing and measure the time at the same time'.
>
> - Vague answers, such as 'systematic error', 'random error', or 'parallax error' do not gain credit without further explanation.
>
> - 'Measurements not repeated' is not a valid answer as, where appropriate, measurements should have been repeated. Similarly, 'zero error in a micrometer' is not valid, as a correction should have been made for this.

↑ Raise your grade

1 **(a)** Set up the circuit shown.

> Make sure the positive terminal of the multimeter is connected to the positive side of the d.c. supply.

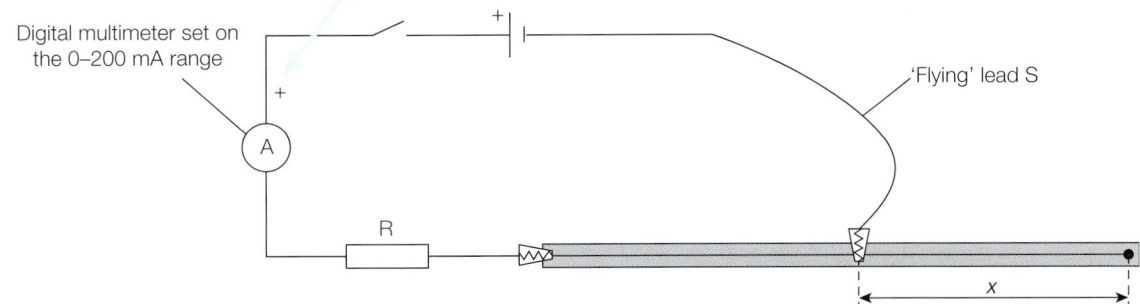

> 'Approximately' means within 0.5 cm.

Attach the flying lead S so that the distance x is approximately 50 cm.

(b) Measure and record x.

> Close enough to 50 cm and has the correct unit.

$x =$ ....49.7 cm.... ✔   [1]

(c) **(i)** Close the switch.

(ii) Measure and record the reading I on the ammeter.

$I =$ ...136 A.... ✗   [1]

(iii) Open the switch.

> I recorded in A not mA.

(d) Repeat (b) and (c) for different values of x until you have six values of x and I.

Include the values of $\dfrac{1}{I}$ in your table.   [10]

✔✔✔✔✔ 6 sets of readings of x and I showing the correct trend (I increasing as x increases)

✔ A good **range** of results, including values of x close to the largest and smallest possible values of x.

x/cm	I/mA	$1/I$ / A^{-1}
⑤	109	9.17
20.3	117	8.55
35.8	126	7.94
58.1	143	6.99
76.2	159	6.28
�95.1	195	5.13

✔ All the column headings have the quantity and an appropriate unit, separated by a /.

✗ This value of $1/I$ has been rounded incorrectly – it should be 6.29.

✗ All the values of x must be recorded to the nearest mm (the precision of the metre rule). No mark for consistency – the first value should be 5.0.

✔ All the values of $1/I$ have been recorded to the same number of sig. figs. as I.

(e) (i) Plot a graph of $\frac{1}{I}$ on the y-axis against x on the x-axis.

(ii) Draw the straight line of best fit.

(iii) Determine the gradient and y-intercept of this line. [8]

Poor choice of scale for the y-axis (the scale is 'compressed' so the points only occupy 3 large squares vertically). ✗

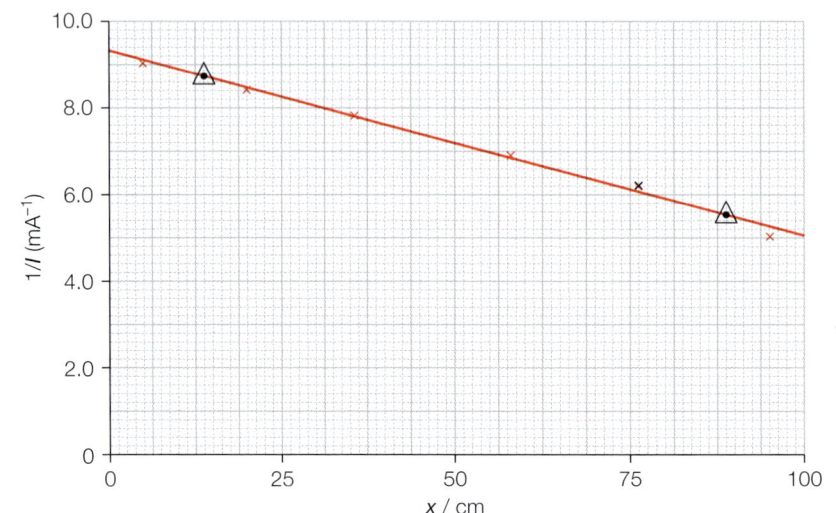

All 6 points plotted correctly with small crosses. ✔

A good 'best line', with some points above and some below the line. ✔

All the points are within ± 0.4 mA⁻¹ of the line. ✔

gradient $= \dfrac{\Delta y}{\Delta x} = \dfrac{5.6 - 8.8}{88.75 - 13.75} = -0.043 \text{ mA}^{-1}\text{cm}^{-1}$ ✔

Using the point (51.25, 7.2) in $y = mx + c$:

y-intercept $= c = 7.2 - (-0.043 \times 51.25) = 9.4 \text{ mA}^{-1}$ ✔

(f) The quantities x and $\frac{1}{I}$ are related by the equation:

$$\frac{1}{I} = -Px + Q$$

Using your answers to (e)(iii), determine the values of P and Q.

$$P = -\text{ gradient} = 0.043$$

$$Q = y\text{-intercept} = 9.4$$

Don't forget to include the units for P and Q – use the equation and the graph axes to work out what they should be.

As a check, see if your calculated value of the y-intercept is the same as the direct read-off from the graph.

One mark for matching P to -gradient and Q to the intercept. ✔

Does not score mark for units (should be mA⁻¹ cm⁻¹ for P and mA⁻¹ for Q). ✗

Exam-style questions

1 Estimate the absolute uncertainty in measuring:

  (a) the diameter of a copper wire of approximate diameter 0.5 mm using a micrometer [1]

  (b) the period of a pendulum of approximate period 2.0 s using a stopwatch with a precision of 0.01 s. [1]

2 Determine the reading on the vernier scale. [1]

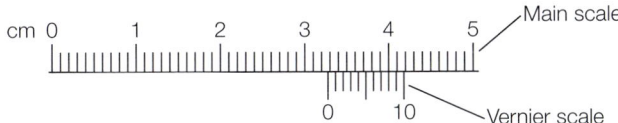

3 The resistance of a length of wire has a percentage uncertainty of 1.0%. The percentage uncertainty in measuring the diameter of the wire is 5% and the percentage uncertainty in measuring the length of the wire is 2%.

  Calculate the percentage uncertainty in calculating the resistivity of the wire. [2]

4 Describe the difference between systematic errors and random errors, and give one example of each. [4]

5 A student carries out an experiment to measure the time t taken for a table-tennis ball to fall to the ground, bounce several times and finally come to rest, when released from a height h.

h / cm	t_1 / s	t_2 / s	t_3 / s	mean t / s	$\sqrt{h}$ / cm$^{1/2}$
96.5	5.2	5.5	5.4		
87.3	5.0	4.9	4.8		
69.5	4.5	4.3	4.3		
54.4	3.6	3.8	3.7		
40.0	3.3	3.3	3.1		
29.8	2.7	2.8	2.9		

  Copy and complete the table. [3]

6 The table shows the results of an experiment in which the period of oscillation T of a compound pendulum is recorded for different values of a variable x.

x / cm	T / s
2.5	2.8
2.3	3.3
1.8	5.7
1.5	7.8
1.3	10.2
1.1	14.7

  (a) Plot a graph of T^2 on the y-axis against x on the x-axis. [2]

  (b) Draw the straight line of best fit. [1]

  (c) Determine the gradient and y-intercept of this line. [2]

7 The table shows the results of an electrical experiment, recording the current I in a circuit for different values of a resistor of resistance R.

R / Ω	I / mA
100	35.4
220	19.0
330	14.4
470	13.7
560	10.7
820	9.0

  (a) Plot a graph of I on the y-axis against $1/R$ on the x-axis. [2]

  (b) Identify the anomalous result and draw a circle around it. Draw the straight line of best fit for the remaining five points. [2]

  (c) Determine the gradient and y-intercept of this line. [2]

8 Measurements of the viscosity η of a gas are made at two different temperatures.

T/K	η/10^{-6} Pa s
323	18.9
373	21.2

  It is suggested that the relationship between T and η is:

  $$T = k\eta^2$$

  where T is the absolute temperature and k is a constant.

  (a) Using the data in the table, calculate two values of k. [2]

  (b) The overall uncertainty in the measurements used to calculate k is estimated to be 10%. State and explain whether the measurements support the suggested relationship. [1]

Paper 5: Planning, Analysis and Evaluation

This paper, together with Paper 4, are A level papers, taken after completing the whole course. There are two questions on Paper 5 *Planning, Analysis and Evaluation*, which lasts 1 hour 15 minutes. There are 15 marks for each question. The first question is designed to test your ability to plan an investigation and is likely to be concerned with a topic that you have not seen before. The second question looks at analysing the results of an experiment, identifying relationships, and evaluating the reliability of experimental evidence.

Question 1

The question will ask you to design a laboratory experiment to test a mathematical relationship between two variables. It may also list some of the apparatus that is available to you. You need to describe the procedure to be followed and the measurements to be taken.

Defining the problem (2 marks)

Start by identifying the **independent variable**, the **dependent variable**, and any other relevant variables that could affect the results, and so need to be kept constant. Some simple examples are given in Table 26.7.

▼ **Table 26.7** Identifying variables

Investigation	Independent variable	Dependent variable	Variables to keep constant
How does the resistance of a metal wire vary with diameter?	diameter	resistance	temperature, length of wire
How does the period of oscillation of a mass on a spring vary with mass?	mass	period	spring stiffness, amplitude

Outlining the procedure (4 marks)

Draw a clear, labelled diagram of the apparatus you intend to use; if the question is an electrical one, draw a circuit diagram. Outline in detail the measurements you intend to make. Draw the column headings (with units) of the table that you will use to record your results

Some key points to consider:

- How will you vary the independent variable?

- How will you **measure** the dependent and independent variables?

- What is the **range** of measurements you intend to make? What are the largest and smallest values of the independent variable?

- What is the **precision** of your measurements?

- Is it necessary to repeat your measurements and calculate an average? If so, when?

Analysing the data (3 marks)

A mathematical relationship between the dependent and independent variables will be suggested in the question. You should state the graph you would plot to test whether the relationship is valid. Normally you should plot the graph that would give you a straight line.

> **Remember**
>
> The **independent variable** is the one **you change and control**; for example the length of a resistance wire or the number of masses hung on a spring.
>
> The **dependent variable** is the one that is **altered** by the change you make; for example the current in a circuit or the extension of a spring.

> ★ **Exam tip**
>
> It is a good idea to leave a couple of 'spare' columns in your table for any calculated quantities you might need to test the relationship.

Additional details, including safety (6 marks)

Up to six marks are awarded for additional details. This might include a description of:

- how specific variables are to be kept constant

- initial experiments to establish a suitable range of values

- the use of an oscilloscope (or storage oscilloscope) to measure voltage, current, time, and frequency

- how to use light gates connected to a data logger to determine time, velocity, and acceleration

- how other sensors, such as motion or pressure sensors, can be used with a data logger.

Additional marks are also awarded for identifying any potential hazards in the experiment and the safety procedures that should be followed. Table 26.8 gives some examples.

▼ Table 26.8 Examples of safety precautions

Activity	Hazard	Safety procedure
Stretching metal wires	Wires break and hit eyes	Wear safety spectacles/goggles
Measuring radioactive decay	Exposure to harmful radiation	Handle the radioactive source using tongs; replace the source in a lead-lined box when not in use
Using heavy weights	Weights fall onto feet/floor	Ensure the weights are as near to the ground as possible; place a sand tray or foam rubber block underneath the weights
Investigating sound waves	Damage to ears	Use ear defenders/switch off loudspeaker(s) when not in use
Investigating diffraction using a laser	Damage to eyes	Do not look directly at the laser

Worked Example

A student is investigating how the resistance of a particular type of thermistor varies with temperature. It is suggested that:

$$R = Ae^{\frac{k}{T}}$$

where R is the resistance of the thermistor at temperature T, and A and k are constants.

Design a laboratory experiment to test the relationship between R and T. Explain how your results could be used to determine values for A and k. You should draw a diagram showing the arrangement of your equipment. In your account pay particular attention to the:

- procedure to be followed,

- measurements to be taken,

- control of variables,

- analysis of the data,

- safety precautions to be taken.

Answer

Defining the problem

The independent variable is the temperature of the thermistor; the dependent variable is the resistance of the thermistor.

Method

- Fill a 250 ml beaker with approximately 200 cm³ of cooking oil. Place a thermometer in the oil and gently heat the oil over a Bunsen burner (or using an electrical immersion heater) until the temperature reaches approximately 120 °C.

- Place the thermistor (supported by a wooden board resting on the top of the beaker) into the oil.

- Use a stirrer to ensure all the oil is at the same temperature.

- After a few minutes (allowing the oil, the thermistor, and the thermometer to reach the same temperature) record the temperature on the thermometer, and the resistance as measured by the digital ohmmeter.

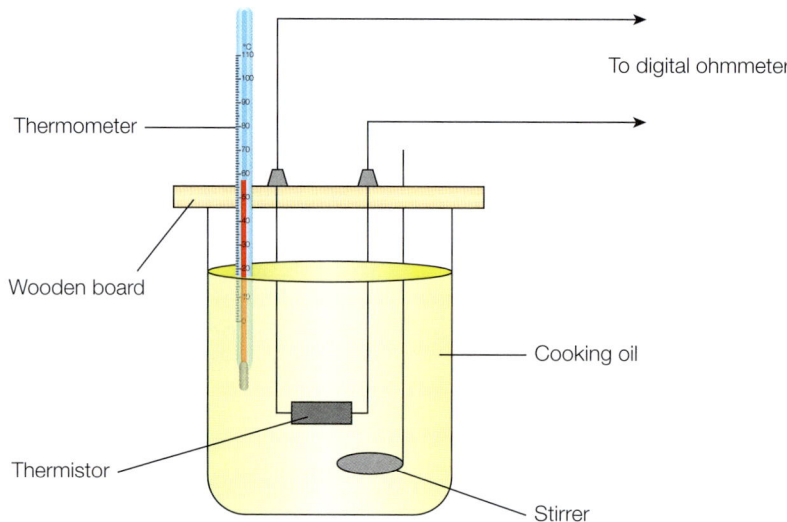

To digital ohmmeter

Thermometer

Wooden board

Cooking oil

Thermistor

Stirrer

▲ **Figure 26.8**

Results

Record the results as shown in Table 26.9.

▼ **Table 26.9**

T/K	R/Ω	$\ln(R/\Omega)$	$\dfrac{1}{T}/s^{-1}$

Include columns for $\ln R$ and $\dfrac{1}{T}$.

Analysis

Plot a graph of $\ln R$ on the y-axis against $1/T$ on the x-axis. The gradient of the graph is k and the y-intercept is $\ln A$.

Safety

Wear safety goggles (in case the hot oil splashes). Remove the heat source before placing the thermistor inside the beaker of oil (to avoid connecting wires being melted).

Question 2

This question provides you with a set of results for analysis and evaluation. A graph is plotted (with error bars) and a line of best fit is drawn together with the worst acceptable straight line so that a calculation of the absolute uncertainty in the gradient or the y-intercept of the graph can be calculated.

Data analysis (1 mark)

Table 26.10 shows the different relationships you will need to be familiar with.

▼ **Table 26.10** Relationships between two variables

Relationship	Graph to plot	Gradient	y-intercept
Linear ($y = mx + c$)	y against x	m	c
Exponential ($y = a\,e^{kx}$)	ln y against x	k	ln a
Power law ($y = ax^n$)	lg y against lg x	n	lg a

> **Maths Skills**
>
> $\lg = \log_{10}$ ('logs to base 10')
>
> $\ln = \log_e$ ('natural logs')
>
> For more about logs see Appendix: *Maths skills*.

Worked Example

When a load is suspended from a metal wire such as copper, the wire stretches. If the load is left on the wire it stretches further over time, a process known as **creep**.

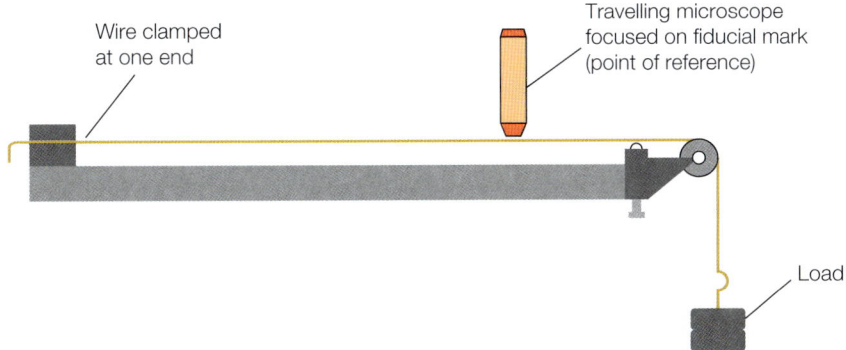

▲ **Figure 26.9** Creep of copper wire

It is suggested that the relationship between the extension x of a wire (excluding the initial extension) and the time t measured from when the load is first placed on the wire, is given by the expression:

$$x = At^n$$

where A and n are constants.

A graph is plotted of lg t on the x-axis against lg x on the y-axis. Determine expressions for the gradient and y-intercept.

Answer

Taking logs of both sides of the equation:

$$\lg x = \lg(At^n) = \lg A + \lg(t^n) = \lg A + n\lg t$$

$$\underset{y}{\lg x} = \underset{m}{n}\ \underset{x}{\lg t}\ +\ \underset{c}{\lg A}$$

The gradient $m = n$; the y-intercept $c = \lg A$.

Table of results (1 mark)

A table of results will need to be completed, following the same guidelines as for Paper 3 *Advanced practical skills*.

Worked Example

The results of the experiment on creep, described earlier are given in **bold** Table 26.11.

Calculate and record values of lg (t/s) and lg (x/mm) in Table 26.11.

Include the absolute uncertainties in lg$x$.

Answer

Units: When the logarithm of a quantity is calculated, the units should be shown with the quantity; for example, lg (t/s) and lg (x/mm). The logarithm itself does not have a unit).

Uncertainty: To calculate the uncertainty (lg x_{max} – lg x_{min})/2

▼ Table 26.11

Time t / s	Extension x / mm	lg (t / s)	lg (x / mm)
10	2.60 ± 0.05	1.00	0.415 ± 0.008
20	3.35 ± 0.05	1.30	0.525 ± 0.006
30	3.90 ± 0.05	1.48	0.591 ± 0.006
40	4.30 ± 0.05	1.60	0.633 ± 0.005
50	4.65 ± 0.05	1.70	0.667 ± 0.005

Decimal places: When calculating the logarithmic value of a quantity, the number of decimal places should be the same as (or one more than) the number of s.f. of the quantity itself. (e.g., if x/mm is 4.65 (3 s.f.) then lg (x/mm) should be 0.667 (3 s.f.) or 0.6675 (4 s.f.).)

Graph (2 marks)

The axes of the graph will already be labelled with the quantities to be plotted, and the scales marked on the axes. You will need to:

- plot the points on the graph correctly, including **error bars**

- draw a straight **line of best fit** and a straight **worst acceptable line** through the points on the graph when the trend on the graph is linear

- draw a curved trend line and a tangent to the curve where appropriate.

The **worst acceptable** line can be either the steepest possible line or the shallowest possible line that still passes through the error bars of all the data points. Draw this as a broken line and label it.

Error bars

When plotting measurements on a graph, the uncertainty in the measurements can be shown by including **error bars** ('uncertainty bars' is a better name for them).

Suppose, for example, you are plotting a value of $y = 6.2 \pm 0.1$ s (see Figure 26.10). The bar should extend 0.1 s (using the scale on the y-axis) either side of the nominal value of 6.2 s (the bar should always be the same height above and below the plotted point). If the uncertainty in both the x values and y values are known, error bars can be drawn both vertically and horizontally.

This question may ask you to plot the log of a quantity. To calculate the uncertainty, calculate the log of the largest value and the log of the measured value – the difference is the uncertainty. Alternatively, calculate:

$$\frac{(\log (\text{max value}) - \log (\text{min value}))}{2}$$

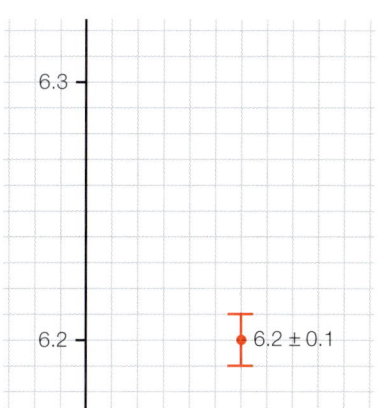

▲ **Figure 26.10** Plotting error bars

A **best fit line** should pass through all the error bars, with an even and balanced distribution of points above and below the line. To estimate the uncertainty in the gradient, a **worst acceptable line** (e.g., the steepest line that *just* passes through all the error bars, should also be drawn). The uncertainty in the gradient is then:

gradient of steepest line – gradient of best fit line

Conclusion (3 marks)

You will need to:

- find the gradient and y-intercept of a straight-line graph or a tangent to a curve,

- derive quantities that equate to the gradient and the y-intercept

- draw conclusions from these quantities.

> 🔗 **Link**
>
> Use the same rules as in Paper 3 *Advanced practical skills*.

Treatment of uncertainties (3 marks)

You need to be able to convert absolute uncertainty estimates of a quantity into fractional or percentage uncertainty estimates and vice versa. A quantity should be expressed as a value, with an uncertainty estimate and a unit (e.g., 6.4 ± 0.2 V).

You may also need to calculate uncertainty estimates in derived quantities.

💡 **Remember**

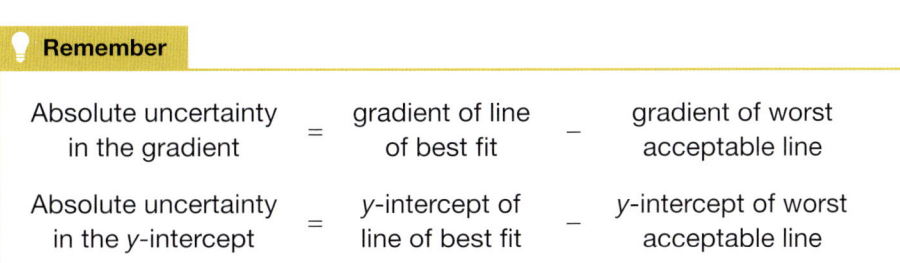

Remaining marks (5 marks)

The remaining five marks for this question are allocated across the different skill areas, and their allocation will vary.

Worked Example

For the data in Table 26.11 (p. 193) plot a graph of lg (x/mm) against lg (t/s). Include error bars for x.

Draw the straight line of best fit and a worst acceptable straight line on your graph. Both lines should be clearly labelled.

Determine the gradient and the y-intercept of the line of best fit.

Use your answers to determine the values of A and n.

Gradient of best fit line

$$= \frac{0.5825 - 0.4375}{1.46 - 1.06}$$

$$= 0.363$$

Gradient of worst acceptable line

$$= \frac{0.6075 - 0.4515}{1.54 - 1.09}$$

$$= 0.347$$

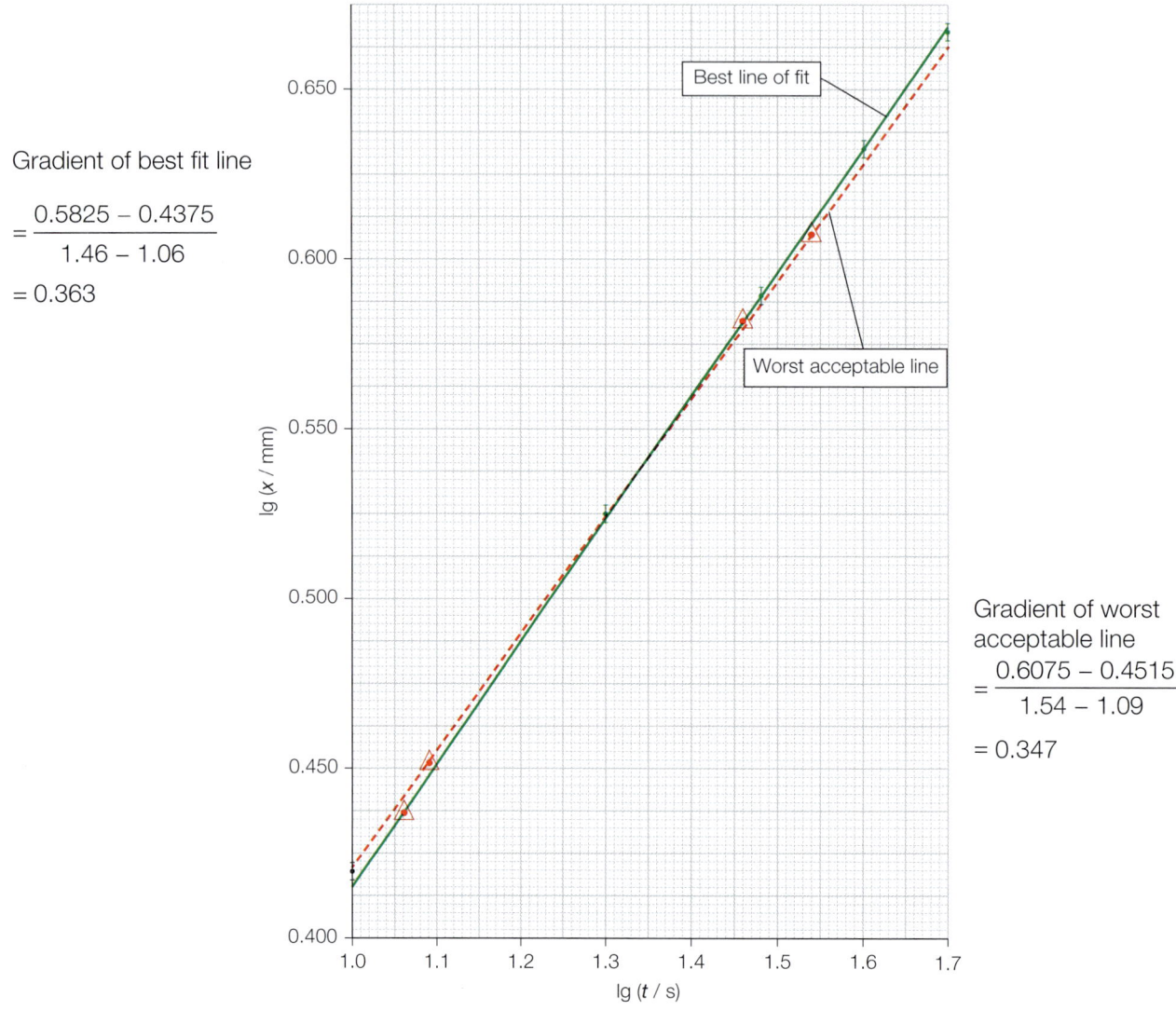

▲ Figure 26.11

Answer

The best fit line is the solid line and the worst acceptable line is the dotted line.

Using the point (1.35, 0.5425) in $y = mx + c$: y-intercept $= c = y - mx$

$$= 0.5425 - 0.363 \times 1.35$$

$$= 5.25 \times 10^{-2}$$

$n = m = 0.363$ lg $A = c = 0.0525$ so $A = 10^c = 1.13$ $x = 1.13\,t^{\,0.363}$

Determine the percentage uncertainty in the value of n.

Since $n = m$, the uncertainty in n is the same as the uncertainty in the gradient.

Absolute uncertainty in the gradient $= 0.363 - 0.347 = 0.016$

% uncertainty in $n = \dfrac{\text{absolute uncertainty in } n}{\text{nominal value of } n} \times 100 = \dfrac{0.016}{0.363} \times 100 = 4.4\%$

Exam-style questions

In the style of Question 1

1 A student is investigating the deflection of a cantilever beam when a load is suspended from one end of the beam.

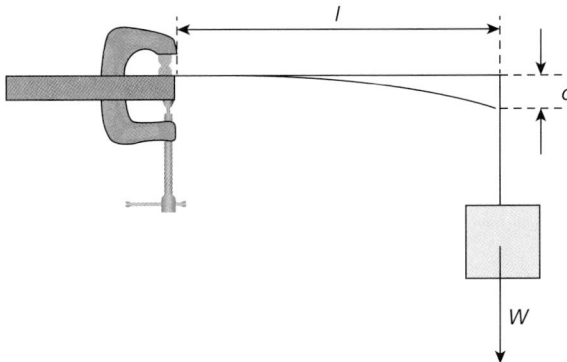

He varies the length l of the beam and measures the deflection d in each case.

(a) State the dependent and independent variables in this experiment. [1]

(b) State any other variables that should be controlled. [1]

(c) It is suggested that:

$$d = kl^3$$

State the graph you would plot to test this relationship. Explain how the value of k could be found from your graph. [3]

(d) State any safety precautions that should be taken when carrying out the experiment. [2]

2 The variables x and y are believed to be related by an equation of the form:

$$y = Ax^n \qquad \text{(a power law)}$$

where A and n are constants.

x	1.0	1.7	2.1	2.8	3.5	3.9
y	3.0	6.6	9.0	14.1	19.5	23.1

(a) Plot a suitable graph using the results shown in the table.

(b) Determine the value of n.

3 A student investigates how the stiffness of a spring varies with the diameter d of the wire.

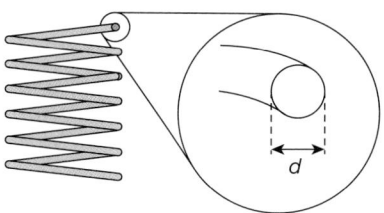

A fixed weight is suspended from a spring and the extension x of the spring is measured. The experiment is repeated using springs with different diameters.

Theory suggests that x and d are related by the equation:

$$x = kd^n$$

where k and n are constants.

(a) A graph is plotted of $\lg x$ on the y-axis against $\lg d$ on the x-axis. Determine expressions for the gradient and intercept. [1]

(b) Values of d and x are given in the table.

d / mm	x / cm	lg (d / mm)	lg (x / cm)
6.4	5.8 ± 0.2		
7.0	4.0 ± 0.2		
7.6	2.9 ± 0.2		
8.2	2.1 ± 0.2		
8.8	1.6 ± 0.2		
9.5	1.2 ± 0.2		

Calculate values of $\lg (x / \text{cm})$ and $\lg (d / \text{mm})$. Record your values on a copy of the table. Include the absolute uncertainties in $\lg (x / \text{cm})$. [2]

(c) (i) Plot a graph of $\lg x$ on the y-axis against $\lg d$ on the x-axis. Include error bars for $\lg (x / \text{cm})$.

(ii) Draw the straight line of best fit and a worst acceptable straight line on your graph.

(iii) Determine the gradient of the line of best fit. Include the absolute uncertainty.

(iv) Determine the y-intercept of the line of best fit. Include the absolute uncertainty. [8]

(d) (i) Using your answers to (c)(iii) and (c)(iv), determine the values of n and k.

(ii) Calculate the percentage uncertainty in your value of n. [4]

Paper 1 style questions: Multiple choice

The actual paper will have 40 multiple choice questions and you will have 1 hour 15 minutes to answer them. You should aim to answer the questions in this sample paper in about 30 minutes. The exam paper will include the standard list of data and formulae, and you will be provided with a separate answer sheet.

1   The Reynolds number Re is a dimensionless constant used in studying the flow of liquids in pipes. It is given by the equation:

$$Re = \frac{\rho v D}{\mu}$$

where ρ is the density of the liquid, v its velocity, and D the diameter of the pipe. What are the SI base units of μ, the viscosity of the liquid?   [1]

A $kg\,m\,s$
B $kg\,m^{-1}\,s$
C $kg\,m\,s^{-1}$
D $kg\,m^{-1}\,s^{-1}$

2   Forces **p** and **q** are represented by two vectors.

Which diagram shows **p**–**q**?   [1]

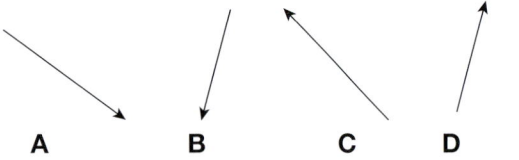

A   B   C   D

3   In an experiment to measure the resistivity of nichrome using a long thin wire, the following measurements were made:

length of wire l	99.5 ± 0.4 cm
diameter of wire d	0.38 ± 0.01 mm
resistance of wire R	$49 \pm 1\,\Omega$

The resistivity ρ was calculated using the equation $\rho = \dfrac{RA}{l}$, where A is the cross-sectional area of the wire.

What is the percentage uncertainty in the value of ρ?   [1]

A   4.2%   B   5.0%   C   6.9%   D   7.7%

4   A stone is thrown vertically down from the top of a tall building. It passes one window travelling at a speed of $7.0\,m\,s^{-1}$. It then passes a lower window travelling at a speed of $11.0\,m\,s^{-1}$.

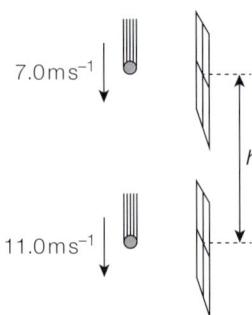

What is the height h between the two windows?   [1]

A   0.82 m   B   1.6 m
C   3.7 m   D   7.3 m

5   A force of 120 N is needed to operate a foot pedal brake, as shown.

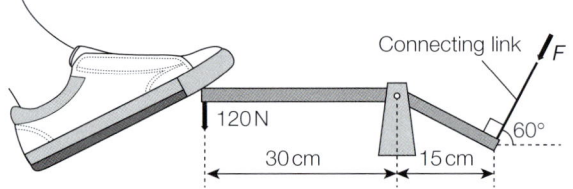

What is the force F in the connecting link?   [1]

A   120 N   B   210 N
C   280 N   D   480 N

6   A lorry of mass 3.0×10^3 kg is moving at a constant speed of $18\,m\,s^{-1}$ up a slope. The slope is inclined at an angle of $10°$ to the horizontal. The frictional forces exert a force of $6.0\,kN$ down the slope.

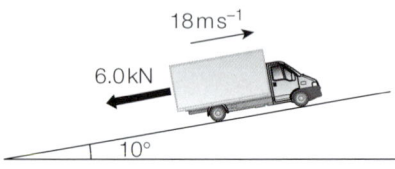

What is the output power of the lorry's engine?   [1]

A   92.0 kW   B   108 kW
C   101 kW   D   200 kW

7   A deep-sea submersible used for exploring the bottom of the deepest oceans can withstand pressures up to150 MPa. What is the maximum depth the submersible can descend?   [1]

A   10 km   B   15 km
C   20 km   D   30 km

8 A loudspeaker connected to a signal generator emits a single frequency of sound waves. It is placed above a closed pipe, as shown. The pipe is filled with water. When a valve is opened the water level gradually falls.

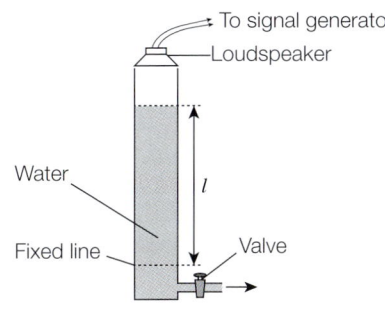

A louder sound is first heard when the length l above a fixed line, drawn on the pipe, is 67.0 cm. A second louder sound is heard when $l = 42.0$ cm. The speed of sound is 330 m s^{-1}.

What is the frequency of the sound emitted from the loudspeaker? [1]

A   330 Hz   B   660 Hz
C   990 Hz   D   1300 Hz

9 An oil drop with charge $-q$ is held stationary in the uniform electric field between two parallel plates. The potential difference between the plates is V and the distance between the two plates is d.

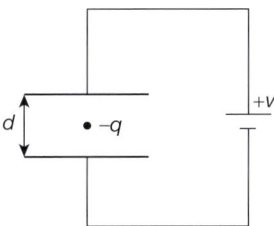

What is the mass m of the drop? [1]

A $\dfrac{dg}{qv}$ B $\dfrac{dq}{gv}$ C $\dfrac{gV}{qd}$ D $\dfrac{qV}{dg}$

10 The siren of an ambulance emits two notes, the higher note having a frequency of 960 Hz. What is the frequency of the higher note heard by a stationary observer when the ambulance is moving directly away from him at a speed of 30 m s^{-1}? [1]

A   870 Hz   B   880 Hz
C   1050 Hz   D   1060 Hz

11 Two cells, with e.m.f.s of 14 V and 10 V and negligible internal resistance, are connected to three 1 Ω resistors, as shown.

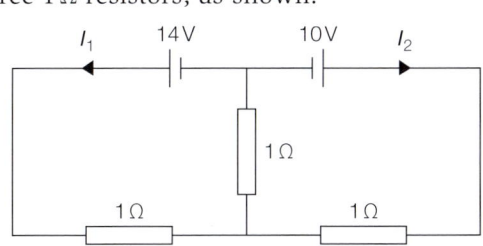

What are the values of I_1 and I_2? [1]

	I_1/A	I_2/A
A	2	10
B	3	8
C	5	4
D	6	2

12 A dry cell, with internal resistance r, is connected to a 5 Ω resistor and an ammeter, as shown.

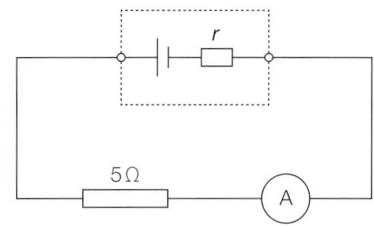

The reading on the ammeter is 2.0 A. When the 5 Ω resistor is replaced by a 7 Ω resistor, the reading on the ammeter falls to 1.5 A.

What is the value of r? [1]

A   0.5 Ω   B   1.0 Ω   C   1.5 Ω   D   2.0 Ω

13 $^{214}_{83}$Bi is a radioactive isotope of bismuth which decays by β$^-$-decay into an isotope of polonium. The polonium decays by α-decay into an isotope of lead.

What are the proton number and the nucleon number of the isotope of lead? [1]

	proton number	nucleon number
A	81	210
B	81	214
C	82	210
D	82	214

14 Which elementary particle is **not** a lepton? [1]

A   an electron   B   a neutrino
C   a positron   D   a quark

Paper 2 style questions: AS structured questions

You should aim to complete this sample paper in 1 hour 15 minutes. In the actual exam, the paper will include the standard list of data and formulae, and there will be spaces in the paper for you to write your answers.

1 **(a)** Explain what is meant by *work done*. [1]

(b) A lorry of total mass 2000 kg is travelling along a road with a constant uphill gradient, as shown. The angle of the road to the horizontal is 6°.

Calculate the component of the weight of the lorry down the slope. [2]

(c) The lorry is travelling at a speed of 20 m s⁻¹ when the driver applies the brakes. The braking force resisting the motion of the lorry is 7200 N.

(i) Show that the deceleration of the lorry when the brakes are applied is 4.6 m s⁻². [2]

(ii) Calculate the distance travelled by the lorry from the moment the brakes are applied until it comes to rest. [2]

(d) (i) Calculate:

1. the kinetic energy lost by the lorry [1]

2. the work done by the braking force. [1]

(ii) Explain why your answers to (d)(i) 1. and (d)(i) 2. are not the same. [1]

2 (a) (i) State the principle of conservation of momentum. [2]

(ii) Explain what is meant by an *elastic* collision. [1]

(b) A snooker ball of mass m, travelling at speed u, collides elastically with a second, stationary ball of equal mass, as shown. The speed of the first ball after the collision is $\frac{u}{2}$. The speed of the second ball after the collision is v at an angle θ to the original direction of the first ball.

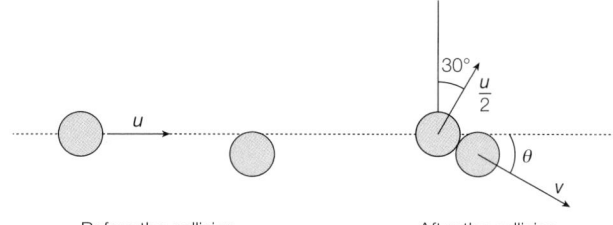

Before the collision   After the collision

(i) State expressions for:

1. the kinetic energy of the first ball before the collision [1]

2. the kinetic energy of the first ball after the collision [1]

3. the kinetic energy of the second ball after the collision. [1]

(ii) Hence show that $v = \dfrac{u\sqrt{3}}{2}$. [1]

(c) Using the principle of conservation of momentum, determine the value of θ. [3]

3 (a) Define, for a metal wire:

(i) *stress* [1]

(ii) *strain* [1]

(iii) *Young modulus.* [1]

(b) An aluminium wire, of diameter 0.28 mm, is fixed at one end and passes over a pulley at the other end. A mark is made on the wire 1.50 m from the fixed end and a travelling microscope is placed above the mark, as shown.

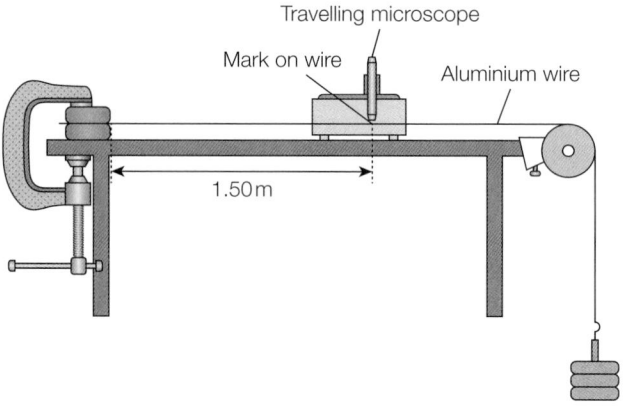

When a load of 3.5 N is hung on the free end of the wire, the mark moves 1.24 mm.

(i) Show that the stress on the wire is 5.7 × 10⁷ Pa. [2]

(ii) Calculate the Young modulus of aluminium. [3]

(c) A student suggests that a wire of aluminium of the same length, but with diameter 0.14 mm, will stretch four times as far with the same load.

State whether, or not, the student is correct. Explain your answer. [2]

4 (a) Explain what is meant by the terms:

(i) *diffraction* [2]

(ii) *interference.* [1]

(b) A student attempts to measure the wavelength of blue light using Young's double-slits experiment, as shown. The two slits are 0.60 mm apart.

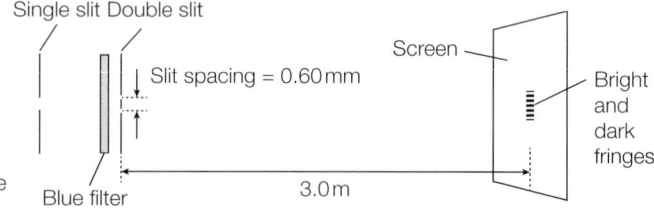

(i) The light emerging from the two slits is coherent. Explain what is meant by *coherent*. [1]

(ii) Bright and dark fringes are observed on a screen placed 3.0 m away from the double slits. Explain why bright and dark fringes are observed. [2]

(c) The distance between adjacent bright fringes is 2.4 mm. Calculate the wavelength of the blue light. [2]

(d) State what would happen to the fringes if the blue filter is replaced by a red filter. Explain your answer. [1]

5 (a) (i) State Kirchhoff's first law. [1]

(ii) Kirchhoff's first law is linked with the conservation of a physical quantity. State this quantity. [1]

(b) A 9.0 V dry cell, with an internal resistance of 0.5 Ω, is connected to a network of resistors, as shown.

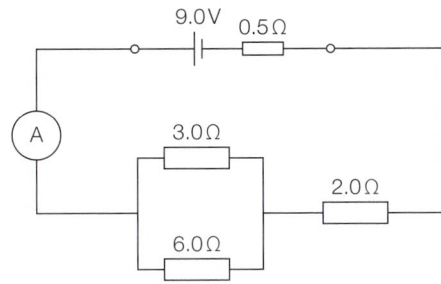

Show that the reading on the ammeter is 2.0 A. [2]

(c) Calculate:

(i) the current in the 6 Ω resistor [1]

(ii) the power dissipated in the 2 Ω resistor [2]

(iii) the potential difference across the terminals of the cell. [2]

6 (a) The radioactive decay of an isotope of cobalt is shown by the nuclear equation:

$$^{60}_{27}\text{Co} \rightarrow {}^{60}_{28}\text{Ni} + \text{X}$$

(i) Explain the meaning of the term *isotope*. [2]

(ii) State

1. the number of protons in the cobalt nucleus, [1]

2. the number of neutrons in the nickel nucleus. [1]

(b) (i) X is a type of radiation. State the name of this radiation. [2]

(ii) Describe **two** properties of this radiation. [2]

(c) The mass of the cobalt (Co) nucleus is greater than the combined mass of the nucleus of nickel (Ni) and X. Use a conservation law to explain how this is possible. [3]

Paper 3 style questions: Advanced practical skills

The actual exam will last for 2 hours, but in this sample paper you are not expected to complete the practical work; you are provided with specimen results. You should aim to complete this sample paper in 1 hour.

1 *Part (a) of this question asks candidates to set up the electrical circuit shown. The movable lead is first placed approximately half-way along the resistance wire. The potential difference V across the fixed resistor and the value of x are recorded. The experiment is then repeated for different values of x – a set of specimen results is provided for you.*

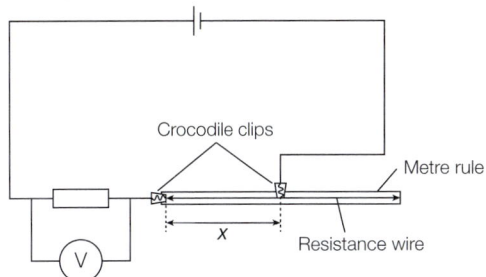

(b) Change x and repeat the experiment until you have six sets of values of x and V.

Record your results in a table. Include values of $\frac{1}{V}$ in your table.

x / cm	V / V	$\frac{1}{V}$ / V⁻¹
5.1	1.43	0.699
16.5	1.24	0.806
36.1	1.01	0.990
59.8	0.87	1.15
72.3	0.80	1.25
96.3	0.72	1.39

(c) (i) Using graph paper, plot a graph of $\frac{1}{V}$ on the y-axis against x on the x-axis. [3]

(ii) Draw the straight line of best fit. [1]

(iii) Determine the gradient and y-intercept of this line. [2]

(d) It is suggested that the quantities $\frac{1}{V}$ and x are related by the equation:

$$\frac{1}{V} = Px + Q$$

where P and Q are constants.

Using your answers in (c)(iii), determine the values of P and Q.

Give appropriate units. [2]

2 In this experiment, you will investigate the deflection of a thin wooden beam when a load is suspended from it. *Specimen results are provided when needed.*

(a) (i) Set up the apparatus as shown.

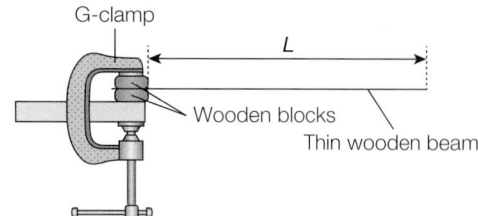
G-clamp
L
Wooden blocks
Thin wooden beam

Adjust the beam so that the length L is approximately 90 cm.

(ii) Measure and record L. $L = \underline{89.7\,cm}$

(b) (i) Suspend the mass M from the beam, as shown below.

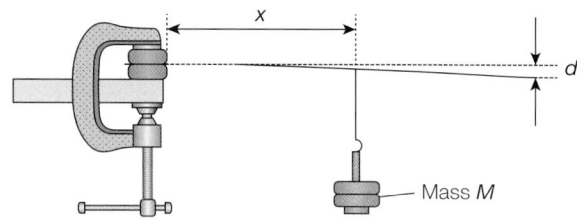
x
d
Mass M

Adjust the position of M so that x is approximately 50 cm.

(ii) Record x. $x = \underline{50.2\,cm}$

(c) (i) Calculate the value of P, where:

$$P = x^2(3L - x) \qquad [1]$$

(ii) Justify the number of significant figures that you have given for your value of P. [1]

(d) (i) Measure and record the deflection d of the end of the beam.

$d = \underline{8\,mm}$

(ii) Estimate the percentage uncertainty in your value of d. [1]

(e) Without changing L, change the position of the mass M so that x is approximately 80 cm.

Repeat (b)(ii), (c)(i) and (d)(i). $x = \underline{79.6\,cm}$

$d = \underline{17\,mm}$ [1]

(f) It is suggested that the relationship between d and P is

$$d = kP$$

where k is a constant.

(i) Using your data, calculate two values of k. [1]

(ii) Explain whether your results in (f)(i) support the suggested relationship. [1]

(iii) Using your first value of k, estimate the deflection of the beam when $x = 65$ cm. [1]

(g) (i) Describe four sources of uncertainty or limitations of the procedure for this experiment. [4]

(ii) Describe four improvements that could be made to this experiment. You may suggest the use of other apparatus or different procedures. [4]

Paper 4 style questions: A Level structured questions

You should aim to complete this sample paper in 2 hours. In the actual exam, the paper will include the standard list of data and formulae, and there will be spaces in the paper for you to write your answers.

1 (a) State Newton's law of gravitation. [2]

(b) A satellite orbits the Earth at a constant speed, at a height of 220 km above the Earth's surface.

A student states that *'the satellite is travelling at constant speed, so the net force on the satellite must be zero'*. Explain why the student is incorrect. [2]

(c) The mass of the satellite is 2.5×10^3 kg. The mass of the Earth is 6.0×10^{24} kg. The radius of the Earth is 6570 km.

Show that:

(i) the gravitational force acting on the satellite is 2.2×10^4 N [2]

(ii) the centripetal acceleration of the satellite is 8.7 m s^{-2}. [1]

(d) Determine:

(i) the speed of the satellite [2]

(ii) the time taken for one complete orbit of the Earth. [1]

(e) Suggest one use of this type of satellite. [1]

2 (a) Define *simple harmonic motion*. [2]

(b) A torsion pendulum consists of a long, thin metal wire supporting a horizontal bar of length 6.0 cm, as shown. The bar is twisted in a horizontal plane through an angle of 20° and released. The bar rotates back and forth with simple harmonic motion. The period T of the oscillation is 1.2 s.

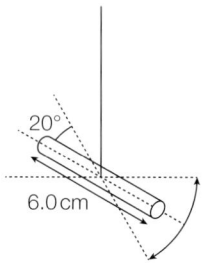
20°
6.0 cm

(i) Calculate the frequency of the oscillation. [2]

(ii) Show that the angular frequency of the oscillation is $5.2 \, \text{rad} \, \text{s}^{-1}$. [1]

(iii) Determine the position of the bar 5.7 s after it is released. [1]

(c) Calculate the maximum speed of one end of the bar. Assume the oscillation is undamped. [3]

(d) The graph shows the variation of the kinetic energy of the bar and wire over half a cycle.

On a copy of the axes, sketch:

(i) the variation of potential energy with time over ½ a cycle [1]

(ii) the variation of total energy of the bar and wire with time over ½ a cycle. [1]

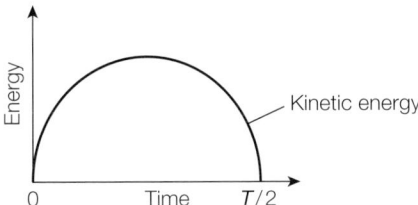

3 (a) Define *capacitance*. [1]

(b) A student says that '*capacitors are for storing charge*' Explain why the student is incorrect. [2]

(c) Three parallel-plate capacitors are connected to a 12 V d.c. supply, as shown.

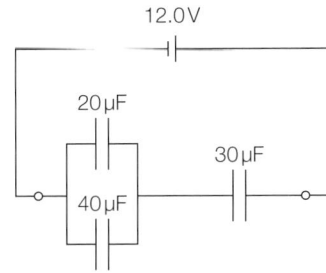

(i) Calculate the combined capacitance of the three capacitors. [2]

(ii) Show that the potential difference across the 30 μF capacitor is 8.0 V. [2]

(iii) Determine the charge on one plate of the 20 μF capacitor. [1]

(d) (i) Calculate the energy stored in the 40 μF capacitor. [2]

(ii) Describe how energy is stored in a capacitor. [1]

(iii) Suggest one use of capacitors as energy storage devices. [1]

4 (a) An experiment is carried out to calibrate a negative temperature coefficient thermistor as a thermometer.

What is meant by the term *negative temperature coefficient*? [1]

(b) The circuit used is shown.

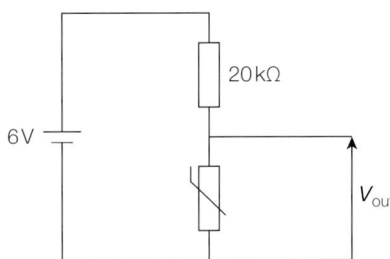

(i) What is the name given to this type of circuit? [1]

(ii) At 30 °C the output p.d. V_{out} is 4 V. Show that the resistance of the thermistor at 30 °C is 40 kΩ. [2]

(c) The calibration curve for the thermistor is shown.

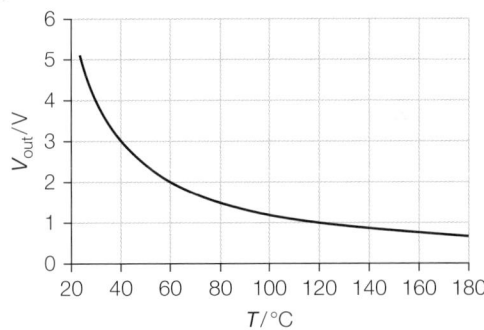

(i) Explain what is meant by *calibration curve*. [1]

(ii) Determine the output p.d. when the temperature is 70 °C. [2]

(iii) Determine the temperature of the thermistor when V_{out} is 0.8 V. [1]

(d) Suggest a reason why the thermistor is a less reliable thermometer when used at high temperatures. [1]

5 (a) (i) State Faraday's law. [1]

(ii) Define *magnetic flux*. [2]

(b) A metal bar of length l is moving at a constant speed v along a frictionless wire loop. The bar is pulled by a weight W connected to the bar by a string passing over a frictionless pulley, as shown. A uniform magnetic field B acts vertically downwards.

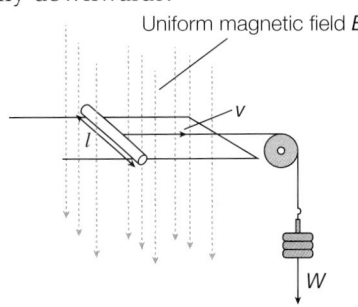

205

(i) State the area 'swept out' by the bar in one second. [1]

(ii) Hence show that the e.m.f. E induced across the ends of the bar is given by the equation $E = Blv$. [1]

(c) The electrical resistance of the bar is R ohms. Determine:

(i) the current I in the bar [1]

(ii) the direction of the current [1]

(iii) the magnitude and direction of the magnetic force F on the bar. [2]

(d) State what would happen to the direction of the magnetic force if the direction of the magnetic field was reversed. [1]

(e) Show that the speed v of the bar is given by the equation: $v = \dfrac{RW}{B^2 l^2}$ [3]

Paper 5 style questions: Planning, analysis, and evaluation

The actual exam will last for 1 hour 15 minutes; you should aim to complete this sample paper in the same time. There will be spaces in the exam paper for you to write your answers.

1 A student is investigating the stiffness of metal springs.

It is suggested that the spring stiffness k of the spring is related to the diameter d of the coils of the spring by the equation:

$$k = \frac{C}{d^3}$$

where C is a constant.

Design a laboratory experiment to test the relationship between k and d. Explain how your results could be used to determine the value of C. You should draw a diagram showing the arrangement of your equipment. In your account you should pay particular attention to:

- the procedure to be followed
- the measurements to be taken
- the control of variables
- the analysis of the data
- any safety precautions to be taken. [15]

2 A student is investigating how the time taken for a water tank to empty is related to the initial height h of water in the tank, as shown.

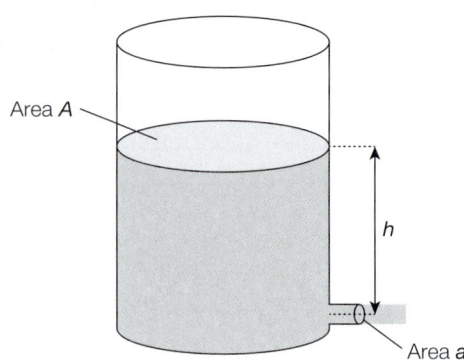

The student records the time t taken for the tank to empty from different heights.

Theory suggests that h and t are related by the equation:

$$t = \frac{A}{a}\sqrt{\frac{2}{g}}\,h^n$$

where A is the cross-sectional area of the tank and a is the area of the water outlet. g is the acceleration of free fall and n is a constant.

(a) A graph is plotted of lg t on the y-axis against lg h on the x-axis.

Determine expressions for the gradient and y-intercept. [2]

(b) Values of h and t are given in the table.

h / cm	t / s	lg (h)	lg(t)
75.4	98 ± 2		
63.6	86 ± 2		
51.5	74 ± 2		
46.9	70 ± 2		
37.1	59 ± 2		
25.7	45 ± 2		

Calculate and record values of lg (h / cm) and lg (t / s).

Include the absolute uncertainties in lg (t / s). [3]

(c) (i) On graph paper, plot a graph of lg (t / s) against lg (h / cm). Include error bars for lg (t / s). [2]

(ii) Draw the straight line of best fit and a worst acceptable straight line on your graph.

Both lines should be clearly labelled. [2]

(iii) Determine the gradient of the line of best fit. Include the absolute uncertainty in your answer. [2]

(iv) Determine the y-intercept of the line of best fit. Include the absolute uncertainty in your answer. [2]

(d) Using your answers to (a), (c)(iii) and (c)(iv), determine the values of n and $\dfrac{A}{a}$. [2]

Take $g = 981$ cm s^{-2}

Index

Headings in **bold** indicate key terms. Page numbers in *italic* refer to figures; page numbers in **bold** refer to tables.